AF721570

Dendrobium nobile

Research and Therapeutic Applications

Dendrobium nobile

Research and Therapeutic Applications

Jingshan Shi, Archana Jain, Surendra Sarsaiya, Qin Wu and Qihai Gong

Zunyi Medical University, Zunyi, Guizhou, China

CABI

CABI is a trading name of CAB International

CABI
Nosworthy Way
Wallingford
Oxfordshire OX10 8DE
UK

Tel: +44 (0)1491 832111
E-mail: info@cabi.org
Website: www.cabi.org

CABI
200 Portland Street
Boston
MA 02114
USA

Tel: +1 (617)682-9015
E-mail: cabi-nao@cabi.org

© Archana Jain, Jiangshan Shi, Qin Wu, Surendra Sarsaiya, Qihai Gong 2025. All rights, including for text and data mining, AI training, and similar technologies, are reserved. No part of this publication may be reproduced in any form or by any means, electronically, mechanically, by photocopying, recording or otherwise, without the prior permission of the copyright owners.

The views expressed in this publication are those of the author(s) and do not necessarily represent those of, and should not be attributed to, CAB International (CABI). Any images, figures and tables not otherwise attributed are the author(s)' own. References to internet websites (URLs) were accurate at the time of writing.

CAB International and, where different, the copyright owner shall not be liable for technical or other errors or omissions contained herein. The information is supplied without obligation and on the understanding that any person who acts upon it, or otherwise changes their position in reliance thereon, does so entirely at their own risk. Information supplied is neither intended nor implied to be a substitute for professional advice. The reader/user accepts all risks and responsibility for losses, damages, costs and other consequences resulting directly or indirectly from using this information.

CABI's Terms and Conditions, including its full disclaimer, may be found at https://www.cabi.org/terms-and-conditions/.

A catalogue record for this book is available from the British Library, London, UK.

ISBN-13: 9781800627963 (hardback)
9781800627970 (ePDF)
9781800627987 (ePub)

DOI: 10.1079/9781800627987.0000

Commissioning Editor: Rebecca Stubbs
Editorial Assistant: Helen Elliott
Production Editor: Rosie Hayden

Typeset by Straive, Pondicherry, India
Printed and bound in the UK by CPI Group (UK) Ltd, Croydon, CR0 4YY

Contents

Foreword

In the rich tapestry of traditional Chinese medicine, few plants have been as extensively studied as *Dendrobium nobile*. Known for its myriad therapeutic properties and historical significance, this extraordinary orchid has captured the interest of both traditional healers and modern scientists. As we stand at the intersection of ancient wisdom and contemporary research, *Dendrobium nobile: Research and Therapeutic Applications* offers a comprehensive exploration of this plant's multifaceted applications and potential.

This book is a testament to the enduring relevance of traditional medicinal practices and their integration into modern science. Each chapter meticulously delves into various aspects of *D. nobile*, from its traditional use in Chinese medicine to cutting-edge research uncovering its bioactive compounds and pharmacological activities. Through this blend of historical context and scientific enquiry, the book provides a holistic understanding of *D. nobile*'s significance and its potential in contemporary healthcare.

Chapter 1 presents an overview of *D. nobile*, emphasizing its importance and outlining its traditional and modern applications. The subsequent chapters build upon this foundation, offering detailed examinations of its use in traditional Chinese medicine, the chemical constituents responsible for its therapeutic effects and the latest findings on its health benefits. Topics range from its role in enhancing immunity, combating oxidative stress and managing diabetes to its potential in treating cardiovascular, respiratory, gastrointestinal, renal and neurodegenerative diseases. The discussion also extends to its applications in skincare, dry eye disease and pain management, highlighting both preclinical and clinical evidence.

One of the strengths of this book is its balanced approach, addressing not only the promising therapeutic potential but also the challenges and limitations associated with *D. nobile*. These chapters underscore the necessity for rigorous research, standardization of extracts and comprehensive safety assessments to ensure effective and safe integration into modern medicine.

The final chapter, dedicated to future research directions, encapsulates the ongoing quest to unlock the full potential of *D. nobile*. Interdisciplinary collaboration, robust funding and ethical considerations are required to advance our understanding and application of this valuable medicinal plant.

As we embark on this journey through the pages of *Dendrobium nobile: Research and Therapeutic Applications*, we are reminded of the profound wisdom embedded in traditional medicinal practices and the transformative potential of scientific exploration. This book is not only a tribute to the legacy of *D. nobile* but also a beacon guiding future research and innovation. I hope that this comprehensive book will inspire and inform researchers, clinicians and herbal enthusiasts, contributing to the continued evolution and integration of traditional and modern medicine.

Professor Jianguo Liu
President, Zunyi Medical University

Preface

Dendrobium nobile: Research and Therapeutic Applications represents a collaborative endeavour to explore and document the multifaceted potential of *Dendrobium nobile*, a traditional medicinal orchid with a rich history in Chinese medicine. This comprehensive volume delves into the traditional uses, chemical composition, pharmacological activities and potential health benefits of this remarkable plant, bridging the gap between ancient wisdom and contemporary scientific research.

D. nobile, known as 'shi hu' or 'shi hu gen' in traditional Chinese medicine, has been revered for its tonic, nourishing and therapeutic properties for centuries. Modern research has begun to unveil the scientific underpinnings of these traditional uses, revealing a plethora of bioactive compounds with significant pharmacological potential. This book aims to provide a holistic view of *D. nobile*, encompassing its traditional applications, scientific discoveries and future research directions.

The book is divided into 15 chapters, each meticulously crafted to offer in-depth insights into various aspects of *D. nobile*. The first chapter sets the stage with an overview of its historical significance, modern uses and economic importance, followed by chapters dedicated to exploring its traditional medicinal applications, chemical constituents and diverse pharmacological activities.

Subsequent chapters delve into the specific health benefits of *D. nobile*, covering its roles in enhancing immunity, providing antioxidant protection, managing diabetes, supporting cardiovascular health and offering analgesic and anti-inflammatory effects. The therapeutic potential of *D. nobile* in treating respiratory diseases, gastrointestinal disorders, renal diseases and neurodegenerative conditions is also thoroughly examined. Additionally, chapters on skin health and the potential treatment of dry eye disease highlight the plant's modern applications in dermatology and ophthalmology.

The final chapter offers a forward-looking perspective, identifying key areas for future research and development. It emphasizes the need for standardized extraction methods, comprehensive safety assessments and well-designed clinical trials to fully realize the therapeutic potential of *D. nobile*.

This book is a testament to the collaborative efforts of researchers, clinicians and traditional medicine practitioners in advancing our understanding of *D. nobile* and its applications in modern healthcare. We hope this volume serves as a valuable resource for researchers, healthcare professionals and anyone interested in the intersection of traditional medicine and modern science.

We express our deepest gratitude to all the contributors in this fascinating field for their expertise and insights, and to the reader for their interest in it. It is our sincere hope that *Dendrobium nobile: Research and Therapeutic Applications* will inspire further research, foster interdisciplinary collaboration and ultimately contribute to the integration of traditional knowledge with modern scientific approaches for the betterment of human health.

EDITORS
Jingshan Shi and Archana Jain

CONTRIBUTORS
Surendra Sarsaiya, Qin Wu and Qihai Gong

Biographies

Professor Shi Jingshan
Key Laboratory of Basic Pharmacology and Joint International Research Laboratory of Ethnomedicine of Ministry of Education, Zunyi Medical University, Zunyi, Guizhou, China; Email: shijs@zmu.edu.cn

Professor Shi Jingshan, MD, PhD is former president of Zunyi Medical College and vice president of the Chinese Pharmacological Society. He is expert with special allowance from the State Council, core expert of Guizhou Province, expert of Guizhou Province Traditional Chinese Medicine Modernization Expert Group, leader of Guizhou Province Science and Technology Innovation Talent Team, review expert of National Natural Science Foundation, expert of National 863 Program Expert Database and review expert of the National Science and Technology Progress Award. He has been successively named the Outstanding Science and Technology Worker of Guizhou Province, Advanced Individual of the Tenth Anniversary of the Construction of Traditional Chinese Medicine Modernization Science and Technology Industrial Base of the Ministry of Science and Technology, and the Outstanding Science and Technology Worker of the Country. He is currently a reviewer for periodicals such as *Aging Cell, Basic & Clinical Pharmacology & Toxicology, Journal of Ethnopharmacology, Pharmacology and Therapeutics, Journal of Functional Food, Behavioural Brain Research* and *Phytomedicine*. He is an editorial board member of *Journal of Alzheimer's Disease, Chinese Journal of Preventive Medicine* and *Chinese Journal of New Drugs and Clinical Practice*. He is mainly engaged in neuropharmacology research; the effective ingredients of local Chinese medicinal materials such as *Epimedium* and *Dendrobium candidum* have been used in the treatment of degenerative diseases of the central nervous system, especially Alzheimer's disease and Parkinson's disease, and phased results have been achieved. He has presided over more than 20 provincial and ministerial projects, including the National Natural Science Foundation, the '973' preliminary special projects and the major projects of the Guizhou Provincial Department of Science and Technology. He has edited and participated in the editing of more than 10 textbooks, won five national invention patents, one second prize of the National Science and Technology Progress Award, one first prize of the Guizhou Provincial Science and Technology Progress Award, three second prizes and four third prizes. To date he has published more than 500 papers worldwide, of which more than 138 have been included in the Science Citation Index. He led the pharmacy discipline of his school to be rated as a first-class discipline in Guizhou Province and was approved for the construction of the International

Cooperation Joint Laboratory of Characteristic Ethnic Medicines of the Ministry of Education, and took the lead in establishing the Dendrobium Special Class of Zunyi Medical University. He has a high academic status in the pharmacology community both at home and abroad.

Professor Gong Qihai

Key Laboratory of Basic Pharmacology and Joint International Research Laboratory of Ethnomedicine of Ministry of Education, Zunyi Medical University, Zunyi, Guizhou, China; Email: gqh@zmu.edu.cn

Professor Gong Qihai, doctor of medicine, second-level professor, doctoral supervisor, is an expert in Guizhou Province, expert with a special allowance from the State Council and director of the Chinese Pharmacological Society. He is currently a member of the Party Committee and vice president of Zunyi Medical University and the leader of the innovation team of the Ministry of Education (including 'rolling' support). His main research direction is the neuropharmacology and toxicology of traditional Chinese medicine. He has presided over more than 20 projects or projects, such as the Key Special Project of the National Key R&D Program Modernization Research of Traditional Chinese Medicine the National Natural Science Foundation and the Ministry of Education. He has obtained five national invention patents and published more than 200 papers at home and abroad, of which more than 100 have been included in the Science Citation Index. He won the first prize for provincial scientific and technological progress and the first prize for provincial graduate teaching achievements. He was selected for the New Century Excellent Talent Support Program of the Ministry of Education, the Excellent Young Science and Technology Talent of Guizhou Province, and the 'Hundred'-level Talent of High-level Innovative People in Guizhou Province. He is an editorial board member of *Pharmacology and Clinical Practice of Traditional Chinese Medicine* and *Journal of Chinese New Drugs and Clinical Practice* and was selected as being in the world's top 2% scientists in the field of pharmacy.

Professor Wu Qin

Key Laboratory of Basic Pharmacology and Joint International Research Laboratory of Ethnomedicine of Ministry of Education, Zunyi Medical University, Zunyi, Guizhou, China; Email: wuqin@zmu.edu.cn

Professor Wu Qin, master supervisor, is currently the deputy director of the Key Laboratory of Basic Pharmacology of the Ministry of Education of Zunyi Medical University. She is mainly engaged in digestive pharmacology, drug metabolism and toxicology; she is employed as an evaluation expert of the National Natural Science Foundation of China and has presided two projects of the National Natural Science Foundation of China as well as five provincial projects, such as social development research projects of the Guizhou Provincial Science and Technology Department. She won second prize of Guizhou Provincial Science and Technology Progress, two first prizes, three third prizes, one second prize of Guiyang Science and Technology Progress Award and one first prize of Zunyi Science and Technology Progress Award. She has four national invention patents and has published more than 30 papers included in the Science Citation Index.

Dr Archana Jain
Key Laboratory of Basic Pharmacology and Joint International Research Laboratory of Ethnomedicine of Ministry of Education, Zunyi Medical University, Zunyi, Guizhou, China; Email: assjain80@gmail.com, assjain@zmu.edu.cn

Dr Archana Jain is a post-doctoral researcher affiliated with the Key Laboratory of Basic Pharmacology and Joint International Research Laboratory of Ethnomedicine of the Ministry of Education at the Zunyi Medical University. Dr Jain completed her PhD in Education at Mansarovar Global University (MGU) in Sehore, Madhya Pradesh, India, in May 2023. With her diverse educational background and research interests in education and microbiology, Dr Jain has contributed to the field of microbiology and education by exploring the dynamics of teaching and learning, particularly in relation to students who require additional support in their academic journey.

Dr Surendra Sarsaiya
Key Laboratory of Basic Pharmacology and Joint International Research Laboratory of Ethnomedicine of Ministry of Education and Bioresource Institute for Healthy Utilization (BIHU), Zunyi Medical University, Zunyi, Guizhou, China; Email: sarsaiya.s@gmail.com, surendra@zmu.edu.cn

Dr Surendra Sarsaiya is a professor and researcher in microbiology. Dr Sarsaiya is currently affiliated with the Key Laboratory of Basic Pharmacology and Joint International Research Laboratory of Ethnomedicine of the Ministry of Education and Bioresource Institute for Healthy Utilization (BIHU) at Zunyi Medical University. Dr Sarsaiya is an accomplished researcher who has published numerous articles in reputable Science Citation Index journals such as *Plant Disease*, *International Journal of Molecular Science (IJMS)*, *Bioengineered*, *Bioresource Technology*, *Journal of Applied Biology and Biotechnology*, *Environmental Science and Pollution Research* and *Renewable and Sustainable Energy Reviews*. His research has contributed significantly to the advancement of the field of biological sciences with total a JCR Impact Factor of approximately 360 with 5600 citations.

1 Overview of *Dendrobium nobile*: Research Perspectives

Archana Jain, Surendra Sarsaiya, Qihai Gong, Qin Wu and Jingshan Shi*
Key Laboratory of Basic Pharmacology and Joint International Research Laboratory of Ethnomedicine of Ministry of Education, Zunyi Medical University, Zunyi, Guizhou, China

Abstract
This chapter delves into various aspects of *Dendrobium nobile*, offering a comprehensive overview of its traditional and modern uses. Initially, it provides background information and highlights the significance of studying this plant. Traditional applications in Chinese medicine are discussed, emphasizing its therapeutic efficacy and safety. Historical uses and specific therapeutic applications, such as antipyretic and analgesic properties, are explored. Modern research, including preclinical studies, clinical trials and potential adverse effects, are reviewed. The chapter also examines the commercial uses and economic importance of *D. nobile*, followed by future research directions and challenges. Tables detail its traditional uses, bioactive compounds and potential health benefits, highlighting its roles in immune enhancement and anti-inflammatory, anti-cancer, anti-obesity, anti-diabetes and neuroprotective activities. The concluding remarks focus on the implications, recommendations and limitations, guiding further research on this valuable medicinal plant.

1.1 Introduction

Over the last eight decades, researchers have delved into over 40 species belonging to the genus *Dendrobium*. Notably, common bibenzyls, such as moscatilin and gigantol, have been identified in nearly 20 different *Dendrobium* species, featuring three to six hydroxyl and methoxy groups. While fluorenone and coumarin are also present, they are comparatively less abundant than bibenzyl and phenanthrene in *Dendrobium* species (Thanh and Yen, 2022). Alkaloids, one of the primary compounds scrutinized and structurally elucidated in *Dendrobium*, exhibit diverse skeletal frameworks, such as sesquiterpene, indolizidine, pyrrolidine, phthalide and imidazole. Sesquiterpenoids are predominantly found in *Dendrobium nobile* Lindl. *D. nobile*, historically used in traditional Chinese medicine (TCM) for its broad therapeutic applications, is a promising candidate for the discovery of novel natural remedies for various ailments.

Known as noble dendrobium, 'shi hu' or 'shi hu gen' in Chinese, *D. nobile* is an orchid species entrenched in the annals of Chinese medicinal practice for millennia. Native to regions encompassing China, Japan, the Korean peninsula and other parts of Asia, this botanical marvel is esteemed not only for its exquisite blossoms, but also for its medicinal attributes (Teoh and Teoh, 2016). In TCM, *D. nobile* is considered to

*Corresponding author: shijs@zmu.edu.cn

© Jingshan Shi, Archana Jain, Surendra Sarsaiya, Qin Wu and Qihai Gong 2025. Dendrobium nobile: *Research and Therapeutic Applications* (J. Shi, A. Jain, S. Sarsaiya, Q. Wu and Q. Gong)
DOI: 10.1079/9781800627987.0001

have cooling and nourishing properties, and is used to nourish yin, replenish fluids and strengthen the stomach and lungs. It is often used to treat conditions such as dry mouth and throat, thirst, cough and fever (Ramesh *et al.*, 2019). Modern research has revealed various activities associated with *D. nobile*. Studies have shown that it possesses anti-inflammatory, antioxidant, immunomodulatory and neuroprotective properties. Additionally, *D. nobile* contains a variety of bioactive compounds including polysaccharides, alkaloids, flavonoids and phenols (Wang *et al.*, 2022). Some of the most interesting recent studies on *D. nobile* have focused on its potential as an anti-cancer agent. *In vitro* and animal studies have shown that it has antitumour effects, and may be capable of triggering apoptosis (programmed cell death) in cancer cells. The potential applications of *D. nobile* appear promising, although further research is necessary to validate these findings in cancer treatment and prevention (Fan *et al.*, 2023). Despite its long history of traditional use and potential medicinal properties, *D. nobile* has become endangered in the wild due to overharvesting and habitat destruction. Consequently, efforts are underway to conserve and cultivate this important plant (Seyler, 2017).

1.1.1 *D. nobile*: background information

D. nobile boasts a plethora of medicinal virtues. Traditionally acknowledged for its capacity to replenish yin and stimulate bodily fluid production, this botanical gem is reputed for its anti-inflammatory, analgesic and antipyretic properties (Paudel *et al.*, 2022). Recently, the potential of *D. nobile* has garnered increasing attention, catalysing numerous investigations into its prospective therapeutic utility for various ailments. Among the observed benefits are its antioxidative properties; it has been shown to shield against oxidative stress (Hsu *et al.*, 2022) and to provide neuroprotection and enhance cognitive faculties (Li *et al.*, 2022).

Besides its medicinal utility, *D. nobile* has extensive applications in the food and cosmetics industries. Abundant in polysaccharides, it provides a spectrum of health benefits, spanning immune modulation, anti-inflammatory properties and antitumour properties (Chakraborty *et al.*, 2019). Polysaccharides derived from *D. nobile* have been used in the formulation of various functional foods and dietary adjuncts. Despite its venerable history in traditional healing practices, the attributes and underlying mechanisms of *D. nobile*'s properties remain unclear. Nonetheless, amid the burgeoning interest in natural remedies and their therapeutic promise, *D. nobile* is on a compelling frontier for the discovery of novel therapeutics and functional nutraceutical constituents (Ding *et al.*, 2020).

1.1.2 The significance of studying *D. nobile*

D. nobile has been traditionally used in Chinese medicine for its various medicinal properties, and recent research has shown promising results in terms of its potential therapeutic applications. Several recent studies have highlighted the potential of *D. nobile* as a source of novel drugs and functional food ingredients. For example, one study found that polysaccharides extracted from *D. nobile* had immune-modulating and anti-inflammatory effects (Seyler, 2017), whereas another study showed that an extract of *D. nobile* had neuroprotective effects and improved cognitive function in mice. Recent research has confirmed its potential therapeutic applications (Li *et al.*, 2022). *D. nobile* has antioxidant properties and can protect against oxidative stress, has neuroprotective effects and can improve cognitive function. A comprehensive review of the therapeutic potential of plants and their identification are potential areas for further research. Moreover, the growing interest in natural products and their potential therapeutic applications make *D. nobile* an important source of new drugs and functional food ingredients. It also highlights the importance of preserving traditional knowledge and practices in the context of modern scientific research (Tiwari *et al.*, 2024). *D. nobile* has a wide range of therapeutic applications, and further research is needed to fully explore its potential. *D. nobile*'s active ingredients are emerging as valuable agents for liver protection, showing potential to prevent liver damage, reduce inflammation and support liver function through various mechanisms (Fu *et al.*, 2023).

1.2 Traditional Use of *D. nobile* in Chinese Medicine

With its extensive heritage in TCM, *D. nobile* is noted for its reputed cooling and nourishing characteristics (Table 1.1 and Fig. 1.1). It has been employed to address an array of maladies, including fever, thirst, dry mouth, cough and lung infections. According to TCM, it is also thought to strengthen the stomach, promote the production of bodily fluids and benefit eyesight and hearing (Fan *et al.*, 2023). One of the most well-known traditional uses of *D. nobile* is for treating a condition called 'yin deficiency'. Yin deficiency is a concept in Chinese medicine that refers to a lack of cooling and nourishing fluids in the body. It is often characterized by symptoms such as dry mouth and throat, night sweats and restlessness. *D. nobile* is believed to help nourish yin and alleviate these symptoms (Hong *et al.*, 2022a).

Numerous studies have explored the traditional applications of *D. nobile* in Chinese medicine. For instance, one study demonstrated its anti-inflammatory effects in rats with lung inflammation, while another highlighted its antioxidant effects in mice (Sun *et al.*, 2021). Traditionally, *D. nobile* has been valued for its ability to nourish yin, promote fluid generation, reduce heat, improve vision, tonify the stomach and support digestion. Numerous bioactive compounds have been identified in *D. nobile*, including polysaccharides, alkaloids, flavonoids and bibenzymes (Ri *et al.*, 2023). These compounds display a range of beneficial activities, such as antioxidant, anti-inflammatory, immunomodulatory and antitumour effects, providing scientific insights into their long-standing use in TCM (Table 1.2) (Zhang *et al.*, 2023). *D. nobile* is often used in TCM to treat various respiratory conditions, such as cough, asthma and bronchitis. In TCM, *D. nobile* is also used to treat stomach disorders such as poor appetite, indigestion and nausea. Its tonifying and digestive properties help strengthen the stomach and improve digestion (Zhou *et al.*, 2022). *Uncaria rhynchophylla* and *Scutellaria baicalensis* have been used to create formulas for treating fever and convulsions. Additionally, in TCM, *D. nobile* serves as a tonic to bolster overall health and promote longevity. It is believed to nourish the yin and blood, strengthen the immune system and improve vitality.

Table 1.1. Traditional uses of *D. nobile* (authors' own table).

Disease/condition	Traditional use of *D. nobile*	Part used
Diabetes	Helps regulate blood sugar levels	Stem, extract
Fever	Antipyretic properties	Whole plant
Cough	Soothes respiratory system	Stem, aerial parts
Asthma	Assists in relieving symptoms	Stem, aerial parts
Sore throat	Relieves throat irritation	Stem, extract
Inflammation	Anti-inflammatory effects	Stem, aerial parts
Pain relief	Analgesic properties	Stem, extract
Digestive issues	Aids in digestion	Stem, aerial parts, extract
Hypertension	Helps regulate blood pressure	Stem, extract
Fatigue	Boosts energy levels	Stem, extract
Insomnia	Promotes better sleep	Stem, extract
Vision impairment	Supports eye health	Stem, aerial parts
Immunomodulation	Enhances immune function	Stem, aerial parts, extract
Cardiovascular health	Supports heart health	Stem, extract
Kidney disorders	Assists in kidney function	Stem, extract
Liver health	Supports liver function	Stem, extract
Anti-ageing effects	Helps combat signs of ageing	Stem, extract
Stress management	Calms the nervous system	Stem, extract
Skin health	Improves skin condition	Leaves, extract
Eye disorders	Treats eye-related issues	Flower, extract
Bone health	Strengthens bones and prevents osteoporosis	Root, extract

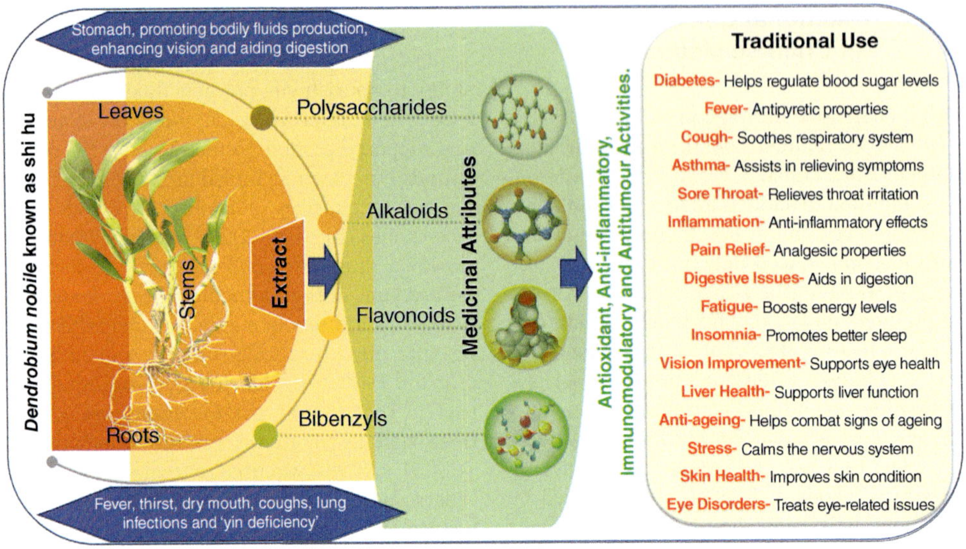

Fig. 1.1. Medicinal attributes and traditional uses of *D. nobile* (authors' own figure).

Table 1.2. Bioactive compounds in *D. nobile* (authors' own table).

Compound	Biological activity
Polysaccharides	Immunomodulatory, anti-inflammatory, anti-cancer, antioxidant, anti-diabetes, neuroprotective
Alkaloids	Anti-inflammatory, anti-cancer, anti-obesity, anti-diabetes, neuroprotective
Phenanthrenes	Anti-inflammatory, anti-cancer, anti-obesity, anti-diabetes, neuroprotective
Bibenzyls	Anti-inflammatory, anti-cancer, anti-obesity, anti-diabetes, neuroprotective
Flavonoids	Anti-inflammatory, anti-cancer, antioxidant, anti-diabetes
Phenolic acids	Anti-inflammatory, anti-cancer, antioxidant, anti-diabetes

1.3 History of Use in Traditional Chinese Medicine

D. nobile boasts a rich historical lineage in TCM, where it has been employed for multifaceted health purposes. In TCM, *D. nobile* is classified as a 'yin-nourishing' herb. The use of *D. nobile* in TCM can be traced back to the Tang Dynasty (618–907 AD), as recorded in the *Tang Bencao*, one of the earliest Chinese materia medica texts. It has been described in the *Tang Materia Medica* as a nourishing tonic for the yin, which helps to moisten the lungs, nourish the stomach and promote the production of body fluids. During the following dynasties, *D. nobile* continued to be used for its health benefits (Shurtleff and Aoyagi, 2021). During the Ming Dynasty (1368–1644 AD), it was included in the *Compendium of Materia Medica* by Li Shizhen, which is still one of the most important Chinese materia medica texts (Chiu, 2023). *D. nobile* is useful to nourish the stomach, promote the production of body fluids, and relieve thirst (Li, 2003). In modern times, *D. nobile* has continued to be an important herb in TCM. It is also used to treat stomach disorders such as poor appetite, indigestion and nausea. In recent years, scientific studies have been conducted on the bioactive compounds found in *D. nobile*, which have been found to have various effects. These studies have provided scientific evidence to support some traditional uses of *D. nobile* in TCM.

1.3.1 Preparation and administration

The preparation and administration of *D. nobile* can vary depending on the specific health conditions being treated and the form in which the herb is used. In TCM, *D. nobile* is often decocted with other herbs and taken orally as tea or soup. They can also be processed into powder, taken in capsules or mixed with water (Teoh, 2019). *D. nobile* can also be prepared as an alcoholic extract. The extract exhibits antioxidant and anti-inflammatory properties both *in vitro* and *in vivo*, indicating its potential utility as a natural antioxidant in functional foods and pharmaceuticals. Impact of various processing techniques on bioactive constituents of *D. nobile* were studied by Ye *et al.* (2023). Steaming of *D. nobile* increased the antioxidant activity of the herb, whereas boiling and frying decreased it. Thus steaming could be the preferred processing method to preserve the health benefits of *D. nobile* (Liu *et al.*, 2023a). Overall, the preparation and administration of *D. nobile* can vary depending on its intended use and form. It is important to follow the instructions of a qualified healthcare practitioner and use high-quality sources of the herb to ensure safety and efficacy.

1.3.2 Therapeutic applications and indications

D. nobile has various therapeutic applications and indications in TCM, some of which have been supported by modern research. *D. nobile* is commonly used in TCM to treat respiratory conditions such as cough, asthma and bronchitis. It has been used to nourish the yin and generate fluids, which helps moisturize the lungs and relieve dry coughs. Studies have shown that *D. nobile* extracts have anti-inflammatory effects on the respiratory system, which may explain its traditional use in treating respiratory conditions (Duan *et al.*, 2022). *D. nobile* is used in TCM to treat digestive disorders such as poor appetite, indigestion and nausea. Studies have found that *D. nobile* extracts can stimulate the secretion of digestive enzymes and improve gastrointestinal motility, which may explain their traditional use in treating digestive disorders (Zhang, 2021). *D. nobile* contains flavonoids and alkaloids, which exhibit antioxidant and anti-inflammatory properties. They may aid in diminishing oxidative stress and inflammation in the body, potentially mitigating the onset of various chronic ailments (Xu *et al.*, 2017).

D. nobile is also used in TCM to treat fever and convulsions. Its cooling and calming effects on the body make it useful in treating these conditions. *D. nobile* extracts have antipyretic and anticonvulsant effects, which may explain its traditional use in treating these conditions. *D. nobile* is also used to treat stomach disorders, such as poor appetite, indigestion and nausea. *D. nobile* has been shown to have antitumour effects in various studies. Its bioactive compounds have been found to inhibit the growth and proliferation of various cancers, including lung, liver and breast cancers (Ganapathy *et al.*, 2023). *D. nobile* has been found to have immunomodulatory effects, which may explain its use as a tonic for overall health and longevity in TCM. Research indicates that extracts from *D. nobile* have the capacity to boost cytokine production and augment the functionality of immune cells, such as natural killer cells and macrophages (Hong *et al.*, 2022b). Overall, *D. nobile* has a range of potential therapeutic applications, based on its traditional use and modern research.

1.3.3 Efficacy and safety

A systematic review and meta-analysis of clinical studies on *D. nobile* found that it may have beneficial effects on cognitive function and gastrointestinal disorders; however, more high-quality studies are needed to confirm these findings. A study on the safety and efficacy of *D. nobile* extract in treating dry eye disease found that it improved tear production and alleviated dry eye symptoms with no significant adverse effects (Huang *et al.*, 2023). A randomized controlled trial on the effects of *D. nobile* extract on postoperative cognitive dysfunction in elderly patients undergoing hip replacement surgery found that it improved cognitive function and reduced inflammation with no significant adverse effects.

A review of the safety of *D. nobile* found that it is generally safe when used appropriately, but high doses or long-term use may cause gastrointestinal discomfort, allergic reactions and liver and kidney damage in some individuals (Hong and Soh, 2019). *D. nobile* extract in rats did not cause any significant toxic effects on the liver, kidney or reproductive system at high doses. A case report of a patient who developed acute liver injury after consuming a dietary supplement containing *D. nobile* highlights the potential for adverse effects with inappropriate use or adulteration of *D. nobile* products (Selvaraj *et al.*, 2021).

1.4 Modern Research on *D. nobile*

D. nobile has been the subject of increasing scientific interest in recent years, with numerous studies investigating its potential effects and clinical applications. *D. nobile* has been used traditionally to treat chronic fatigue syndrome (Zhang *et al.*, 2021). Gogoi and Sen (2023) found that *D. nobile* polysaccharides had beneficial effects on gut microbiota and neuroinflammation in mice, suggesting a potential mechanism and a therapeutic application for chronic fatigue syndrome. The anti-cancer properties of *D. nobile* polysaccharides in lung cancer cells were investigated. Researchers observed that these polysaccharides hindered the growth of A549 cancer cells and prompted apoptosis, potentially by influencing the phosphoinositol 3-kinase (PI3K)/Akt signalling pathway (Tian *et al.*, 2023).

There has also been assessment of the antidepressant-like effects of *D. nobile* alkaloids in a rat model of depression induced by chronic unpredictable stress. Researchers have explored the antidepressant effects of *D. nobile* alkaloids, revealing their positive impact on depressive-like behaviour in rats subjected to chronic unpredictable stress, likely through modulation of the serotonergic system (Xiong *et al.*, 2021). The anti-inflammatory effects of *D. nobile* extracts on RAW264.7 cells stimulated with lipopolysaccharide and in mice with lipopolysaccharide-induced acute lung injury were investigated. These findings suggest that these extracts possess anti-inflammatory effects on RAW264.7 cells and mice, potentially by modulating signalling pathways (Li *et al.*, 2024).The potential impact of *D. nobile* alkaloids on gene expression associated with metabolism in HepG2 liver cancer cells suggested that the alkaloids positively influenced gene expression related to metabolism, potentially through modulation of the AMP-activated protein kinase (AMPK) signalling pathway. Additionally, alkaloids exhibit anti-inflammatory effects in immune cells stimulated with lipopolysaccharides, indicating potential therapeutic applications in inflammatory diseases (Xu *et al.*, 2017).

1.5 Studies on Traditional Uses of *D. nobile*

1.5.1 Preclinical studies

D. nobile polysaccharides were found to enhance cognitive function in rats with vascular dementia, while *Dendrobium* alkaloids ameliorated Aβ-induced spatial memory deficits in mice. Additionally, *D. nobile* exhibited memory-improving effects in sleep-deprived mice. These findings corroborate ancient records suggesting the traditional intelligence-boosting properties of *D. nobile*, further supported by scientific observations highlighting its advantages in neurodegenerative diseases, age-related cognitive decline and overall cognitive enhancement (Zhang *et al.*, 2024). Identification using advanced analytical techniques such as ultra performance liquid chromatography coupled to mass spectrometry–ion trap–time of flight (UPLC–IT–TOF) and liquid chromatography–mass spectrometry (LC–MS) revealed a total of 20 chemical constituents in *D. nobile*, including phenanthrenes, bibenzyls and various other compounds. Notably, phenanthrenes, which exhibit potent antitumour activity, were prominent. Extracts from *D. nobile* exhibit robust antitumour activity against A549 cells, with pathway enrichment analysis indicating significant involvement in the non-small cell lung cancer pathway (Liu *et al.*, 2023b).

The neuroprotective effects of *D. nobile* alkaloids were explored in rat primary cultured neurons exposed to oxygen-glucose deprivation/reperfusion (OGD/RP) *in vitro*, with the goal of alleviating neuronal injury in ischaemic brain vascular disorders. Treatment with *D. nobile* alkaloids significantly attenuated neuronal damage, as evidenced by enhanced cell viability, reduced apoptosis and mitigated morphological impairment. *D. nobile* alkaloids exerted protective effects against OGD/RP-induced neuronal damage by stabilizing mitochondrial membrane potential. Moreover, *D. nobile* alkaloids have been explored for their potential in mitigating neuroinflammation, a mechanism implicated in Alzheimer's disease (Cakova *et al.*, 2017). *D. nobile* has several beneficial properties including anti-ageing, antioxidant, antitumour and immunoregulatory activities. The abundant polysaccharides, flavones, alkaloids and polyphenols found in *D. nobile* exhibit strong biological activity and have significant medicinal value. Additionally, polysaccharides have been discovered to possess the ability to form polysaccharide–zinc chelates, transforming inorganic zinc into an organic form that reduces damage induced by zinc stress (Fan *et al.*, 2021).

1.5.2 Clinical trials

Recent research aimed to explore the impact of *Dendrobium* polysaccharides on the gastrointestinal system and gut microbiota in individuals with notable changes in consciousness (Zhang *et al.*, 2022). The potential of *Sarcodon aspratus* polysaccharides (SAFPs) to mitigate gastric ulcers caused by water immersion and restraint stress (WIRS) was explored. These findings suggest that SAFPs effectively countered stress-induced damage, inflammation, pathological changes and lesions in the gastric mucosa of rats. They displayed protective effects against WIRS-induced gastric ulcers by obstructing the TLR4 signalling pathway and activating the Nrf2 signalling pathway. Additionally, SAFPs exhibited the ability to regulate WIRS-induced dysbiosis in the gut microbiota provoked by WIRS. Consequently, SAFPs have emerged as a promising natural agent for fortifying the gastric mucosa and averting gastric ulcers and related ailments in both food and pharmaceutical domains (Zhang *et al.*, 2022).

In a separate investigation, scientists delved into the variations in secondary metabolites in *D. nobile* at different growth phases (Luo *et al.*, 2024). Alkaloids and sesquiterpene glycosides in *D. nobile* exhibit growth stage-dependent patterns, with 1-year-old specimens showing relatively higher levels of alkaloids and 3-year-old specimens displaying elevated levels of sesquiterpene glycosides. These variations significantly shaped the overall secondary metabolite profile in *D. nobile*, implying potential differences in pharmacological activity between specimens in different stages of growth. Moreover, researchers plan to employ an alloxan-induced diabetic mouse model to evaluate the hypoglycaemic effects of *D. nobile* at different growth stages. Advanced analytical techniques, such as ultra performance liquid chromatography (UPLC)–Q Exactive/Orbitrap/MS, will be utilized for subsequent analysis of the overall secondary metabolite profile in *D. nobile*, with different components screened and identified. Their effects on key diabetes-related targets have been predicted using network pharmacology to identify potential quality markers (Luo *et al.*, 2024).

In the context of lung cancer, Su *et al.* (2020) reviewed studies focusing on the clinical application of TCM as a supplementary treatment. Dendrobine, a colourless solid alkaloid derived from *D. nobile*, has the potential to trigger apoptosis in cancer cells via a mitochondrially mediated pathway. The combination of cisplatin and dendrobine is promising for lung cancer therapy (Su *et al.*, 2020). Targeting osteoclast differentiation has emerged as a therapeutic strategy for osteolytic conditions characterized by bone loss, fractures and pain. Dendrobine has analgesic and anti-inflammatory properties and inhibits osteoclastogenesis by impeding the reactive oxygen species, p38-c-Fos and NFATc1-MMP9 pathways *in vitro*. This finding suggests the potential of dendrobine as an osteoclast inhibitor for mitigating bone erosion (Deng *et al.*, 2022). *D. nobile* is effective in alleviating menstrual pain and enhancing quality of life compared to placebo (Fan *et al.*, 2020). Abundant in biologically active polysaccharides, *Dendrobium* plants show significant anti-ageing, antioxidant, hypoglycaemic, anti-cancer and immunomodulatory activities. However, despite the notable immune activity displayed by polysaccharides extracted from *D. nobile*, their efficacy is hindered by weak binding affinity with receptors and limited capacity to attach to target organs (Fan *et al.*, 2020). Ongoing endeavours will focus on exploring safe and efficient methods to enhance the hydrophilicity and biological activity of *D. nobile* polysaccharides.

1.5.3 Adverse effects and safety issues

D. nobile is a species of orchid commonly used in TCM as a dietary supplement. Although it has been used for centuries in various medicinal preparations, there are potential adverse effects and safety issues associated with its consumption, particularly when used in high doses or for prolonged periods. Like many plants, *D. nobile* may cause allergic reactions in sensitive individuals. Symptoms can range from mild skin irritation to severe allergic responses such as difficulty breathing or anaphylaxis. In some cases, consumption of *D. nobile* may lead to digestive issues such as nausea, vomiting, diarrhoea or stomach discomfort. These symptoms are often mild and transient but can be problematic for some individuals (Chen *et al.*, 2020). *D. nobile* contains diverse compounds that can potentially interact with specific medications. It is crucial for patients to seek guidance from a healthcare provider before incorporating *Dendrobium* supplements into one's regimen, particularly if currently taking prescription medications (Teoh, 2019). While *D. nobile* has a long history of traditional use, excessive consumption or use of high-potency extracts may lead to toxicity. This can manifest as symptoms such as dizziness, headache, elevated heart rate or liver damage in severe cases (Chung and Kim, 2022). The quality and purity of *D. nobile* supplements vary significantly among products and brands. Contamination with pesticides, heavy metals or other impurities is possible, particularly in products sourced from unreliable or unregulated suppliers (Cheng *et al.*, 2019).

Despite its traditional use, there is limited scientific research on the efficacy and safety of *D. nobile*. Many of its purported health benefits lack robust clinical evidence, making it challenging to accurately assess its overall safety profile (Lu, 2022). Due to the limited safety data available, pregnant and breastfeeding women are advised to avoid using *D. nobile* supplements to prevent any potential risks to themselves and their babies. As with any supplement or botanical product, individual responses to *D. nobile* vary widely. What may be safe for one person may cause adverse effects for another person. It is essential to start with low doses and monitor for adverse reactions (Chen *et al.*, 2023). Overall, whereas *D. nobile* may offer some potential health benefits, it is imperative for patients to exercise caution when considering its usage and to seek advice from a healthcare professional before using it, particularly in those with underlying health conditions or who are taking medications. Additionally, choosing high-quality supplements from reputable sources can help minimize the risk of adverse effects and ensure product safety.

1.6 Potential Applications and Future Directions

Dendrobium is known for its potential therapeutic benefits in various medical conditions such as cancer, diabetes and neurodegenerative diseases. Studies have suggested that *Dendrobium* alkaloids and polysaccharides have anti-cancer, anti-diabetes and neuroprotective properties. *Dendrobium* has been used as a sports nutritional supplement because of its potential ergogenic effects. Studies have suggested that *Dendrobium* extracts may enhance physical performance and improve recovery after exercise. *Dendrobium* has been used as an ingredient in various food and beverage products such as teas, juices and candies, owing to its potential health benefits and pleasant taste. *Dendrobium* is widely cultivated as an ornamental plant because of its beautiful flowers and long blooming seasons. Further research is ongoing to improve the cultivation and breeding of *Dendrobium* for commercial purposes.

Dendrobium is a rich source of bioactive compounds such as alkaloids and polysaccharides, which have potential applications in biotechnology. Studies have suggested that *Dendrobium* extracts may have antimicrobial, antioxidant and immunomodulatory properties. *D. nobile* has been studied for its potential applications in medicine and health. *D. nobile* has been identified to possess antioxidative properties, potentially serving as a defence against oxidative stress and associated ailments such as cancer, cardiovascular disease and neurodegenerative disorders. Furthermore, research has indicated that *D. nobile* exhibits anti-inflammatory effects, offering promise in alleviating conditions such as arthritis, inflammatory bowel disease and asthma. Studies have also suggested that it has potential antitumour activity, which is attributed to its ability to induce cell cycle arrest and apoptosis in cancer cells. Moreover, *D. nobile* has shown neuroprotective effects, suggesting implications for the prevention and management of neurodegenerative conditions, such as Alzheimer's and Parkinson's diseases. TCM has long employed *D. nobile* for diabetes management, with research indicating its ability to regulate blood glucose levels and enhance insulin sensitivity. Additionally, its potential for wound healing has been recognized, offering prospects for the treatment of various types of wounds (Table 1.3).

1.7 Commercial Uses and Economic Importance

D. nobile has been used in TCM for centuries, and its demand has increased over the years due to its perceived health benefits. It is used to treat various ailments such as fever, cough and diabetes. *D. nobile* extract is used as an ingredient in functional foods and beverages because of its potential health benefits. It is also used as a natural sweetener and flavour enhancer. *D. nobile* extract is used in skincare products owing to its anti-ageing and skin-lightening effects. *D. nobile* is a popular ornamental plant and its cultivation is a source of income for many farmers and nurseries. *D. nobile* has a complex genome, and genetic research has the potential to uncover the mechanisms behind its properties and aid in the development of new *D. nobile*-based products. The economic importance of *D. nobile* has led to overharvesting and illegal harvesting, resulting in the depletion of wild populations and endangerment of the species. Sustainable cultivation methods and conservation efforts are required to protect the species and ensure its sustainable use (Fig. 1.2).

Table 1.3. Potential health benefits of *D. nobile* extract (author's own table).

Health benefit	Potential mechanism
Immune system enhancement	Activation of immune cells, increase in cytokine production
Anti-inflammatory	Inhibition of pro-inflammatory cytokines
Anti-cancer	Induction of apoptosis, inhibition of cell proliferation, induction of cell cycle arrest
Anti-obesity	Inhibition of adipogenesis, increase in lipolysis
Anti-diabetes	Activation of insulin signalling pathways, regulation of glucose metabolism
Neuroprotection	Inhibition of oxidative stress, regulation of neurotransmitter levels

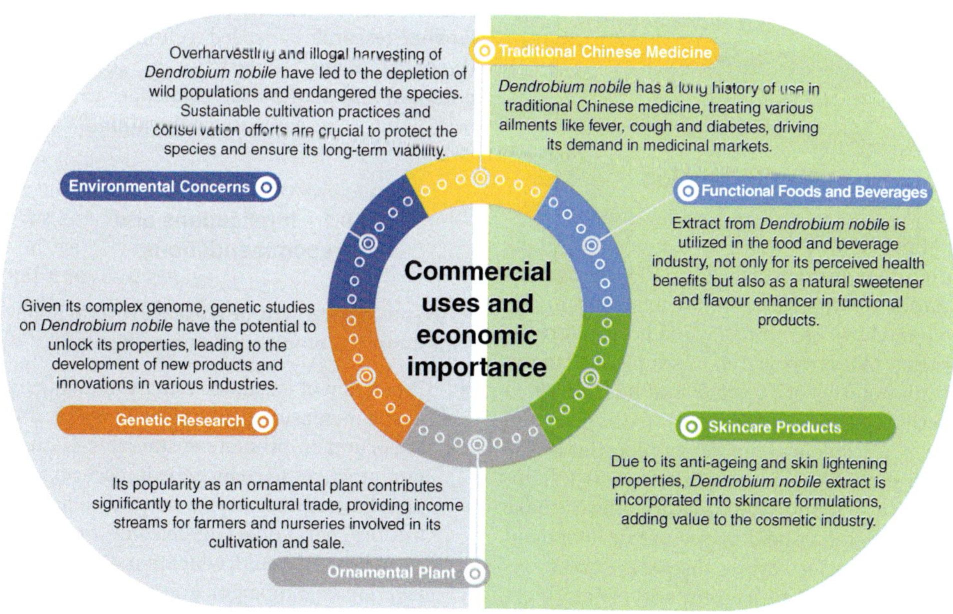

Fig. 1.2. Commercial uses and economic importance of *D. nobile* (authors' own figure).

1.8 Future Research Directions and Challenges

Despite the identification of several bioactive compounds in *D. nobile*, there is still a need to identify and isolate additional compounds to fully understand their therapeutic potential. Further mechanistic studies are required to understand the mechanisms underlying the effects of *D. nobile*. This can help in the development of new drugs based on *D. nobile* or its bioactive compounds. Although *D. nobile* has shown promising health benefits, there is a need for more extensive toxicity studies to ensure its safety for human consumption. As the demand for *D. nobile* increases, there is a need for sustainable cultivation methods to reduce its impact on wild populations and ensure its availability for future generations. Some of the challenges that may arise in future research and cultivation of *D. nobile* indicate that the availability of *D. nobile* may be limited due to the depletion of wild populations and challenges in cultivating it. Legal and ethical issues may exist surrounding the use of *D. nobile*, particularly in cases of overharvesting or illegal harvesting. Standardization and quality control of *D. nobile* products can be challenging owing to variations in cultivation practices and processing methods. Lack of funding may limit the scope and depth of research on *D. nobile*, particularly in areas such as toxicity and mechanistic studies.

1.9 Conclusion

In conclusion, *D. nobile* is a medicinal plant with a long history of use in TCM. Its use has expanded beyond traditional medicine to the food and beverage, cosmetic and horticulture industries. Modern research has supported many traditional uses of *D. nobile* and has identified several bioactive compounds with potential therapeutic applications. The properties of the plant, including its antioxidant, anti-inflammatory and immunomodulatory effects, make it a promising candidate for the development of new drugs for various diseases. Further research is needed to identify and isolate more bioactive compounds and understand the underlying mechanisms of the properties of *D. nobile*. Toxicity studies are necessary to ensure the safety of human consumption. Despite the challenges in the research and cultivation of *D. nobile*, its potential health benefits and economic importance make it a valuable research area. Continued research and sustainable cultivation practices can ensure the availability and therapeutic potential of this plant in future generations.

Recent scientific investigations have unveiled a multitude of potential health advantages attributed to its bioactive constituents, including anti-inflammatory, anti-cancer and anti-ageing properties. *D. nobile* extract is used in various industries, including TCM, food and beverages, cosmetics, horticulture and biotechnology.

Future research directions include identification of more bioactive compounds, mechanistic studies, toxicity studies and sustainable cultivation methods. Challenges in future research and cultivation of *D. nobile* include limited availability, legal and ethical issues, standardization and quality control, and a lack of funding. Overall, the traditional use and modern research of *D. nobile* have shown promising results in its potential health benefits; however, sustainable cultivation methods and conservation efforts are necessary to ensure its availability for future generations. Further research is needed to fully understand the mechanisms underlying its therapeutic potential and address the challenges that may arise in future research and cultivation.

1.9.1 Implications and recommendations

Based on the overview of *D. nobile*, several implications and recommendations can be made.

- Promotion of sustainable cultivation: owing to the increasing demand for *D. nobile*, it is important to promote sustainable cultivation practices to reduce the impact on wild populations and ensure its availability for future generations. This objective can be accomplished through educational initiatives, awareness campaigns and enforcement of cultivation regulations and guidelines.

- Encouragement of further research: despite promising findings on the potential health benefits of *D. nobile*, there is still a need for further research to fully understand its therapeutic potential. Mechanistic studies, toxicity studies and identification of bioactive compounds are areas that require further research. Encouraging and supporting such research can lead to the development of new drugs based on *D. nobile* or its bioactive compounds.
- Standardization and quality control: the standardization and quality control of *D. nobile* products can be challenging owing to variations in cultivation practices and processing methods. The development of standardized protocols for cultivation and processing, as well as the implementation of quality control measures, can ensure the consistency and quality of *D. nobile* products.
- Preservation of wild populations: overharvesting and illegal harvesting have led to the depletion of wild *D. nobile* populations, endangering the species. Efforts to preserve wild populations through conservation measures and regulations can help to protect it and ensure its sustainable use.

In general, *D. nobile* exhibits promising potential as a reservoir for therapeutic agents; however, additional research is warranted to comprehensively comprehend its mechanisms of action and ascertain its safety and efficacy.

1.9.2 Limitations and suggestions for further research

Some limitations of the current research on *D. nobile* include the lack of standardization in the cultivation, processing and preparation of *D. nobile* products, which makes it difficult to compare results across studies. Although preclinical studies have shown promising results, more rigorous clinical trials are needed to confirm the therapeutic effects of *D. nobile* in humans. Although *D. nobile* is generally considered safe for consumption, there is a need for more extensive toxicity studies to ensure its safety and potential side effects. Although several bioactive compounds have been identified in *D. nobile*, there is still much to learn about their mechanisms of action and potential therapeutic applications.

To address these limitations, future research on *D. nobile* should focus on standardizing cultivation, processing and preparation methods and performing more extensive toxicity studies. Additionally, further research should aim to identify and isolate more bioactive compounds and investigate their mechanisms of action in order to develop new drugs based on *D. nobile* or its derivatives. *D. nobile* has the potential to be used in a variety of products, including functional foods, cosmetics and pharmaceuticals. Further research can aid in the development of new *D. nobile*-based products. As the demand for *D. nobile* increases, there is a need for sustainable cultivation methods to reduce its impact on wild populations and ensure its availability for future generations. Research on sustainable cultivation practices can help achieve this goal. Standardization and quality control of *D. nobile* products can be challenging owing to variations in cultivation practices and processing methods. Further research can help develop standardized cultivation and processing methods to ensure the consistency and quality of *D. nobile* products.

Acknowledgements

The authors express their gratitude for the financial support received through the Distinguished High-Level Talents Research Grant from the Guizhou Science and Technology Corporation Platform Talents Fund (grant nos. [2017]5733-001 and CK-1130-002), the National Natural Science Foundation of China (U1812403, 82373981 and 82060750), 2011 Collaborative Innovation Centre of Traditional Chinese Medicine in Guizhou Province (no. [2022]026) and the support provided by the Zunyi Medical University, China. Special appreciation is extended to all laboratory colleagues and research staff members for their valuable insights, constructive guidance and assistance with the contents of this chapter.

References

Cakova, V., Bonte, F. and Lobstein, A. (2017) *Dendrobium*: Sources of active ingredients to treat age-related pathologies. *Aging and Disease* 8(6), 827–849.

Chakraborty, I., Sen, I.K., Mondal, S., Rout, D., Bhanja, S.K. *et al.* (2019) Bioactive polysaccharides from natural sources: A review on the antitumor and immunomodulating activities. *Biocatalysis and Agricultural Biotechnology* 22, 101425.

Chen, H., Wang, C., Bai, J., Song, J., Bu, L. *et al.* (2023) Targeting microbiota to alleviate the harm caused by sleep deprivation. *Microbiological Research* 275, 127467.

Chen, Z., Ye, S.Y. and Zhu, R.G. (2020) The extraordinary transformation of traditional Chinese medicine: Processing with liquid excipients. *Pharmaceutical Biology* 58(1), 561–573.

Cheng, J., Dang, P.-P., Zhao, Z., Yuan, L.-C., Zhou, Z.-H. *et al.* (2019) An assessment of the Chinese medicinal *Dendrobium* industry: Supply, demand and sustainability. *Journal of Ethnopharmacology* 229, 81–88.

Chiu, P. (2023) From Hai Yao, Yang Yao to Xi Yao: Sinification of material medical from the West. *Chinese Medicine and Culture* 6(4), 319–329.

Chung, G. and Kim, S.K. (2022) Therapeutics for chemotherapy-induced peripheral neuropathy: approaches with natural compounds from traditional eastern medicine. *Pharmaceutics* 14(7), 1407.

Deng, W., Ding, Z., Wang, Y., Zou, B., Zheng, J. *et al.* (2022) Dendrobine attenuates osteoclast differentiation through modulating ROS/NFATc1/ MMP9 pathway and prevents inflammatory bone destruction. *Phytomedicine* 96, 153838.

Ding, Z., Zhong, R., Xia, T., Yang, Y., Xing, N. *et al.* (2020) Advances in research into the mechanisms of Chinese Materia Medica against acute lung injury. *Biomedicine & Pharmacotherapy* 122, 1–18.

Duan, F.P., Li, Y.S., Hu, T.Y., Pan, X.Q. and Ma, F. (2022) *Dendrobium nobile* protects against ovalbumin-induced allergic rhinitis by regulating intestinal flora and suppressing lung inflammation. *Chinese Journal of Natural Medicines* 20(6), 443–457.

Fan, C., Sun, X., Wang, X. and Yu, H. (2023) Therapeutic potential of the chemical composition of *Dendrobium nobile* Lindl. *Frontiers in Pharmacology* 14, 1–15.

Fan, Y., Jiang, T., Chun, Z., Wang, G., Yang, K. *et al.* (2021) Zinc affects the physiology and medicinal components of *Dendrobium nobile* Lindl. *Plant Physiology and Biochemistry* 162, 656–666.

Fan, Y., Yu, Q., Wang, G., Tan, J., Liu, S. *et al.* (2020) Effects of non-thermal plasma treatment on the polysaccharide from *Dendrobium nobile* Lindl. And its immune activities *in vitro*. *International Journal of Biological Macromolecules* 153, 942–950.

Fu, X., Chen, S., Xian, S., Wu, Q., Shi. J. *et al.* (2023) *Dendrobium* and its active ingredients: Emerging role in liver protection. *Biomedicine & Pharmacotherapy* 157, 1–13.

Ganapathy, A.A., Haripriya, V.M., Acharya, N., Somappa, S.B. and Kumaran, A. (2023) Ethnobotanical significance of medicinal plants: Beta-amyloid and tau aggregation inhibitors against Alzheimer's disease. *Journal of Biochemical and Molecular Toxicology* 37(6), 23339.

Gogoi, P. and Sen, S. (2023) Orchids of Dibru-Saikhowa: A systematic review on their traditional use, pharmacological activity and phytochemistry. *Current Topics in Medicinal Chemistry* 23(24), 2277–2299.

Hong, H. and Soh, S.B. (2019) *Healing Orchids*. World Scientific Publishing Co. Pte. Ltd, Singapore.

Hong, H., Wee, K. and Soh, S.B. (2022a) *Chinese Medicine for Health: Holistic Healing, Inner Harmony and Herbal Recipes*. World Scientific Publishing Co Pte Ltd, Singapore.

Hong, S., Kim, E.Y., Lim, S.E., Kim, J.H., Sohn, Y. *et al.* (2022b) *Dendrobium nobile* Lindley administration attenuates atopic dermatitis-like lesions by modulating immune cells. *International Journal of Molecular Sciences* 23(8), 1–22.

Hsu, W.H., Chung, C.P., Wang, Y.Y., Kuo, Y.H., Yeh, C.H. *et al.* (2022) *Dendrobium nobile* protects retinal cells from UV-induced oxidative stress damage via nrf2/HO-1 and MAPK pathways. *Journal of Ethnopharmacology* 288, 114886.

Huang, Q., Zhan, M. and Hu, Z. (2023) Auricular acupressure for dry eye disease: A systematic review and meta-analysis of randomized controlled trials. *Medicina* 59(1), 1–18.

Li, S. (2003) *Compendium of Materia Medica* (Bencao Gangmu). Foreign Languages Press, China. Vol. 6, pp. 1–4397.

Li, D.D., Zheng, C.Q., Zhang, F. and Shi, J.S. (2022) Potential neuroprotection by *Dendrobium nobile* Lindl alkaloid in Alzheimer's disease models. *Neural Regeneration Research* 17(5), 972–977.

Li, L., Chen, H., Huang, G., Lv, Y., Yao, L., Guo, Z., Qiu, S., Wang, X., Wei, C. (2024) Structure of Polysaccharide from *Dendrobium nobile* Lindl. and Its Mode of Action on TLR4 to Exert Immunomodulatory Effects. *Foods* 13(9),1356.

Liu, C., Zuo, Z., Xu, F. and Wang, Y. (2023a) Authentication of herbal medicines based on modern analytical technology combined with chemometrics approach: A review. *Critical Reviews in Analytical Chemistry* 53(7), 1393–1418.

Liu, J., Li, Z., Zeng, M., Liang, Y., Liang, B. *et al.* (2023b) Rapid comparison of antitumor chemical constituents and mechanisms between *Dendrobium nobile* and *Dendrobium officinale* by UPLC-IT-TOF, network pharmacology and experimental verification. *Acta Chromatographica* 35(2), 123–138.

Lu, Z.F. (2022) *Authentication of by Multiplex Polymerase Chain Reaction. Authentication Of Chinese Medicinal Materials by DNA Technology: Techniques and Applications*, 2nd edn. World Scientific Publishing Co Pte Ltd, Singapore.

Luo, Y., Yang, D., Xu, Y., Wu, D., Tan, D. *et al.* (2024) Hypoglycemic effects and quality marker screening of *Dendrobium nobile* Lindl. at different growth years. *Molecules* 29(3), 1–23.

Paudel, M.R., Bhattarai, H.D. and Pant, B. (2022) Traditionally used medicinal *Dendrobium*: A promising source of active anticancer constituents. In: Mérillon, JM. and Kodja, H. (eds) *Orchids Phytochemistry, Biology and Horticulture*. Reference Series in Phytochemistry, Springer, Cham, pp. 389–414.

Ramesh, T., Koperuncholan, M., Praveena, R., Ganeshkumari, K., Vanithamani, J. *et al.* (2019) Medicinal properties of some *Dendrobium* orchids – A review. *Journal of Applied and Advanced Research* 4(4), 119–128.

Ri, M.H., Xing, Y., Zuo, H.X., Li, M.Y., Jin, H.L. *et al.* (2023) Regulatory mechanisms of natural compounds from traditional Chinese herbal medicines on the microglial response in ischemic stroke. *Phytomedicine* 116, 154889.

Selvaraj, C., Yogeswari, C. and Singh, S.K. (2021) Molecular features of potential herbal products in enhancing human immune system. In: Selvaraj, C., Yogeswari, C. and Singh, S.K. (eds) *Traditional Herbal Therapy for the Human Immune System*. CRC Press, Boca Raton, Florida, p. 301.

Seyler, B.C. (2017) The role of botanical gardens in the conservation of orchid biocultural diversity in Sichuan Province, China. PhD Thesis, University of Hawai'i at Manoa, USA.

Shurtleff, W. and Aoyagi, A. (2021) *History of Soy Nutritional Research (200 BCE to 1945)*. Soyinfo Center, USA 1945.

Su, X.L., Wang, J.W., Che, H., Wang, C.F., Jiang, H. *et al.* (2020) Clinical application and mechanism of traditional Chinese medicine in treatment of lung cancer. *Chinese Medical Journal* 133(24), 2987–2997.

Sun, Y., Geng, J. and Wang, D. (2021) Cardioprotective effects of Ginsenoside compound-Mc1 and *Dendrobium nobile* Lindl against myocardial infarction in an aged rat model: Involvement of TLR4/NF-κB signaling pathway. *European Journal of Inflammation* 19, 1–9.

Teoh, E.S. (2019) *Orchids as Aphrodisiac, Medicine or Food*. Springer, pp. 1–85.

Teoh, E.S. and Teoh, E.S. (2016) Sources of medicinal orchids and conservation. In: *Medicinal Orchids of Asia*. Springer, Cham, pp. 691–727.

Thanh, N.T.V. and Yen, P.H. (2022) Phenolic compounds from *Dendrobium nobile* Lindl. *JST: Engineering and Technology for Sustainable Development* 32(3), 034–041.

Tian, W., Huang, J., Zhang, W., Wang, Y., Jin, R. *et al.* (2023) Harnessing natural product polysaccharides against lung cancer and revisit its novel mechanism. *Pharmacological Research* 199, 1–11.

Tiwari, P., Sharma, A., Bose, S.K. and Park, K.I. (2024) Advances in orchid biology: Biotechnological achievements, translational success, and commercial outcomes. *Horticulturae* 10(2), 1–28.

Wang, G., Wang, J., Deng, Y., Qin, L., He, Y. *et al.* (2022) Chemical constituents and nutritional health functions of *Dendrobium nobile*: A review. *Food Science and Technology* 42, 1–19.

Xiong, T.-W., Liu, B., Wu, Q., Xu, Y.-Y., Liu, P. *et al.* (2021) Beneficial effects of *Dendrobium nobile* Lindl. Alkaloids (DNLA) on anxiety and depression induced by chronic unpredictable stress in rats. *Brain Research* 1771, 147647.

Xu, Y.Y., Xu, Y.S., Wang, Y., Wu, Q., Lu, Y.F. *et al.* (2017) *Dendrobium nobile* Lindl. Alkaloids regulate metabolism gene expression in livers of mice. *Journal of Pharmacy and Pharmacology* 69(10), 1409–1417.

Ye, G., Zhang, J., Xu, X., Zeng, C., Ye, Q. *et al.* (2023) Comparative analysis of water-soluble polysaccharides from *Dendrobium* second love "Tokimeki" and *Dendrobium nobile* in structure, antioxidant, and anti-tumor activity *in vitro*. *International Journal of Molecular Sciences* 24(12), 1–17.

Zhang, D., Xiang, M., Jiang, Y., Wu, F., Chen, H. *et al.* (2022) The protective effect of polysaccharide SAFP from *Sarcodon aspratus* on water immersion and restraint stress-induced gastric ulcer and modulatory effects on gut microbiota dysbiosis. *Foods* 11(11), 1–25.

Zhang, Q., Fu, X., Liu, H., Chen, Y., Chen, S. *et al.* (2023) A systematical review on ethnobotanical, phytochemical and pharmacological aspects of *Dendrobium nobile* Lindl. *Phytochemistry Reviews* 22, 743–780.

Zhang, Q., Li, Y., Fan, B., Wang, F., Li, Z. *et al.* (2024) *Dendrobium nobile* Lindl ameliorates learning and memory deficits in scopolamine-treated mice. *Journal of Ethnopharmacology* 324, 117416.

Zhang, X., Wang, M., Zhang, C., Liu, Z. and Zhou, S. (2021) Clinical study of *Dendrobium nobile* Lindl intervention on patients with metabolic syndrome. *Medicine* 100(12), 1–6.

Zhang, Y. (2021) Diet according to traditional Chinese medicine for health and longevity. *Nutrition, Food and Diet in Ageing and Longevity* 14, 331–356.

Zhou, W., Zhang, H., Wang, X., Kang, J., Guo, W. *et al.* (2022) Network pharmacology to unveil the mechanism of Moluodan in the treatment of chronic atrophic gastritis. *Phytomedicine* 95, 153837.

2 Traditional Uses of *Dendrobium nobile* in Chinese Medicine

Archana Jain, Surendra Sarsaiya, Qihai Gong, Qin Wu and Jingshan Shi*
Key Laboratory of Basic Pharmacology and Joint International Research Laboratory of Ethnomedicine of Ministry of Education, Zunyi Medical University, Zunyi, Guizhou, China

Abstract

This chapter explains the traditional uses of *Dendrobium nobile* in Chinese medicine, detailing its historical, cultural and medicinal significance. Known as 'shi hu' or 'shi hu gen', this herb is used because of its tonic, nourishing and replenishing properties. It is used to treat various health conditions such as respiratory and digestive disorders, providing immune support and helping with physical endurance and skin health. *D. nobile* is associated with the Stomach, Kidney and Lung meridians, and its pharmacological potential is attributed to its diverse chemical constituents. This chapter outlines the traditional methods of processing and formulation, and the safety and toxicity considerations, and highlights the current research. Although its therapeutic properties are supported by preclinical and clinical studies, further research is needed to understand its mechanisms of action, establish quality control measures and integrate traditional knowledge with modern research. This chapter provides a comprehensive overview emphasizing the need for continued research and development in the field of *D. nobile*.

2.1 Introduction

Dendrobium nobile Lindl. has been used in traditional Chinese medicine (TCM) for centuries. *D. nobile* has a long history of use in Chinese medicine and is highly esteemed for its medicinal properties. It has been incorporated into various Chinese herbal formulations and is known to have diverse therapeutic effects. *D. nobile* has been traditionally used in Chinese medicine for a wide range of health conditions. It is believed to possess tonic, nourishing and replenishing properties, and is often associated with the nourishment of yin and the promotion of bodily fluids (Lee *et al.*, 2018). In the Chinese medical theory, *D. nobile* is considered to have a sweet and slightly cold nature and is associated with the Stomach, Kidney and Lung meridians. It is commonly used in Chinese herbal formulations to tonify yin and replenish the body's vital essence. The medicinal history of *Dendrobium* can be traced back to the Qin and Han dynasties. *Shen Nong Ben Cao Jing*, also known as *Shennong's Classic of Materia Medica*, is a foundational Chinese text on medicinal plants and agriculture, and is regarded as a cornerstone of TCM that *Dendrobium* is recognized as a top-quality herbal medicine that strengthens yin and tonifies the five viscera. Among many precious Chinese herbal medicines, it is ranked as "the first of the nine immortal herbal medicines" (Li *et al.*, 2023b).

*Corresponding author: shijs@zmu.edu.cn

© Jingshan Shi, Archana Jain, Surendra Sarsaiya, Qin Wu and Qihai Gong 2025. Dendrobium nobile: *Research and Therapeutic Applications* (J. Shi, A. Jain, S. Sarsaiya, Q. Wu and Q. Gong)
DOI: 10.1079/9781800627987.0002

D. nobile has traditionally been used in Chinese medicine to support respiratory health. It is believed to have expectorant properties and can help clear heat and phlegm from the respiratory system (Zhou *et al.*, 2020). It is commonly used in formulas for cough, bronchitis and other respiratory conditions. *D. nobile* has also been traditionally used in Chinese medicine to promote healthy digestive function. It is believed to have digestion-stimulating properties and can help tonify the Stomach and Spleen meridians (Nie *et al.*, 2020). It is commonly used in formulas to treat indigestion, poor appetite and other digestive disorders. *D. nobile* is also believed to have immunomodulatory properties and can help support the immune system. It is commonly used in Chinese herbal formulations to enhance defence mechanisms and boost immunity (Fan *et al.*, 2020). It is also believed to have anti-inflammatory properties, and can help reduce inflammation in the body. *D. nobile* has been traditionally used in Chinese medicine because of its potential benefits in enhancing physical endurance. It is believed to have tonifying properties and can help nourish the body and improve stamina (Gao *et al.*, 2022). It is commonly used by athletes and individuals engaged in strenuous physical activity to support endurance and performance. *D. nobile* is also believed to have analgesic and antioxidant properties. It is believed to relieve pain and protect against oxidative stress (Natta *et al.*, 2022). These properties make it useful in the treatment of inflammatory conditions and pain management, and as an antioxidant to protect against cellular damage. *D. nobile* has been traditionally used in Chinese medicine for its potential benefits in promoting healthy ageing and improving skin health. It is believed to have nourishing properties and can help replenish the body's vital essence, the decline of which is associated with ageing (Long *et al.*, 2023). It is also believed to help improve skin complexion, promote skin elasticity and reduce the signs of ageing. *D. nobile* is believed to have potential benefits in enhancing cognitive function. It is commonly used in Chinese herbal formulations owing to its potential to support memory, concentration and overall cognitive health. It is believed to have neuroprotective properties, and may help improve brain function and prevent age-related cognitive decline (Li *et al.*, 2022).

2.1.1 Brief overview of *D. nobile*, its botanical characteristics and its historical use in Chinese medicine

D. nobile is an orchid species traditionally used in Chinese medicine for its therapeutic properties. It is known by various names: 'shi hu' or 'shi hu gen' in Chinese, which translates to 'stone fungus' (Wang, 2021b), and noble dendrobium or noble orchid in English (Teoh, 2016a). Botanically, *D. nobile* belongs to the family Orchidaceae and is a sympodial epiphytic or lithophytic orchid species. It is native to various regions of Asia, including China, India, Nepal, Bhutan, Myanmar, Japan, the Korean peninsula, Hong Kong, Taiwan and other South-east Asian regions, and is widely cultivated for medicinal and ornamental purposes (Teoh, 2019). Historically, *D. nobile* has been used in Chinese medicine for its health benefits. It has been documented in ancient Chinese texts, such as the *Shen Nong Ben Cao Jing*, as a medicinal plant with properties such as tonifying the yin, nourishing the stomach and promoting the production of body fluids (Peng *et al.*, 2022). *D. nobile* has also been used in TCM to tonify the kidneys, improve vision, promote digestion and enhance overall vitality.

2.1.2 Importance and relevance of *D. nobile* in TCM

D. nobile holds significant importance and relevance in TCM because of its various medicinal properties and therapeutic applications. The use of *D. nobile* in TCM is based on its traditional applications, as well as its reported pharmacological activities supported by scientific research. *D. nobile* is highly regarded in TCM for its nourishing and tonifying properties, particularly in supporting the health of the kidneys, lungs and stomach. They are believed to have properties such as nourishing yin, clearing heat, replenishing essence and promoting physical strength (Ramesh *et al.*, 2019). *D. nobile* is also used in TCM formulations to support the immune system, improve digestion, promote healthy ageing and manage certain health conditions (Wu *et al.*, 2023a). Moreover, scientific studies have reported various pharmacological activities of

D. nobile, such as antioxidant, anti-inflammatory, immunomodulatory, anti-tumour, anti-diabetes, neuroprotective and hepatoprotective activities. These findings provide scientific evidence for the traditional use of *D. nobile* in TCM and highlight its potential therapeutic applications.

2.2 Botanical Description and Taxonomy of *D. nobile*

D. nobile is an epiphytic or lithophytic orchid in the family Orchidaceae, native to various regions of Asia, including China, India, Nepal, Bhutan, Myanmar, and adjacent areas in Vietnam (Knapp, 2021). Known for its showy, fragrant flowers that bloom in winter or early spring and growing up to 40–50 cm in height, *D. nobile* exhibits a sympodial growth habit with cane-like pseudobulbs, which grow up to 60 cm, covered by papery sheaths. Its leaves are generally thin and oblong, and alternate along the pseudobulb, growing up to 20 cm long and 2 cm wide. The flowers, large and often colourful, vary from white to pink, lavender and bi-coloured forms, featuring a distinctive, often fringed three-lobed lip with prominent callus or contrasting markings (Chase *et al.*, 2020; Moore, 2021). The plant is adapted to tolerate drought, storing water and nutrients in its pseudobulbs, and its aerial roots, covered by velamen, help absorb moisture from the air. Morphologically, it grows in clusters with cylindrical pseudobulbs, typically lanceolate leaves, and flowers arising from the nodes of pseudobulbs. *D. nobile* thrives in high altitudes, forests and rocky cliffs and has been widely cultivated in parts of Asia, such as Japan, South Korea and Taiwan, for both ornamental and medicinal uses (Buyun *et al.*, 2021; Fei, 2021). Taxonomically, *D. nobile* belongs to the kingdom Plantae, phylum Magnoliophyta, class Liliopsida, order Asparagales, family Orchidaceae, subfamily Epidendroideae, tribe Dendrobieae and genus *Dendrobium* (Song *et al.*, 2022). It is placed within the subsection *nobile* of the genus *Dendrobium*, which is one of the largest genera in Orchidaceae, comprising over 1500 species. The classification is based on the Angiosperm Phylogeny Group (APG) system, though it may evolve with ongoing research (Zhang *et al.*, 2016; Rentsch *et al.*, 2020). This orchid is notable in TCM and holds cultural significance due to its unique botanical traits (Fan *et al.*, 2023) (Fig. 2.1).

2.3 Traditional Medicinal Uses of *D. nobile*

D. nobile has a long history of traditional use in Chinese medicine. It has been used for various therapeutic purposes, owing to its potential health benefits. *D. nobile*, a staple in TCM, is revered for its multifaceted medicinal properties that are deeply rooted in historical practices and cultural beliefs (Zhou *et al.*, 2020). Believed to tonify yin and nourish the kidney, it is frequently employed to address conditions associated with yin deficiency such as dry mouth, dry throat and dizziness (Hsu *et al.*, 2022). Furthermore, its traditional use extends to treating issues such as blurred vision, dim vision and dry eyes, which are often linked to liver and kidney deficiencies (Cakova *et al.*, 2017). *D. nobile* is also believed to have immune-boosting properties, acting as a tonic herb to bolster the body's defence mechanisms and enhance overall health (Wu *et al.*, 2023b). Additionally, it is traditionally used to support digestive health by soothing the digestive system, alleviating stomach discomfort and promoting healthy digestion (Chen *et al.*, 2022). Moreover, its anti-inflammatory and analgesic effects make it a good remedy for sore throat, swollen gums and mouth ulcers (Li *et al.*, 2022). Notably, research suggests it has potential for addressing cognitive dysfunction in Alzheimer's disease through various mechanisms, such as regulating tau protein hyperphosphorylation and enhancing synaptic connections (Li *et al.*, 2022). However, it is crucial to emphasize that while these traditional uses are deeply ingrained, further scientific investigation is imperative to comprehensively understand their medicinal properties and mechanisms of action (Zhou *et al.*, 2020). It is always advisable for patients to seek guidance from qualified healthcare professionals before integrating *D. nobile* or any herbal remedy into medicinal practices (Table 2.1 and Table 2.2).

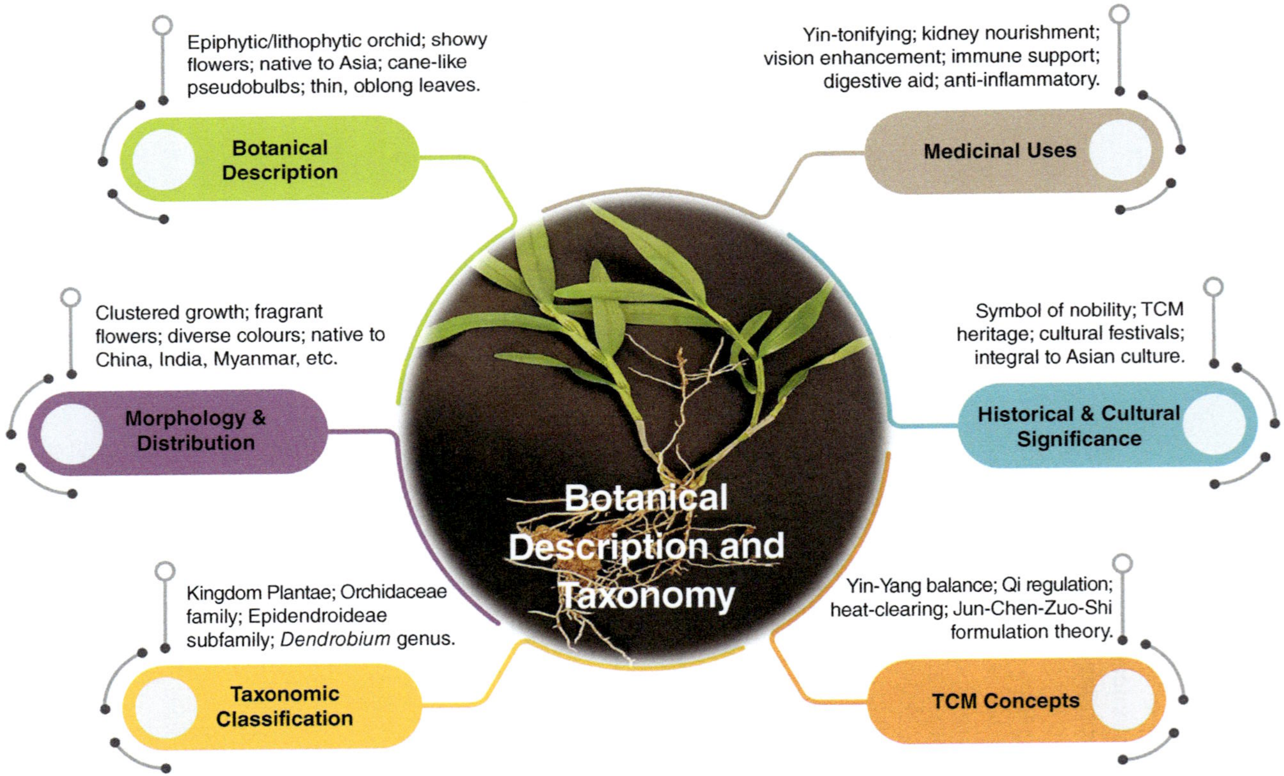

Fig. 2.1. Botanical description and taxonomy (authors' own figure).

Table 2.1. Traditional uses of *D. nobile* in Chinese medicine (authors' own table).

Health condition	Traditional use
Yin deficiency and heat syndrome	Nourishes yin, clears heat and promotes body fluids
Lung heat and cough	Clears lung heat, stops cough and moistens dryness
Stomach heat and thirst	Clears stomach heat, quenches thirst and promotes production of body fluids
Sore throat and mouth ulcers	Relieves sore throat and mouth ulcers
Blurred vision and eye disorders	Improves vision, clears heat and nourishes liver and kidney
Weakness and fatigue	Tonifies qi, strengthens spleen and improves physical strength
Kidney deficiency and impotence	Tonifies kidney, strengthens yang and improves sexual function
Ageing-related conditions	Nourishes yin and tonifies kidney essence
Diabetes	Helps regulate blood sugar levels
Rheumatoid arthritis	Alleviates joint pain and inflammation
Skin disorders	Clears heat, detoxifies and promotes skin health
Gastrointestinal disorders	Relieves symptoms such as nausea, vomiting and diarrhoea
Cardiovascular disorders	Promotes cardiovascular health and improves blood circulation
Immune system support	Enhances immune function and strengthens bodily defences
Anti-inflammatory and antioxidant properties	Reduces inflammation and oxidative stress

Table 2.2. Traditional uses of *D. nobile* in Chinese medicine categorised by TCM syndrome (authors' own table).

TCM syndrome	Traditional use
Yin deficiency	Nourishes yin and replenishes body fluids
Heat syndrome	Clears heat, reduces inflammation and cools the body
Qi deficiency	Tonifies qi, boosts energy and strengthens the body
Blood deficiency	Nourishes blood and promotes blood circulation
Stagnation syndrome	Breaks up stagnation, promotes qi and blood circulation
Phlegm-dampness syndrome	Resolves phlegm, promotes digestion and clears dampness
Wind cold syndrome	Dispels wind and cold, warms the body and promotes circulation
Wind-heat syndrome	Clears wind-heat, reduces fever and resolves inflammation
Kidney deficiency	Tonifies kidney, strengthens yang and nourishes essence
Liver stagnation	Soothes liver, promotes Qi flow and alleviates stress
Spleen deficiency	Tonifies spleen, promotes digestion and strengthens the body

2.4 Historical and Cultural Significance of *D. nobile* in TCM

D. nobile holds significant historical and cultural significance in TCM. It has been used for centuries as a valuable medicinal herb, and its importance in TCM can be attributed to several factors. *D. nobile* boasts a rich heritage entrenched in TCM, spanning over two millennia, as documented in ancient Chinese medicinal texts such as the *Shennong Bencao Jing* (Xu *et al.*, 2022). Revered as a yin-tonifying herb in TCM, its capacity to replenish yin energy, which is pivotal for maintaining health equilibrium, is characterized by qualities of moisture, coolness and nourishment (Zhou *et al.*, 2020). Beyond its medicinal attributes, *D. nobile* has profound symbolic significance in Chinese culture, symbolizing nobility, purity and longevity, and is thus frequently incorporated into traditional festivities as a harbinger of good fortune and blessings (Wang *et al.*, 2020). Embedded deeply within the cultural fabric of China and other Asian nations, where TCM is deeply rooted, *D. nobile* serves as an integral component of their cultural heritage, perpetuated by traditional healers, practitioners and herbalists over many generations (Heubl, 2022). Despite the advancements in modern medicine, *D. nobile* maintains its relevance in contemporary TCM practices, often prescribed for a spectrum of health conditions ranging from yin deficiency-related ailments to digestive health, vision and immune support (Wu *et al.*, 2023b). The historical and cultural significance of *D. nobile* in TCM is a testament to its long-standing use as a valued medicinal herb in traditional Chinese herbal medicine practices. However, it is important to note that while *D. nobile* has rich historical and cultural significance, further scientific research is needed to fully understand its medicinal properties and mechanisms of action in the context of modern evidence-based medicine (Table 2.2).

2.5 TCM Theories and Concepts Related to *D. nobile*

In TCM, the use of *D. nobile* is based on several theories and concepts that are fundamental to TCM's understanding of health and disease. Some of the key TCM theories and concepts related to *D. nobile* indicate that, central to TCM theory, the concept of yin and yang delineates the balance and interplay of opposing yet complementary forces within the body, with *D. nobile* identified as a yin-tonifying herb crucial for nourishing and replenishing yin energy, which is essential for overall health and well-being (Rosenthal *et al.*, 2022). Moreover, TCM underscores the significance of qi, often translated as vital energy or life force, the balanced flow of which through the body's meridians is deemed imperative for health maintenance. *D. nobile* is believed to regulate qi flow and foster harmony (Zhaoguo *et al.*, 2019). In the context of health equilibrium, TCM recognizes disruptions in the yin/yang balance, with excess heat and dryness regarded as potential disruptors, a concern that *D. nobile* addresses by clearing excess heat and moistening dryness, thus facilitating balance restoration and health promotion (Liu *et al.*, 2024). Adhering to traditional Chinese herbal medicine principles, *D. nobile* is frequently incorporated into TCM formulations, typically alongside other herbs, in accordance with the 'Jun-Chen-Zuo-Shi' theory, which dictates the strategic use of herbs in synergistic roles, with specific herbs designated as 'emperor', 'minister', 'assistant' or 'guide' to optimize therapeutic efficacy while minimizing adverse effects (Wang *et al.*, 2022).

These are some of the key TCM theories and concepts relevant to the traditional use of *D. nobile*. It is important to note that TCM is a holistic system of medicine that takes into account the individual's constitution, symptoms and overall health condition, and the use of *D. nobile* in TCM formulations is based on a comprehensive assessment by a qualified TCM practitioner (Table 2.3).

2.5.1 Formulations and preparation of *D. nobile* in Chinese medicine

D. nobile is commonly used in TCM and is available in various formulations and preparations. In TCM, various methods are employed to prepare *D. nobile* for therapeutic use, including

Table 2.3. Traditional uses of *D. nobile* in Chinese medicine by dosage form (authors' own table).

Dosage form	Traditional use
Decoction	Used as a base herb or combined with other herbs in TCM formulas for various health conditions
Powder	Mixed with other powdered herbs or used in topical applications for skin disorders
Pill/capsule	Formulated as oral supplements for convenience and ease of use
Extract	Concentrated forms for more potent effects
Topical application	Used externally for skin conditions or applied to acupoints for targeted effects
Syrup	Formulated for respiratory or digestive issues
Plaster	Used externally for localized pain or inflammation

decoction ('shuiyin' in Chinese), where the herb is boiled in water to create a concentrated liquid extract suitable for oral consumption (Wang, 2021a). Alternatively, *D. nobile* can be ground into a fine powder ('san') for making herbal teas or to be orally ingested with water or other liquids (Teoh, 2016b). For convenience, *D. nobile* can also be formulated into pills or capsules ('wan'), sometimes combined with other herbal ingredients to serve specific therapeutic purposes, offering ease of oral administration (Han *et al.*, 2018). Additionally, *D. nobile* can be prepared as a tincture ('jiu') by soaking the herb in alcohol or another solvent to create a liquid extract, commonly taken orally with water or other fluids (Stéphane *et al.*, 2021). Beyond oral consumption, *D. nobile* finds application in topical forms such as creams, ointments or lotions, catering to skin health and dermatological conditions (Wang, 2021b). Furthermore, it is frequently integrated into combination formulas alongside other herbs in TCM, available in various preparations, such as decoctions, powders, pills, capsules or tinctures, capitalizing on synergistic effects to address specific health concerns (Hong and Soh, 2019). Notably, the dosage, preparation and administration of *D. nobile* in Chinese medicine may vary depending on the specific health condition, individual patient characteristics and expertise of the TCM practitioner. It is recommended to consult a qualified TCM practitioner or healthcare professional for proper diagnosis, dosage and administration of *D. nobile* or any other herbal medicine (Fig. 2.2).

2.5.2 Traditional methods of processing, preparing and formulating *D. nobile* in Chinese medicine

Traditional methods of processing, preparing and formulating *D. nobile* in Chinese medicine may include various techniques and practices, depending on the specific application and intended use. *D. nobile* undergoes a series of processing steps designed to enhance its medicinal efficacy and usability. Upon harvesting, stems are typically subjected to traditional drying methods such as sun drying, air drying or oven drying to reduce the moisture content and preserve the integrity of the botanical (Li *et al.*, 2023b). Following drying, *D. nobile* can be utilized in various preparations, including decoctions, where the dried stems are boiled in water to extract medicinal properties, resulting in a potent liquid that can be consumed as a tea or serve as a base for other formulations (Yang *et al.*, 2022). Additionally, the dried stems may be ground into a fine powder, facilitating their incorporation into herbal formulas or other preparations and expanding their applications in traditional medicine (Jia *et al.*, 2023). Furthermore, to obtain concentrated forms of bioactive compounds, *D. nobile* can be subjected to extraction using various solvents. These extracts, which are rich in medicinal constituents, can be further utilized in tinctures, capsules or other formulations to maximize their therapeutic potential (Gutiérrez, 2018).

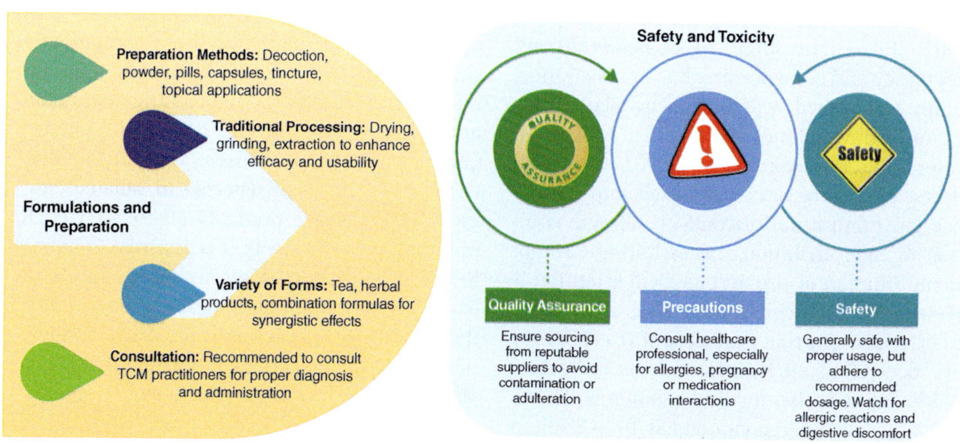

Fig. 2.2. Formulations, preparation, safety and toxicity of *D. nobile* (authors' own figure).

It is important to note that the specific methods of processing, preparing and formulating *D. nobile* may vary depending on the tradition, school of thought or individual practitioner in the field of Chinese medicine. It is always recommended to consult reputable sources, experienced practitioners or authoritative texts for accurate and up-to-date information on traditional methods of processing, preparing and formulating *D. nobile* or any other herbal medicine in Chinese medicine.

2.6 Safety and Toxicity of *D. nobile*

D. nobile is generally considered safe when used appropriately and in accordance with traditional usage. When using *D. nobile*, it is essential to adhere to the recommended dosage instructions provided by a qualified healthcare professional or as indicated on the product label, as excessive use may increase the risk of adverse effects (Marra and Bailey, 2018). Ensuring the quality and source of *D. nobile* is crucial for minimizing the potential for contamination or adulteration, emphasizing the importance of sourcing from reputable suppliers (Li *et al.*, 2019). Individuals with known allergies to plants or flowers, particularly those from the Orchidaceae family, should exercise caution when using *D. nobile* to mitigate the risk of allergic reactions (Teoh, 2019).

Given its potential interactions with medications or other herbal supplements, consultation with a healthcare professional is imperative, especially for individuals with existing health conditions or those taking medications (Teng *et al.*, 2015). The safety of *D. nobile* during pregnancy and breastfeeding has not been thoroughly established, warranting avoidance unless under the guidance of a qualified healthcare professional (Fauzdar *et al.*, 2020). Although adverse effects are rare, individuals may experience gastrointestinal discomfort, allergic reactions or skin irritation, necessitating prompt discontinuation of use and medical attention if severe symptoms arise (Kumar *et al.*, 2022). It is important to note that the safety of *D. nobile* may vary depending on factors such as individual health, age and other medical conditions. Therefore, it is always recommended to consult a qualified healthcare professional before using any herbal medicine or dietary supplement, especially if a patient has any underlying health conditions or is taking medications (Fig. 2.2).

2.6.1 Toxicity of *D. nobile*

D. nobile is generally considered safe when used in TCM practices. However, similar to any herbal medicine, it may carry potential risks and side effects, especially when used improperly or in excessive doses. Limited studies have been conducted on the toxicity of *D. nobile*, and available data suggest that it has a relatively low toxicity profile. There have been reports of allergic reactions, such as skin rashes and itching, associated with the use of *D. nobile* in some individuals, although these cases are rare (Hong and Soh, 2019). In addition, there have been isolated cases of adverse effects on the digestive system, such as nausea, vomiting and diarrhoea, reported with the use of products containing *D. nobile* (Teoh, 2019). However, the adverse effects are typically mild and transient. It is important to note that the safety of *D. nobile* may vary depending on factors such as the quality, dosage, duration of use and individual health conditions.

2.6.2 Summary of the safety profile of *D. nobile*, including potential adverse effects, contraindications and precautions in its use

D. nobile is regarded as safe within TCM practices, with a generally low toxicity profile when used appropriately. Despite this, there are potential risks to consider. Allergic reactions, although rare, have been observed, including skin rashes and itching (Hong and Soh, 2019). Mild digestive discomfort, such as nausea, vomiting or diarrhoea, has also been reported in isolated cases involving *D. nobile* products (Teoh, 2019). To minimize adverse effects, it is essential to adhere to the recommended doses and usage durations, as excessive intake or prolonged use may increase the risk of side effects (Liu *et al.*, 2023). Individuals with known allergies or sensitivities to *D. nobile* should use it cautiously or avoid it altogether. Consultation with a qualified healthcare professional is advised for those with existing health conditions, as well as for pregnant or breastfeeding individuals, to ensure safe and appropriate use (Crislip, 2018). Additionally, ensuring the quality and authenticity of *D. nobile* products is critical because adulteration or contamination can present further health risks (Liu *et al.*, 2023).

2.6.3 Evaluation of the toxicological data and potential interactions with other medications or herbs

As stated, *D. nobile* has a relatively low toxicity profile and is well-tolerated when used as directed. However, further studies are needed to fully understand its long-term safety and potential toxic effects, especially with prolonged use or higher doses (Belwal *et al.*, 2020). *D. nobile* may interact with other medications or herbs, and caution should be exercised when using it concomitantly with other substances. *D. nobile* may have potential interactions with specific health conditions, and caution should be exercised in individuals with pre-existing health conditions, such as liver or kidney diseases, to avoid potential adverse effects (Crislip, 2018). Patients should consult a qualified healthcare professional or a licensed herbalist before using *D. nobile* or any other herbal medicine to assess potential interactions and ensure safe and appropriate use (Mileva *et al.*, 2015). The potential interactions of *D. nobile* with other medications or herbs may vary depending on the specific formulation, dosage and individual health conditions. It is recommended to consult with qualified healthcare professionals or licensed herbalists for comprehensive and up-to-date information on potential interactions with medications or herbs when using *D. nobile* or any other herbal medicine (Huang *et al.*, 2019). A subchronic toxicity study on the toxicological effects of the aqueous extract of *D. nobile* in rats reported that the extract did not cause acute toxicity but caused mild hepatotoxicity at high doses (Wu *et al.*, 2023b). *D. nobile* contains alkaloids and flavonoids that may interact with other medications or herbs (Mou *et al.*, 2021).

2.7 Current Research and Future Perspectives

There has been increasing interest in the pharmacological properties of *D. nobile*, with studies investigating its potential antioxidant, anti-inflammatory, immunomodulatory, anti-cancer, anti-diabetes and neuroprotective effects, among others. These studies used various *in vitro*, *in vivo* and clinical models to evaluate the therapeutic potential of *D. nobile* (Mou *et al.*, 2021). Researchers identified and characterized the chemical constituents of *D. nobile*, including alkaloids, flavonoids, polysaccharides and other bioactive compounds. These studies provide insights into the chemical composition and variability of *D. nobile* from different sources as well as their potential contributions to the medicinal properties of the herb (Gong *et al.*, 2023).

Other studies focused on optimizing the formulation and dosage of *D. nobile* preparations in Chinese medicine, including decoctions, powders, pills and other formulations. They aimed to enhance the therapeutic efficacy, safety and standardization of *D. nobile*-based products, as well as to explore novel delivery systems for better bioavailability and clinical outcomes (Song *et al.*, 2022). Research on the safety profile and quality control of *D. nobile* has also been a focus of attention. Studies have evaluated the potential toxicity, adverse effects and interactions with other medications or herbs, as well as establishing quality control methods to ensure the authenticity, purity and consistency of *D. nobile* products (Peng *et al.*, 2021). *D. nobile* has significant historical and cultural significance in TCM, and research has also explored its ethnopharmacological and cultural aspects, including its traditional uses, processing methods, cultural significance and conservation strategies (Cheng *et al.*, 2019).

2.8 Potential Future Directions and Challenges in the Research and Development of *D. nobile* in Chinese Medicine

Further research is imperative to unravel the phytochemical composition of *D. nobile* comprehensively, including alkaloids, flavonoids, polysaccharides and other bioactive compounds, thus shedding light on its mechanisms of action and synergistic effects (Gong *et al.*, 2023). Delving deeper into mechanistic studies is essential to unveil the cellular and molecular mechanisms underlying the pharmacological activities of *D. nobile* and to provide a scientific foundation for its traditional applications (Li *et al.*, 2023a). Robust large-scale clinical trials are warranted to ascertain the safety, efficacy and optimal dosages of *D. nobile* in human subjects, thereby substantiating its therapeutic potential across diverse health conditions (Mou *et al.*, 2021). Given the chemical complexity and variability of *D. nobile*, establishing quality control measures and standardization protocols is imperative for ensuring product consistency and reproducibility (Song *et al.*, 2022). Addressing concerns regarding sustainability is paramount, necessitating research efforts toward sustainable sourcing practices to safeguard *D. nobile* and its natural habitat (Huang *et al.*, 2019). Moreover, comprehensive safety assessments, including toxicity studies and evaluation of potential drug interactions, are crucial to ascertain the safety profile of *D. nobile* (Peng *et al.*, 2021). Integrating traditional knowledge with modern research methodologies can synergistically validate *D. nobile*'s traditional uses and facilitate its further development, highlighting the importance of collaboration between traditional practitioners and modern researchers (Gong *et al.*, 2023).

2.9 Conclusion

In conclusion, *D. nobile* has a long history of traditional use in Chinese medicine to treat various health conditions. It is known for its potential therapeutic properties, including immunomodulatory, anti-inflammatory, anti-cancer, anti-diabetes and anti-ageing effects. Traditional formulations and preparations of *D. nobile* include decoctions, powders, pills and other formulations, which are often used in combination with other herbs to enhance its efficacy. While *D. nobile* has been used for centuries in Chinese medicine and has shown promising results in preclinical and clinical studies, it is important to note that more research is needed to further understand its mechanisms of action, evaluate its safety and efficacy in larger clinical trials, establish quality control measures and ensure sustainable sourcing practices. Additionally, it is important to integrate traditional knowledge with modern research to validate its traditional use and guide further development. As with any herbal medicine, it is recommended patients consult a qualified healthcare professional or a licensed TCM practitioner before using *D. nobile* or any other herbal remedy, especially for those with underlying health conditions, who are pregnant or breastfeeding, or who are taking medications or other herbs, to ensure safe and appropriate use. Overall, *D. nobile* holds promise as a traditional herbal remedy in Chinese medicine, and further research and development in this field may contribute to its potential therapeutic applications for various health conditions.

2.9.1 Summary of the main findings and key points

This chapter provides a comprehensive overview of the historical, cultural and medicinal significance of *D. nobile*. *D. nobile*, revered as the 'noblest of all orchids', has been an integral part of TCM for centuries, embodying profound historical and cultural significance deeply ingrained in Chinese folklore, symbolizing longevity, beauty and health. Its utilization aligns with fundamental theories and concepts of TCM, notably its classification as a yin-tonifying herb, its meridian actions and its properties according to TCM principles, such as nourishing yin and clearing heat (Chen *et al.*, 2021). Its pharmacological potential stems from its diverse chemical constituents, including alkaloids, flavonoids, polysaccharides and other bioactive compounds, which underpin its immunomodulatory, anti-inflammatory, anti-cancer, anti-diabetes and anti-ageing effects, as elucidated by preclinical and clinical studies. Traditional methods of processing and formulation *of D. nobile*, including decoctions, powders and pills, are detailed, reflecting its versatility in Chinese medicine applications.

Considerations regarding safety and toxicity are paramount, with the chapter outlining potential adverse effects, contraindications, precautions and interactions based on available toxicological data and traditional knowledge. Current scientific research endeavours focus on unravelling the herb's mechanisms of action, establishing quality control measures and integrating traditional knowledge with modern research, paving the way for future advancements and challenges in the realm of *D. nobile* research and development (Li *et al.*, 2023a). The main findings and key points discussed provide a comprehensive overview of the traditional uses of *D. nobile* in Chinese medicine, highlighting its historical and cultural significance, TCM theories and concepts related to its use, chemical constituents, pharmacological activities, formulations and preparations, safety and toxicity, as well as current research and future perspectives.

2.9.2 Concluding remarks and recommendations for further research

The traditional use of *D. nobile* in Chinese medicine has a long history and cultural significance and its potential therapeutic properties have been supported by preclinical and clinical studies. However, further research is required to fully understand the mechanisms of action, establish quality control measures and integrate traditional knowledge with modern research. Based on the findings discussed here, more comprehensive mechanistic studies are imperative to unravel the intricate pathways and molecular mechanisms underlying the pharmacological activities of *D. nobile*, shedding light on its bioactive compounds and their interactions with cellular and molecular targets. Furthermore, well-designed clinical trials, including randomized controlled trials and long-term safety studies, are necessary to substantiate the safety and efficacy of *D. nobile* in various health conditions and formulations (Zhang *et al.*, 2021). Establishing stringent quality control measures, such as standardization of bioactive compounds, extraction methods and dosage determination, is crucial to ensure the consistency and safety of *D. nobile* products. Integrating traditional knowledge with modern research will facilitate a holistic understanding of the therapeutic potential of *D. nobile*, including its synergistic effects with other herbs and medications (Chen *et al.*, 2019).

Continuous evaluation of its safety profile and potential interactions with other substances is paramount to mitigate any risks associated with its use. Additionally, concerted efforts towards sustainable harvesting, conservation and cultivation methods are essential to safeguard the long-term availability and preservation of *D. nobile*. In conclusion, although *D. nobile* has shown promising traditional uses in Chinese medicine, further research is needed to fully understand its mechanisms of action, establish quality control measures, integrate traditional knowledge with modern research, evaluate its safety and efficacy in clinical settings and ensure sustainable harvesting and conservation practices. These research endeavours will contribute to the advancement of our knowledge of *D. nobile* and its potential therapeutic applications in Chinese medicine.

Acknowledgements

The authors express their gratitude for the financial support received through the Distinguished High-Level Talents Research Grant from the Guizhou Science and Technology Corporation Platform Talents Fund (grant nos. [2017]5733-001 and CK 1130-002), the National Natural Science Foundation of China (U1812403, 82373981 and 82060750), 2011 Collaborative Innovation Centre of Traditional Chinese Medicine in Guizhou Province (no. [2022]026) and the support provided by the Zunyi Medical University, China. Special appreciation is extended to all laboratory colleagues and research staff members for their valuable insights, constructive guidance and assistance with the contents of this chapter.

References

Belwal, T., Nabavi, S.M., Nabavi, S.F., Dehpour, A.R. and Shirooie, S. (2020) *Naturally Occurring Chemicals Against Alzheimer's Disease*. Academic Press, Cambridge, Massachusetts.

Buyun, L., Ivannikov, R., Batsmanova, L., Taran, N., Kovalska, L. *et al.* (2021) Structural adaptation and antioxidant response of *Guarianthe bowringiana* (O'Brien) Dressler & W. E. Higgins (Orchidaceae juss.) seedlings during ex vitro acclimatization. *Acta Agrobotanica* 74(1), 1–16.

Cakova, V., Bonte, F. and Lobstein, A. (2017) *Dendrobium:* sources of active ingredients to treat age-related pathologies. *Aging and Disease* 8(6), 827–849.

Chase, M., Christenhusz, M. and Mirenda, T. (2020) *The Little Book of Orchids: Gems of Nature*. Ivy Press, Lewes.

Chen, H., Shi, X., Zhang, L., Yao, L., Cen, L, *et al.* (2022) Ultrasonic extraction process of polysaccharides from *Dendrobium nobile* Lindl.: optimization, physicochemical properties and anti-inflammatory activity. *Foods* 11(19), 1–16.

Cheng, J., Dang, P.P., Zhao, Z., Yuan, L.C., Zhou, Z.H. *et al.* (2019) An assessment of the Chinese medicinal dendrobium industry: supply, demand and sustainability. *Journal of Ethnopharmacology* 229, 81–88.

Crislip, M. (2018) *Flies in the Ointment*. Bitingduck Press LLC, California.

Fan, C., Sun, X., Wang, X. and Yu, H. (2023) Therapeutic potential of the chemical composition of *Dendrobium nobile* Lindl. *Frontiers in Pharmacology* 14, 1–15.

Fan, Y., Yu, Q., Wang, G., Tan, J., Liu, S. *et al.* (2020) Effects of non-thermal plasma treatment on the polysaccharide from *Dendrobium nobile* Lindl. and its immune activities *in vitro*. *International Journal of Biological Macromolecules* 153, 942–950.

Fauzdar, M., Pandey, P.P. and Kaul, V. (2020) Dendrobium: Vital element for im-munomodulatory activities. *Biotica Research Today* 2(9), 933–937.

Fei, B. (2021) Economic value and research significance of Moso bamboo. In: *The Moso Bamboo Genome*. Springer, Cham, pp. 1–12.

Gao, X., Liu, J., Luo, Y., Lei, Y., Long, W. *et al.* (2022) Various fractions of alcoholic extracts from *Dendrobium nobile* functionalized antioxidation and antiaging in D-galactose-induced aging mice. *Frontiers in Bioscience-Landmark* 27(11), 315.

Gong, D., Li, B., Wu, B., Fu, D., Li, Z. *et al.* (2023) The integration of the metabolome and transcriptome for *Dendrobium nobile* Lindl. in response to methyl jasmonate. *Molecules* 28(23), 1–27.

Gutiérrez, R.M.P. (2018) Orchids: a review of uses in traditional medicine, its phytochemistry and pharmacology. *Journal of Medicinal Plants Research* 4(8), 592–638.

Han, K., Wang, M., Zhang, L. and Wang, C. (2018) Application of molecular methods in the identification of ingredients in Chinese herbal medicines. *Molecules* 23(10), 2728.

Heubl, G. (2022) DNA-based authentication of TCM-plants: current progress and future perspectives. In: Wagner, H. and Ulrich-Merzenich, G. (eds) *Evidence and Rational Based Research on Chinese Drugs*. Springer, Vienna.

Hong, H. and Soh, S.B. (2019) *Healing Orchids*. World Scientific, Singapore.

Hsu, W.H., Chung, C.P., Wang, Y.Y., Kuo, Y.H., Yeh, C.H. *et al.* (2022) *Dendrobium nobile* protects retinal cells from UV-induced oxidative stress damage via nrf2/HO-1 and MAPK pathways. *Journal of Ethnopharmacology* 288, 114886.

Huang, W.Y., Cai, Y.Z. and Zhang, Y. (2019) Natural phenolic compounds from medicinal herbs and dietary plants: potential use for cancer prevention. *Nutrition and Cancer* 62(1), 1–20.

Jain, A., Sarsaiya, S., Chen, J., Wu, Q., Lu, Y. *et al.* (2021) Changes in global Orchidaceae disease geographical research trends: recent incidences, distributions, treatment, and challenges. *Bioengineered* 12(1), 13–29.

Jia, X., Li, L., Tan, D., Wu, F., He, Y. *et al.* (2023) Effect of superfine-grinding on the physicochemical and antioxidant properties of *Dendrobium nobile* powders. *Food Science and Technology* 43, e117322.

Knapp, S. (2021) *Extraordinary Orchids*. University of Chicago Press, Chicago, USA.

Kumar, P., Rai, S., Verma, S.K., Prakash, P.S. and Chitara, D. (2022) Classification, mode of action and uses of various immunomodulators. In: *Immunomodulators and Human Health*. Springer Nature, Singapore, pp. 3–38.

Larridon, I., Semmouri, I., Léveillé-Bourret, É., Brewer, G., Epitawalage, N. *et al.* (2018) Changing tribal and generic concepts in cyperaceae, new insights from phylogenomics. In: *6th International Conference on Comparative Biology of Monocotyledons (Monocots VI)*, EDUFRN, pp. 61–61.

Lee, C.T., Kuo, H.C., Chen, Y.H. and Tsai, M.Y. (2018) Current advances in the biological activity of polysaccharides in *Dendrobium* with intriguing therapeutic potential. *Current Medicinal Chemistry* 25(14), 1663–1681.

Li, D.D., Zheng, C.Q., Zhang, F. and Shi, J.S. (2022) Potential neuroprotection by *Dendrobium nobile* Lindl alkaloid in Alzheimer's disease models. *Neural Regeneration Research* 17(5), 972–977.

Li, M., Jiang, C., Pui-Hay, P., Shaw, P.C. and Yuan, Y. (2019) Molecular identification of traditional medicinal materials. In: Huang, Lq. (ed.) *Molecular Pharmacognosy*. Springer, Singapore.

Li, M., Li, J., Deng, L., Zhao, Z., Luo, C. *et al.* (2023a) Species of associated bryophytes and their effects on the yield and quality of *Dendrobium nobile*. *BMC Plant Biology* 23(1), 516.

Li, P.Y., Li, L. and Wang, Y.Z. (2023b) Traditional uses, chemical compositions and pharmacological activities of *Dendrobium*: a review. *Journal of Ethnopharmacology* 310, 116382.

Liu, C., Zuo, Z., Xu, F. and Wang, Y. (2023) Authentication of herbal medicines based on modern analytical technology combined with chemometrics approach: a review. *Critical Reviews in Analytical Chemistry* 53(7), 1393–1418.

Liu, X., Shi, H. and Li, W. (2024) Review on the potential roles of traditional Chinese medicines in the prevention, treatment, and postoperative recovery of age-related cataract. *Journal of Ethnopharmacology* 324, 117786.

Long, Y., Wang, W., Zhang, Y., Zhang, S., Li, Z. *et al.* (2023) *Dendrobium nobile* Lindl polysaccharides attenuate UVB-induced photodamage by regulating oxidative stress, inflammation and mmps expression in mice model. *Photochemistry and Photobiology* 99(5), 1269–1281.

Marra, M.V. and Bailey, R.L. (2018) Position of the academy of nutrition and dietetics: micronutrient supplementation. *Journal of the Academy of Nutrition and Dietetics* 118(11), 2162–2173.

Mileva, M., Nikolova, I., Nikolova, N., Mukova, L., Georgieva, A. *et al.* (2015) Investigation of antioxidant and antiviral properties of geraniol. *Acta Microbiologica Bulgarica* 31(1), 48–53.

Moore, T. (2021) *The Orchid Album, Comprising Coloured Figures and Descriptions of New, Rare, and Beautiful Orchidaceous Plants*, Vol. 1. Nabu Press, India.

Mou, Z., Kennelly, E.J., Zhao, D, *et al.* (2021) Identification, biological activities and biosynthetic pathway of *Dendrobium* alkaloids. *Frontiers in Pharmacology* 12, 605994.

Natta, S., Mondol, M.S.A., Pal, K., Mandal, S., Sahana, N. *et al.* (2022) Chemical composition, antioxidant activity and bioactive constituents of six native endangered medicinal orchid species from north-eastern Himalayan region of India. *South African Journal of Botany* 150, 248–259.

Nie, X., Chen, Y., Li, W. and Lu, Y. (2020) Anti-aging properties of *Dendrobium nobile* Lindl.: from molecular mechanisms to potential treatments. *Journal of Ethnopharmacology* 257, 112839.

Peng, L.-Q., Zhang, Y., Yan, T.-C., Gu, Y.-X., Zi-XuanYue *et al.* (2021) Carbonized biosorbent assisted matrix solid-phase dispersion microextraction for active compounds from functional food. *Food Chemistry* 365, 130545.

Peng, X., Tang, F., Yang, Y., Li, T., Hu, X. *et al.* (2022) Bidirectional effects and mechanisms of traditional Chinese medicine. *Journal of Ethnopharmacology* 298, 115578.

Pradhan, B.M. (2021) Taxonomy of subtribe dendrobiinae lindley (orchidaceae) of Nepal Himalaya. PhD. Thesis, Tribhuvan University, Nepal.

Ramesh, G., Ramudu, J., Khasim, S.M. and Thammasiri, K. (2020) Structural adaptations of *Bulbophyllum* and *Dendrobium* (Orchidaceae) to the epiphytic habitat and their phylogenetic implications. In: Khasim, S., Hegde, S., González-Arnao, M. and Thammasiri, K. (eds) *Orchid Biology: Recent Trends & Challenges*. Springer, Singapore.

Ramesh, T., Koperuncholan, M., Praveena, R., Ganeshkumari, K., Vanithamani, J. *et al.* (2019) Medicinal properties of some dendrobium orchids – a review. *Journal of Applied and Advanced Research* 4(4), 119–128.

Rentsch, J.D., Hardee, L.J., Shelley, C.E. and Williams, M.T. (2020) The complete plastid genome of *Neottia bifolia* (raf.) Baumbach (Orchidaceae): insights into chlorophyllous and achlorophyllous plastid genomes. *Castanea* 85(2), 285–295.

Rosenthal, L., Hernandez, P. and Vaamonde, D. (2022) Traditional Chinese medicine, ayurveda, and fertility. In: Vaamonde, D., A.C., H. and Garcia-Manso, J.M. (eds) *Fertility, Pregnancy, and Wellness*. Elsevier, pp. 209–247.

Song, C., Wang, Y., Manzoor, M.A., Mao, D., Wei, P. *et al.* (2022) In-depth analysis of genomes and functional genomics of orchid using cutting-edge high-throughput sequencing. *Frontiers in Plant Science* 13, 1018029.

Stéphane, F.F.Y., Jules, B.K.J., Batiha, G.E., Ali, I. and Bruno, L.N. (2021) Extraction of bioactive compounds from medicinal plants and herbs. *Natural Medicinal Plants*.

Teng, L., Shaw, D. and Barnes, J. (2015) Characteristics and practices of traditional Chinese medicine retail shops in London, UK: a cross-sectional study using an observational approach. *Journal of Ethnopharmacology* 173, 318–329.

Teoh, E.S. (2016a) Genus: *Dactylorhiza* to *Dipodium*. In: *Medicinal Orchids of Asia*. Springer, Cham.

Teoh, E.S. (2016b) Processing of medicinal herbs. In: *Medicinal Orchids of Asia*. Springer, Cham.

Teoh, E.S. (2019) *Orchids as Aphrodisiac, Medicine or Food*. Singapore, Singapore.

Wang, Y., Feng, Y., Li, M., Yang, M., Shi, G. *et al.* (2022) Traditional Chinese medicine in the treatment of chronic kidney diseases: theories, applications, and mechanisms. *Frontiers in Pharmacology* 13, 917975.

Wang, Y.H. (2021a) Traditional uses, chemical constituents, pharmacological activities, and toxicological effects of dendrobium leaves: a review. *Journal of Ethnopharmacology* 270, 113851.

Wang, Y.H. (2021b) Traditional uses and pharmacologically active constituents of *Dendrobium* plants for dermatological disorders: a review. *Natural Products and Bioprospecting* 11, 465–487.

Wang, Z., Zhao, M., Cui, H., Li, J. and Wang, M. (2020) Transcriptomic landscape of medicinal *Dendrobium* reveals genes associated with the biosynthesis of bioactive components. *Frontiers in Plant Science* 11, 391.

Wu, J., Cao, M., Peng, Y., Dong, B., Jiang, Y. *et al.* (2023a) Research progress on the treatment of epilepsy with traditional Chinese medicine. *Phytomedicine* 120, 155022.

Wu, W., Lin, Y., Farag, M.A., Li, Z. and Shao, P. (2023b) *Dendrobium* as a new natural source of bioactive for the prevention and treatment of digestive tract diseases: a comprehensive review with future perspectives. *Phytomedicine* 114, 154784.

Xu, Q., Niu, S.-C., Li, K.-L., Zheng, P.-J., Zhang, X.-J. *et al.* (2022) Chromosome-scale assembly of the *Dendrobium nobile* genome provides insights into the molecular mechanism of the biosynthesis of the medicinal active ingredient of *Dendrobium*. *Frontiers in Genetics* 13, 844622.

Yang, C.C., Chien, J.Y., Chou, Y.Y., Ciou, J.W. and Huang, S.P. (2022) The effects of *Lycium chinense*, *Cuscuta chinensis*, *Senna tora*, *Ophiopogon japonicus*, and *Dendrobium nobile* decoction on a dry eye mouse model. *Medicina* 58(8), 1134.

Zhang, G.Q., Xu, Q., Bian, C., Tsai, W.-C., Yeh, C.-M. *et al.* (2016) The *Dendrobium catenatum* Lindl. genome sequence provides insights into polysaccharide synthase, floral development and adaptive evolution. *Scientific Reports* 6(1), 19029.

Zhaoguo, L., Qing, W., Yurui, X., Zhaoguo, L., Qing, W. *et al.* (2019) *Key concepts in traditional Chinese medicine*. Springer Singapore, pp. 1–80.

Zhou, J., Zhang, Y., Li, S., Zhou, Q., Lu, Y. *et al.* (2020) *Dendrobium nobile* Lindl. alkaloids-mediated protection against ccl4-induced liver mitochondrial oxidative damage is dependent on the activation of nrf2 signaling pathway. *Biomedicine and Pharmacotherapy* 129, 110351.

3 Chemical Constituents and Pharmacological Activities of *Dendrobium nobile*

Archana Jain, Surendra Sarsaiya, Qihai Gong, Qin Wu and Jingshan Shi*
Key Laboratory of Basic Pharmacology and Joint International Research Laboratory of Ethnomedicine of Ministry of Education, Zunyi Medical University, Zunyi, Guizhou, China

Abstract

This chapter explores the chemical constituents and pharmacological activities of *Dendrobium nobile*, a plant renowned in traditional Chinese medicine. The plant contains various bioactive compounds such as alkaloids, flavonoids, phenols and polysaccharides. These constituents exhibit a wide range of pharmacological properties, including immunomodulatory, antioxidant, anti-inflammatory, neuroprotective, anti-cancer and anti-diabetes effects. This chapter delves into the mechanisms of these effects and discusses potential drug interactions and safety considerations. Additionally, it addresses the therapeutic potential of *D. nobile* for neurodegenerative diseases, metabolic disorders and skin health. Future research directions should emphasize the need for comprehensive studies to elucidate the mechanisms of action, optimize pharmacological activities and ensure long-term safety and efficacy. The chapter underscores the promising prospects of *D. nobile* as a natural source for developing novel therapeutics while highlighting the importance of standardizing its use in modern medicine.

3.1 Introduction

For centuries, *Dendrobium nobile* Lindl., an orchid, has been used in traditional medicine in China and in various other regions of Asia. It is known for its unique botanical characteristics and has been extensively studied for its chemical constituents and pharmacological activity. Recently, there has been growing interest in exploring the bioactive compounds found in *D. nobile* and their potential pharmacological properties (Fan *et al.*, 2023). *D. nobile*, commonly known as the noble dendrobium or orchid dendrobium, is an orchid species that is native to China and South-east Asia. In traditional Chinese medicine, various species of *Dendrobium*, including *D. nobile*, have been used for centuries for their purported medicinal properties. With its widespread distribution in regions of China, and in the Himalayas, as well as in other parts of Asia including Japan, the Korean peninsula and India, *D. nobile* is commonly referred to as 'shi hu' or 'shi hu gen' in Chinese medicine, and is often used in various formulations and preparations for its reported health benefits (Zhang *et al.*, 2023a).

The chemical constituents of *D. nobile* are diverse and complex, with alkaloids, flavonoids, polysaccharides and other bioactive compounds being identified in this plant species. The chemical

*Corresponding author: shijs@zmu.edu.cn

© Jingshan Shi, Archana Jain, Surendra Sarsaiya, Qin Wu and Qihai Gong 2025. Dendrobium nobile: *Research and Therapeutic Applications* (J. Shi, A. Jain, S. Sarsaiya, Q. Wu and Q. Gong)
DOI: 10.1079/9781800627987.0003

structures, sources and potential pharmacological activities of these compounds have been the focus of extensive research. Studies have shown that *D. nobile* exhibits a diverse array of other properties (Li *et al.*, 2023). It has been a staple in traditional Chinese and Asian medicine for centuries. Recently, interest in the discovery of bioactive compounds has surged. This orchid possesses a broad spectrum of pharmacological activities, including antioxidant, anti-inflammatory, immunomodulatory, anti-cancer, anti-diabetes, neuroprotective and other properties (Fan *et al.*, 2023). These effects are believed to be facilitated through various cellular and molecular mechanisms, such as interactions with cellular targets and modulation of signalling pathways. The traditional use of *D. nobile* in Chinese medicine has provided valuable insights into its potential therapeutic effects and safety considerations (Cakova *et al.*, 2017). However, despite the significant progress made in research, there are still gaps in our knowledge, and there is need for further studies on the bioavailability, pharmacokinetics and pharmacodynamics of its constituents, as well as standardized preparations or formulations for therapeutic use. This chapter provides a comprehensive overview of the current knowledge on the chemical constituents and pharmacological activities of *D. nobile*, highlighting its potential as a source of novel therapeutic agents in natural products and traditional medicine research.

D. nobile, deeply rooted in traditional Chinese medicine for its therapeutic benefits, harbours a plethora of chemical constituents that are crucial for its pharmacological actions. *D. nobile* is rich in bioactive compounds, including alkaloids such as dendrobine, nobilin and dendramine, which have been found to possess anti-inflammatory, antitumour and anti-cancer activities (Wang *et al.*, 2022). Additionally, phenanthrenes such as dendrocandins A–E, nobiletin and moscatilin present in *D. nobile* exhibit antioxidant, anti-inflammatory and anti-cancer properties (Wang *et al.*, 2022). Furthermore, polysaccharides, including dendronan, dendrofucoidan and dendropanaxan, contribute to the immunomodulatory, antitumour and anti-diabetes activities of *D. nobile* (Wang *et al.*, 2022). Moreover, the bibenzyl compounds such as dendrocandins A–C have antioxidant, anti-inflammatory and antimicrobial activities (Wang *et al.*, 2022). Overall, these diverse bioactive constituents highlight the potential therapeutic benefits of *D. nobile* for various health conditions.

D. nobile's anti-inflammatory properties suggest its potential in mitigating inflammation and alleviating the symptoms associated with inflammatory conditions (Li *et al.*, 2023). Moreover, the presence of phenanthrenes and bibenzyls confers robust antioxidant activity to *D. nobile*, aiding the neutralization of free radicals and protection against oxidative damage (Li *et al.*, 2023). The immunomodulatory activity of polysaccharides enhances the immune function and promotes overall health (Li *et al.*, 2023). Additionally, compounds such as alkaloids and phenanthrenes contribute to their antitumour activity by inhibiting tumour cell proliferation and inducing apoptosis in cancerous cells (Li *et al.*, 2023). The polysaccharides found in *D. nobile* also exhibit anti-diabetes activity by regulating blood glucose levels and enhancing insulin sensitivity (Li *et al.*, 2023). Furthermore, its neuroprotective properties suggest potential benefits in protecting against neurodegenerative diseases and in enhancing cognitive function (Li *et al.*, 2023). Moreover, certain compounds, such as bibenzyls, demonstrate antimicrobial activity against various microorganisms, including bacteria and fungi, while their reported anti-ageing properties may help mitigate oxidative stress, promote collagen synthesis and improve skin health (Fan *et al.*, 2023). These multifaceted pharmacological activities underscore need for ongoing research to delve deeper into the potential therapeutic applications of *D. nobile* across a spectrum of health conditions. Nevertheless, it is paramount to emphasise that *D. nobile* should be used for medicinal purposes under the supervision of a qualified healthcare professional (Zhang *et al.*, 2023a).

The chemical constituents and pharmacological activities of *D. nobile* are important because of their potential medicinal properties and health benefits. *D. nobile* holds a prominent place in traditional Chinese medicine because of its multifaceted medicinal properties, which include antioxidant, immunomodulatory and neuroprotective effects, among others, attributed to its diverse array of chemical constituents (Wang, 2021). Beyond traditional use, research indicates that *D. nobile* and its chemical constituents offer a spectrum of health benefits, including immune-boosting properties, enhancement of physical performance, cognitive improvement, support for gastrointestinal health and overall promotion of well-being, making it a valuable resource in both traditional and modern medicine, as well as in functional foods and dietary supplements (Zhang *et al.*, 2021a). Moreover, the chemical constituents of *D. nobile* present opportunities for drug discovery and development, serving as promising leads for investigating potential therapeutic applications, and offering natural sources of novel drugs or drug prototypes for various diseases and conditions (Teixeira da Silva and Ng, 2017). Expanding its utility, *D. nobile* finds application in nutraceuticals and functional foods, where it contributes to the development of health-promoting formulations, dietary supplements and functional food products, providing consumers with natural and safe options to improve their health and well-being (Zhang *et al.*, 2019). The chemical constituents and pharmacological activities of *D. nobile* are of substantial significance owing to their potential medicinal properties, health advantages and utilization in functional foods and nutraceuticals. Continued research and exploration into the imperative can offer valuable insights into its potential therapeutic uses and aid in the formulation of safe and efficacious treatments for a myriad of health conditions.

3.2 Chemical Constituents of *D. nobile*

The chemical composition of *D. nobile* is subject to variation influenced by factors such as geographical location, growth stage and processing techniques. Current research endeavours seek to comprehensively understand the diverse chemical constituents of *D. nobile* and their corresponding pharmacological impacts (Table 3.1 and Fig. 3.1). It is intricate and varied, and their interactions and synergistic effects collectively contribute to the plant's overall pharmacological properties. However, given the evolving nature of scientific studies in this domain, it is essential to understand the chemical composition and pharmacological attributes of *D. nobile*. The chemical constituents of *D. nobile* are outlined in the following sections.

3.2.1 Alkaloids

D. nobile contains various alkaloids, which are organic compounds that contain basic nitrogen atoms. Some alkaloids identified in *D. nobile* include dendrobine, dendramine and nobiline. These alkaloids have been documented to have analgesic and anti-cancer properties, and *D. nobile* shows noteworthy pharmacological properties.

Table 3.1. Major chemical constituents of *D. nobile* (authors' own table).

Constituent	Classification	Biological activity
Polysaccharides	Water-soluble	Immunomodulatory, antitumour, anti-inflammatory
Alkaloids	Alkaloid	Antitumour, anti-inflammatory, antiviral,
Flavonoids	Flavonoid	Anti-inflammatory, anti-cancer, antiviral, antibacterial
Bibenzyls	Phenanthrene	Antioxidant, anti-inflammatory, antitumour
Phenols	Phenol	Antioxidant, anti-inflammatory, antitumour
Acetylenes	Acetylene	Anti-inflammatory, antibacterial
Terpenoids	Terpenoids	Anti-inflammatory, antitumour, immunomodulatory
Sterols	Sterol	Antitumour, anti-inflammatory
Anthocyanins	Anthocyanin	Antioxidant, anti-inflammatory
Hydroquinones	Hydroquinone	Anti-inflammatory, antitumour
Catechins	Catechin	Antioxidant, anti-inflammatory
Orchinols	Bibenzyl	Antitumour, anti-inflammatory
Phenanthrenes	Phenanthrene	Antitumour, anti-inflammatory
Phenanthrene derivatives	Phenanthrene derivative	Antitumour, anti-inflammatory
Glycosides	Glycoside	Antitumour, anti-inflammatory, antibacterial
Organic acids	Organic acid	Antitumour, anti-inflammatory, antiviral

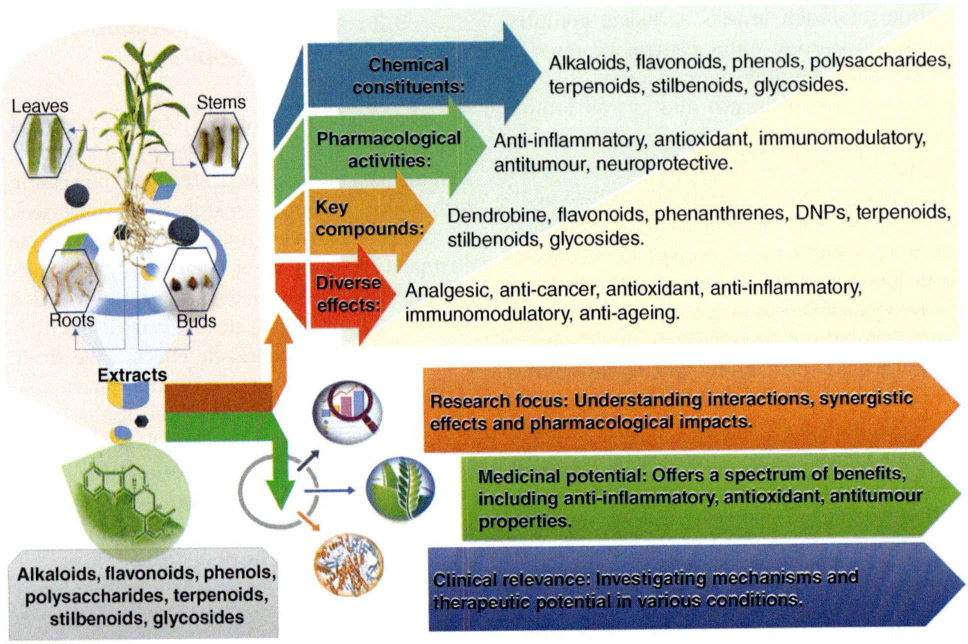

Fig. 3.1. Major chemical constituents, pharmacological activities, key compounds and diverse effects of *D. nobile* (authors' own figure). DNP, *D. nobile* polysaccharides.

Dendrobine has been shown to reduce inflammation by inhibiting the generation of proinflammatory cytokines. In contrast, dendramine exhibits anti-cancer effects through mechanisms involving cell cycle arrest and apoptosis induction in cancerous cells (Zhang *et al.*, 2021b).

3.2.2 Flavonoids

Flavonoids are a class of polyphenolic compounds that are extensively distributed in plants, including *D. nobile*. *D. nobile* harbours diverse types of flavonoids, such as flavones, flavonols and flavanones. Renowned for their antioxidant, anti-inflammatory and antitumour properties, flavonoids found in *D. nobile* exemplify these pharmacological properties. For instance, certain flavonoids in *D. nobile* have been documented to demonstrate potent antioxidant effects in addition to scavenging free radicals, and flavonoids have also been shown to mediate immune responses. Moreover, flavonoids affect cancer cells and hinder angiogenesis (Wu *et al.*, 2023).

3.2.3 Phenols

Phenols constitute a category of organic compounds that are extensively present in *D. nobile*, including phenanthrenes, bibenzyls and phenolic acids. Renowned for their manifold pharmacological activities, phenols found in *D. nobile* demonstrate antioxidant, anti-inflammatory and anti-cancer properties. For instance, certain phenanthrenes identified in *D. nobile* have shown antioxidant properties by neutralizing reactive oxygen species, thereby shielding cells from oxidative harm. Phenolic acids display anti-inflammatory properties by inhibiting the synthesis of proinflammatory agents and regulating immune reactions. Furthermore, phenols have been shown to have anti-cancer effects, inhibit cancer cell growth, promote apoptosis and impede angiogenesis (Liu *et al.*, 2023a).

3.2.4 Polysaccharides

Polysaccharides, which are intricate carbohydrates abundant in *D. nobile*, constitute a notable component with diverse pharmacological attributes.

D. nobile polysaccharides (DNPs) have garnered attention because of their immunomodulatory, antitumour and anti-ageing effects. DNPs are reported to regulate immune responses by augmenting the activity of immune cells, such as macrophages and natural killer (NK) cells, as well as by promoting cytokine production. Moreover, DNPs demonstrate antitumour activities by impeding cancer cell proliferation, fostering apoptosis and inhibiting angiogenesis. Additionally, DNPs exhibit anti-ageing properties and enhance cellular antioxidant defence mechanisms (Xie *et al.*, 2016).

3.2.5 Other bioactive compounds

D. nobile also encompasses a plethora of other bioactive compounds, including terpenoids, stilbenoids and glycosides, which likely contribute to its pharmacological repertoire. Terpenoids are known for their diverse pharmacological properties. Stilbenoids are another class of polyphenolic compounds with anti-inflammatory and anti-cancer activities. Glycosides are organic compounds that contain sugar molecules. These compounds collectively confer a spectrum of beneficial effects, including anti-inflammatory, antioxidant, antitumour, immunomodulatory and anti-ageing properties. Continuing research efforts are focused on fully understanding the bioactive compounds from *D. nobile*.

3.3 Pharmacological Activities of *D. nobile*

D. nobile, a widely utilized traditional Chinese medicinal plant, has garnered increasing attention owing to its myriad pharmacological properties (Table 3.2). Here we summarize recent progress in unravelling the phytochemical components and pharmacological properties of *D. nobile*. The phytochemical profile of *D. nobile* includes alkaloids, flavonoids, phenols, polysaccharides and various other bioactive compounds. Pharmacological investigations have revealed a diverse array of activities associated with *D. nobile*, including anti-inflammatory, antioxidant, immunomodulatory, antitumour, neuroprotective, anti-diabetes and anti-ageing effects. Pharmacological activities of some bioactive compounds present in *D. nobile* exert their effects through various mechanisms, such as inhibiting inflammatory pathways, scavenging reactive oxygen species, modulating immune responses, inducing apoptosis, inhibiting angiogenesis, protecting nerve cells, increasing glucose metabolism and enhancing antioxidant defence systems. The underlying mechanisms and clinical

Table 3.2. Pharmacological activities of *D. nobile* (authors' own table).

Activity/area of action	Constituents
Immunomodulatory	Polysaccharides, terpenoids
Antitumour	Alkaloids, bibenzyls, phenanthrenes, phenanthrene derivatives, polysaccharides, sterols, orchinols, organic acids
Anti-inflammatory	Flavonoids, bibenzyls, phenanthrenes, hydroquinones, terpenoids, organic acids
Antiviral	Alkaloids, flavonoids, organic acids
Antibacterial	Acetylenes, glycosides
Antioxidant	Phenols, anthocyanins, catechins
Hepatoprotective	Polysaccharides, phenanthrenes
Neuroprotective	Phenanthrenes, polysaccharides
Cardiovascular-protective	Flavonoids, phenols, polysaccharides
Anti-diabetes	Polysaccharides, flavonoids
Anti-obesity	Polysaccharides, flavonoids
Anti-ageing	Polysaccharides, phenols
Bone health	Polysaccharides, flavonoids
Skin health	Polysaccharides, flavonoids, phenols

potential of *D. nobile*'s pharmacological activities are outlined in Fig. 3.2.

3.3.1 Sports performance enhancement

D. nobile has been traditionally used in Chinese medicine to enhance physical performance and has gained attention for its potential sports performance-enhancing properties. Nevertheless, it is crucial to acknowledge that scientific evidence in this domain remains limited, necessitating further research to comprehensively understand the impact of *D. nobile* on sports performance. Some studies have suggested that *D. nobile* may enhance sports performance through various mechanisms, such as increasing energy metabolism, improving endurance and reducing fatigue. It has been reported that *D. nobile* extract may increase the release of adrenaline and noradrenaline, which are hormones involved in energy metabolism and the 'fight or flight' response (Cervia and Perrotta, 2019). This may lead to increased energy levels, alertness and endurance during physical activities. Additionally, *D. nobile* has been shown to possess anti-fatigue properties, possibly by reducing the accumulation of lactic acid in the muscles and improving recovery from exercise-induced fatigue (Zhou and Jiang, 2019).

Furthermore, *D. nobile* has been reported to have potential anabolic effects that may contribute to its sports performance-enhancing properties (Conley and Campisi, 2017). This may be beneficial for athletes and individuals engaged in strength training or resistance exercise.

It is important to note that the evidence on the sports performance-enhancing effects of *D. nobile* is limited and conflicting. Moreover, when using any dietary supplements, including *D. nobile* or its extracts, for sports performance enhancement, safety and efficacy considerations are important. Athletes should also be aware of the regulations and restrictions regarding the use of dietary supplements in their respective sports, as some substances may be prohibited by anti-doping agencies.

3.3.2 Skin health

Skin health encompasses the holistic state and condition of the skin, the body's largest organ, and is essential for shielding against external environmental factors and maintaining overall well-being. Maintaining healthy skin is important for both cosmetic and medical reasons because the skin serves as a barrier against harmful microorganisms, regulates body temperature and plays a role in sensory perception (Conley and Campisi, 2017).

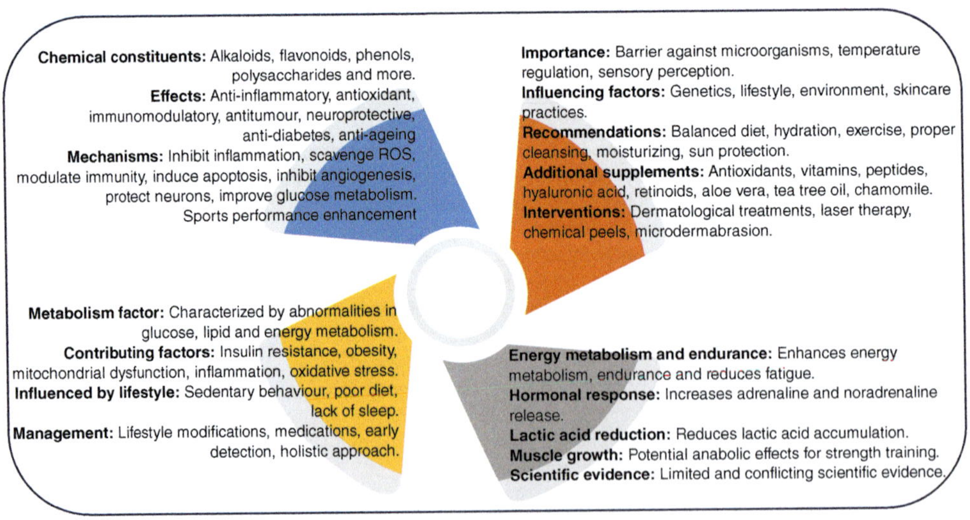

Fig. 3.2. Pharmacological activities of *D. nobile* (authors' own figure).

Various factors can affect skin health, including genetics, lifestyle, environmental factors and skincare practices. Prioritizing a wholesome lifestyle encompassing a well-rounded diet, consistent physical activity, proper hydration and ample sleep can enhance overall skin well-being. Additionally, practising good skincare habits, such as proper cleansing, moisturizing and sun protection, can help maintain the integrity and health of the skin (Knaggs and Lephart, 2023). There are also numerous skincare products and treatments that are specifically formulated to promote skin health, such as moisturisers, sunscreens, exfoliators, serums and masks. These products may contain various active ingredients that provide specific benefits to the skin, such as antioxidants, vitamins, peptides, hyaluronic acid and retinoids. Moreover, some natural ingredients, such as aloe vera, tea tree oil, chamomile, green tea extract and calendula, have traditionally been used to promote skin health. These constituents may offer anti-inflammatory, antioxidant and moisturizing attributes, aiding in the soothing and nourishment of the skin.

In addition to lifestyle choices and skincare products, certain medical and cosmetic interventions, such as dermatological treatments, laser therapies, chemical peels and microdermabrasion, may be employed to address specific skin health issues, including acne, hyperpigmentation, scars and signs of ageing. It is essential to acknowledge that individual skin types and conditions can vary significantly, and what proves effective for one individual may not yield similar results for another. Seeking guidance from a dermatologist or a skincare professional can provide personalized advice and recommendations tailored to individual needs and concerns, ensuring optimal skin health maintenance.

3.3.3 Metabolic disorders

Metabolic disorders are a cluster of medical conditions characterized by abnormalities in typical metabolic functions, including those related to glucose, lipid and energy metabolism (Aron-Wisnewsky *et al.*, 2021). The pathophysiology of metabolic disorders is complex and involves multiple genetic, environmental and lifestyle-related factors. Insulin resistance is often associated with obesity, particularly visceral adiposity, and can lead to a cascade of metabolic abnormalities including dyslipidaemia, hyperglycaemia and systemic inflammation (Rani *et al.*, 2016). In addition to insulin resistance, other mechanisms involved in the development of metabolic disorders include impaired mitochondrial function, chronic low-grade inflammation, oxidative stress, altered gut microbiota and hormonal dysregulation. These mechanisms interact with each other and contribute to the dysfunction of various metabolic pathways, resulting in metabolic imbalances and related health issues (Golonka *et al.*, 2020). Lifestyle factors, such as sedentary behaviour, poor diet quality and inadequate sleep, can also play a significant role in the development and progression of metabolic disorders. These factors can disrupt normal metabolic processes such as glucose homeostasis, lipid metabolism and energy expenditure, leading to metabolic dysfunction.

The diagnosis and management of metabolic disorders often involves a multifaceted approach, including lifestyle modifications (e.g. diet, physical activity, sleep), medications (e.g. insulin sensitizers, lipid-lowering drugs) and other targeted interventions. Early detection and intervention, as well as a holistic approach addressing various contributing factors, are important for effective prevention and management of metabolic disorders (Phillips, 2022). It is worth noting that our understanding of metabolic disorders is still evolving, and additional research is required to comprehensively understand the underlying mechanisms and devise more efficient strategies for the prevention, diagnosis and management of these conditions. Qualified healthcare professionals should seek proper medical evaluation and management for individuals with suspected or diagnosed metabolic disorders.

3.4 Neurodegenerative Diseases

3.4.1 Alzheimer's disease

Alzheimer's disease is a devastating neurological condition marked by the aggregation of beta-amyloid plaques and tau protein tangles in the brain, leading to a gradual decline in cognitive abilities and memory function. *D. nobile*, which has purported neuroprotective properties, has garnered attention for its potential relevance in management of Alzheimer's disease. Studies suggest that *D. nobile* extracts contain antioxidant compounds that may help mitigate oxidative stress, a key contributor to Alzheimer's pathology. Furthermore, its anti-inflammatory effects could potentially reduce neuroinflammation, which is increasingly being recognized as a contributing factor in Alzheimer's disease progression. While preliminary research hints at the therapeutic potential of *D. nobile*, further investigations, including clinical trials, are necessary to validate its efficacy and safety as a treatment or adjunct therapy for Alzheimer's disease (Rabaneda-Bueno *et al.*, 2021).

3.4.2 Parkinson's disease

Parkinson's disease is a progressive neurodegenerative disorder characterized by the degeneration of dopamine-producing neurons in the brain, resulting in motor symptoms, such as tremors, stiffness and impaired balance, alongside non-motor symptoms, such as cognitive impairment and mood disturbances. Research suggests that *D. nobile* may have the potential for managing Parkinson's disease. Its antioxidant compounds can help to reduce oxidative stress and protect neurons from damage. Additionally, its neurotrophic effects might support the survival and function of dopaminergic neurons, which are particularly affected in Parkinson's. However, further studies are required to explore the specific mechanisms and efficacy of *D. nobile* in Parkinson's disease management (Huang *et al.*, 2022).

3.4.3 Huntington's disease

Huntington's disease is an inherited neurodegenerative condition caused by a mutation in the huntingtin gene that leads to the accumulation of abnormal proteins in the brain, particularly affecting the basal ganglia, resulting in progressive motor dysfunction, cognitive decline and psychiatric symptoms. While there is currently no cure for Huntington's disease, research into potential therapeutic approaches includes investigating natural compounds with neuroprotective properties, such as those found in *D. nobile*. Although specific studies linking *D. nobile* to Huntington's disease are limited, its antioxidant and neurotrophic properties make it a potential candidate for exploration in Huntington's disease management. Antioxidants can help mitigate oxidative stress, which contributes to neuronal damage in Huntington's disease, whereas neurotrophic factors may support the survival and function of neurons affected by the disease. Further research is needed to elucidate the efficacy and safety of *D. nobile* in the context of Huntington's disease and to determine its potential as a therapeutic agent or adjunct therapy for Huntington's disease treatment (Ciurea *et al.*, 2023).

3.4.4 Amyotrophic lateral sclerosis

Amyotrophic lateral sclerosis (ALS), also known as Lou Gehrig's disease, is a relentless neurodegenerative disorder characterized by degeneration of motor neurons in the brain and spinal cord, causing muscle weakness, paralysis and eventual respiratory failure. Although ALS primarily affects motor function, it can also involve cognitive and behavioural changes. Research on *D. nobile* suggests that it may offer avenues for exploration of ALS management. The antioxidant compounds found in *D. nobile* extracts could potentially counteract oxidative stress, which plays a role in ALS pathogenesis. Additionally, its neurotrophic effects may support the survival and function of motor neurons affected by the disease. However, the specific mechanisms and efficacy of *D. nobile* in ALS warrant further investigation through preclinical and clinical studies to determine its potential as a therapeutic intervention for this debilitating condition (Mead *et al.*, 2023).

3.4.5 Frontotemporal dementia

Frontotemporal dementia (FTD) encompasses a group of neurodegenerative disorders characterized by gradual deterioration of the frontal and temporal lobes of the brain, leading to changes in behaviour, personality and language abilities. FTD is a significant cause of dementia, particularly in young individuals. Research into potential treatments for FTD has led to investigations into natural compounds with neuroprotective properties, such as those found in *D. nobile*. Although direct studies linking *D. nobile* to FTD are limited, its antioxidant and neurotrophic properties make it a promising candidate for exploration in FTD management. Antioxidants can help mitigate oxidative stress, which is implicated in neuronal damage in FTD, whereas neurotrophic factors may support the survival and function of neurons affected by the disease. Further research is necessary to elucidate the specific mechanisms and efficacy of *D. nobile* in the context of FTD and to determine its potential as a therapeutic intervention or adjunct therapy for FTD (Panza *et al.*, 2020).

3.4.6 Multiple sclerosis

Multiple sclerosis is a chronic autoimmune disease characterized by inflammation and damage to the myelin sheath surrounding the nerve fibres in the brain and spinal cord, resulting in symptoms such as muscle weakness, fatigue and impaired coordination. Although multiple sclerosis primarily affects the central nervous system, it can also involve the degeneration of neurons over time. Research on potential treatments for multiple sclerosis has led to interest in natural compounds with neuroprotective properties, including those from *D. nobile*. Although direct studies linking *D. nobile* to multiple sclerosis are limited, its antioxidant and anti-inflammatory properties suggest potential benefits in mitigating the oxidative stress and inflammation associated with multiple sclerosis pathology. Additionally, its neurotrophic effects may support the preservation of neuronal integrity and function, potentially alleviating symptoms and slowing disease progression. Further research is needed to elucidate the specific mechanisms and efficacy of *D. nobile* in disease management and to determine its potential as a therapeutic agent or adjunct therapy for multiple sclerosis treatment (Eva *et al.*, 2023).

3.4.7 Prion diseases

Prion diseases represent a rare group of progressive neurodegenerative disorders triggered by abnormal infectious proteins called prions, which lead to irreversible damage to the brain and nervous system. These proteins induce abnormal folding of other normal proteins in the brain, leading to neuronal damage and formation of protein aggregates. While prion diseases are not as common as other neurodegenerative disorders, they are often fatal and include conditions such as Creutzfeldt–Jakob disease and variant Creutzfeldt–Jakob disease. Research into potential treatments for prion diseases has led scientists to explore natural compounds with neuroprotective properties, such as those found in *D. nobile*. Although direct studies linking *D. nobile* to prion diseases are limited, their antioxidant and anti-inflammatory properties suggest potential benefits in mitigating the oxidative stress and inflammation associated with prion pathology. Additionally, its neurotrophic effects may support the preservation of neuronal integrity and function, potentially slowing the progression of prion diseases. Further research is needed to elucidate the specific mechanisms and efficacy of *D. nobile* in prion disease management and to determine its potential as a therapeutic agent or adjunct therapy for treating these devastating conditions (Thellung *et al.*, 2022).

3.5 Anti-inflammatory and Immunomodulatory Effects

D. nobile encompasses diverse chemical constituents, including alkaloids, flavonoids, phenols and polysaccharides, which have been scrutinized for their pharmacological attributes, including anti-inflammatory and immunomodulatory effects. Multiple studies have underscored the anti-inflammatory effects of *D. nobile* and its chemical constituents. Notably, alkaloids derived from *D. nobile* have demonstrated substantial anti-inflammatory activity both *in vitro* and *in vivo*. They have been shown to inhibit production of proinflammatory cytokines, pivotal signalling molecules in inflammation, and mitigate inflammation in animal models, thereby suggesting their potential as anti-inflammatory agents (Li *et al.*, 2022a). Additionally, polysaccharides extracted from *D. nobile* have demonstrated immunomodulatory activity. Polysaccharides, large carbohydrate entities renowned for their immune-regulatory traits, have been shown to bolster immune function by upregulating cytokine production and fostering lymphocyte proliferation, pivotal constituents of the immune system (Xie *et al.*, 2023).

Moreover, other chemical constituents of *D. nobile*, such as flavonoids and phenols, have been shown to possess anti-inflammatory and immunomodulatory properties. Flavonoids are plant-derived compounds known for their antioxidant and anti-inflammatory properties, along with phenols and aromatic compounds capable of eliciting anti-inflammatory effects. Nonetheless, further research is required to comprehensively elucidate the mechanisms of action and to ascertain the efficacy and safety of these chemical constituents in *D. nobile* for anti-inflammatory and immunomodulatory purposes (Fan *et al.*, 2023). It is crucial to underscore that while these studies suggest potential anti-inflammatory and immunomodulatory ramifications of *D. nobile* and its chemical constituents, additional research is imperative to validate these findings, delineate optimal dosage and administration, and ensure safety. It is strongly advised to consult a qualified healthcare professional before utilizing any natural product or its chemical constituents for purported health benefits.

3.6 Digestive Disorders

Digestive disorders encompass a broad spectrum of conditions affecting organs within the digestive tract, including the oesophagus, stomach, intestine, liver, gall bladder and pancreas. These disorders may arise from various causes including genetic predispositions, infections, dietary patterns, lifestyle choices, autoimmune reactions and environmental factors. There may be an array of symptoms, such as abdominal discomfort, bloating, diarrhoea, constipation, nausea, vomiting and alterations in bowel habits.

Gastro-oesophageal reflux disease (GORD) is characterized by frequent backflow of stomach acid into the oesophagus, leading to heartburn, regurgitation and chest discomfort (Gyawali and Fass, 2018). Inflammatory bowel disease, including Crohn's disease and ulcerative colitis, involves chronic inflammation of the intestine, resulting in symptoms such as abdominal pain, diarrhoea and rectal bleeding (Gajendran *et al.*, 2019). Irritable bowel syndrome (IBS) is a functional disorder of the digestive tract that causes symptoms such as abdominal discomfort, bloating, irregular bowel habits, and diarrhoea without structural abnormalities (Talley, 2020). Coeliac disease, an autoimmune disorder triggered by gluten consumption, damages the lining of the small intestine and leads to symptoms such as abdominal pain, diarrhoea and malabsorption (Kumar *et al.*, 2024). Gallstones and solid deposits in the gall bladder can cause abdominal pain, nausea, vomiting and jaundice when obstructing bile ducts (Sivakrishnan and Pharm, 2019). Hepatitis, characterized by liver inflammation due to viral infections, alcohol consumption or autoimmune reactions, presents with symptoms such as fatigue, jaundice, abdominal discomfort and impaired liver function (Bergasa, 2022).

The treatment of digestive disorders varies depending on the specific condition and severity of the symptoms. This may encompass lifestyle adjustments, dietary modifications, medications, surgical procedures and other therapeutic interventions. Accurate diagnosis and expert management by qualified healthcare practitioners are essential to ensure the effective treatment and control of digestive disorders. It is imperative to recognise that this list does not encompass all digestive disorders as numerous conditions can affect the digestive system. Patients suspecting they have a digestive disorder or experiencing symptoms related to digestion should seek guidance from a qualified healthcare professional for thorough evaluation, diagnosis and appropriate treatment.

3.7 Mechanisms of Action

D. nobile exhibits a range of pharmacological activities attributed to its various chemical constituents, including polysaccharides, alkaloids, phenols and flavonoids. *D. nobile* demonstrates a diverse array of pharmacological activities, making it a promising candidate for various therapeutic applications. Its immunomodulatory activity, evidenced by augmented lymphocyte proliferation and increased cytokine production, is likely attributable to its polysaccharide and alkaloid constituents (Fan *et al.*, 2020). Additionally, *D. nobile* exhibits antioxidant properties, scavenging free radicals and reducing lipid peroxidation, mainly owing to its flavonoid and phenolic constituents (Lei *et al.*, 2022). Its neuroprotective effects, which protect against oxidative stress, apoptosis and inflammation in the brain, are associated with its alkaloids, polysaccharides and phenolic components (Zhang *et al.*, 2023b). Moreover, *D. nobile* displays anti-inflammatory activity by inhibiting the production of inflammatory cytokines and enzymes, which is likely mediated by its alkaloid and phenolic constituents (Chen *et al.*, 2022). Its anti-cancer potential, attributed to the inhibition of cancer cell proliferation and induction of apoptosis in cancer cells, is linked to its polysaccharide, alkaloid and phenolic constituents (Liu *et al.*, 2023b). Furthermore, *D. nobile* exhibits anti-diabetes activity by regulating glucose metabolism, enhancing insulin sensitivity and mitigating oxidative stress, largely owing to its polysaccharide and alkaloid constituents (Li *et al.*, 2019). Additionally, its antihyperlipidaemic and anti-obesity activities, as evidenced by reduced serum lipid levels, improved lipid metabolism and inhibited adipocyte differentiation, are associated with its alkaloid and polysaccharide components (Hei *et al.*, 2021; Zhang *et al.*, 2023b). These findings underscore the potential of *D. nobile* as a valuable therapeutic agent for various health conditions.

3.8 Drug Interactions and Safety Considerations

D. nobile is generally considered safe for human consumption. However, as with any natural product or medication, potential drug interactions and safety considerations should be considered (Fu *et al.*, 2023). It is crucial to acknowledge the scarcity of studies on the potential drug interactions in *D. nobile*. As such, prudence is advised when combining *D. nobile* with other medications or supplements. Consulting a healthcare professional before initiating a new supplement or medication regimen is strongly recommended (Pawar and Grundel, 2017). *D. nobile* has been reported to cause hypoglycaemia in individuals with diabetes who are taking insulin or other diabetes medications. Therefore, individuals with diabetes should monitor their blood sugar levels closely while taking *D. nobile* and consult with their healthcare provider (Li *et al.*, 2022b). Additionally, *D. nobile* may interact with medications metabolized by the liver. This is because *D. nobile* contains alkaloids, which may induce or inhibit liver enzymes involved in drug metabolism. Therefore, individuals taking medications that are metabolized by the liver should seek guidance from their healthcare provider before using *D. nobile* (Zhang *et al.*, 2023b).

There is also a potential risk of allergic reactions to *D. nobile*, especially in individuals who are allergic to other members of the Orchidaceae family. If symptoms such as itching, rash, swelling or difficulty breathing manifest, they could indicate an allergic reaction. Individuals should discontinue the use of *D. nobile* and seek medical attention (Crislip, 2018). Finally, it is important to note that *D. nobile* may have a stimulant effect and can increase heart rate and blood pressure. Hence, individuals with heart conditions or high blood pressure should exercise caution when taking *D. nobile* and consult their healthcare provider (Zhang *et al.*, 2023b). Although *D. nobile* is widely regarded as safe, it remains crucial to acknowledge potential drug interactions and safety precautions. It is advisable for individuals to seek guidance from their healthcare provider before incorporating *D. nobile* into their regimen, particularly if they have underlying health conditions or are concurrently taking medications (Ramesh *et al.*, 2019).

3.8.1 Potential drug interactions

There is limited information available regarding potential drug interactions with *D. nobile*. However, as with any herbal supplement or medication, it is important to exercise caution and consult with a healthcare professional before using *D. nobile* in combination with other drugs or supplements. One case report described a potential drug interaction between *D. nobile* and warfarin, a blood-thinning medication (Fu *et al.*, 2023). This report describes a patient who experienced an increase in international normalized ratio (INR), which measures blood clotting time, while taking both *D. nobile* and warfarin. The authors suggest that the interaction may be due to the anticoagulant properties of *D. nobile*, which may increase the risk of bleeding when used in combination with warfarin.

3.8.2 Safety considerations

While *D. nobile* is generally considered safe when consumed in moderate amounts, there are some safety considerations to be considered. One study reported that high doses of *D. nobile* extract may cause liver toxicity in rats (Zhang *et al.*, 2021b). However, further research is needed to determine whether these findings are applicable to humans. Individuals with a history of allergies or hypersensitivity to orchids or other plants in the Orchidaceae family may be at an increased risk of allergic reactions to *D. nobile*. Pregnant and breastfeeding women should exercise caution and consult a healthcare professional before using *D. nobile*, as there is limited information available regarding its safety in these populations (Jędrejko *et al.*, 2021). It is also important to exercise caution when purchasing *D. nobile* supplements, as some products may be adulterated with other substances or contain inaccurate labelling. Overall, it is recommended to consult a healthcare professional before using *D. nobile* supplements or incorporating them into a health regimen.

3.8.3 Safety considerations and adverse effects

D. nobile appears to be safe when consumed in recommended amounts. However, there have been reports of adverse effects associated with its use, including gastrointestinal discomfort, headaches, dizziness and allergic reactions (Pawar and Grundel, 2017). The safety of *D. nobile* supplements has not been extensively studied, and there is limited information available on their long-term use. Therefore, caution should be exercised, and a healthcare provider consulted before using *D. nobile* supplements, particularly for individuals with pre-existing medical conditions or those on medication (Rahman and Aziz, 2020). Additionally, there have been some concerns about the potential contamination of *D. nobile* supplements with heavy metals or other harmful substances due to the plant's ability to absorb pollutants from the soil. Hence, acquiring supplements from reputable sources that have undergone thorough quality and safety tests is crucial. Additionally, it is imperative to recognise that *D. nobile* supplements should not replace medical treatment, and individuals should always follow the recommended dosage instructions provided by the manufacturer. If any adverse effects are experienced, use should be discontinued immediately and medical attention should be sought if necessary (Zhang *et al.*, 2023c). Overall, while *D. nobile* appears to have several potential health benefits, caution should be exercised when using supplements.

3.9 Conclusion

D. nobile, a plant steeped in the annals of traditional Chinese medicine for centuries, contains a diverse array of chemical compounds, including polysaccharides, alkaloids, flavonoids and phenolics. Collectively, these constituents underpin their multifaceted pharmacological properties. Research has shown that *D. nobile* exhibits a range of beneficial effects, including immunomodulatory, antioxidant, neuroprotective, anti-inflammatory, anti-cancer, anti-diabetes, antihyperlipidaemic and anti-obesity activities. These effects are attributed to the various chemical constituents of *D. nobile* and their ability to interact with different cellular and molecular targets in the body. Although *D. nobile* is generally considered safe and well tolerated, some potential drug interactions and safety considerations should be taken into account. Patients should consult with a healthcare professional before using *D. nobile* in combination with other medications or if they have any underlying medical conditions. Overall, *D. nobile* is a promising natural source for the development of novel therapeutics for various diseases. Further research is needed to elucidate its mechanisms of action, optimise its pharmacological activities and evaluate its long-term safety and efficacy in humans.

3.9.1 Summary of chemical constituents and pharmacological activities

D. nobile, an esteemed botanical remedy deeply entrenched in traditional Chinese medicine for centuries, harbours a rich assortment of chemical constituents, notably alkaloids, phenols, flavonoids and polysaccharides, each showcasing a diverse spectrum of pharmacological virtues. The alkaloids and phenols in *D. nobile* have been shown to exhibit antioxidative, anti-inflammatory and neuroprotective properties, whereas flavonoids exhibit antitumour, antiviral and immunomodulatory properties. Furthermore, the polysaccharides present in this botanical marvel display immunomodulatory, anti-inflammatory and antitumour potential. Emerging research underscores the therapeutic versatility of *D. nobile*, hinting at its potential to address an array of maladies, including Alzheimer's disease, Parkinson's disease, cancer, diabetes and cardiovascular disorders. Moreover, it may augment physical performance and fortify the immune system. However, prudent consideration must be given to safety profiles and possible drug interactions. Although studies suggest its overall safety when consumed judiciously, excessive intake should be avoided. Notably, caution is advised concerning potential interactions, particularly with medications prescribed for diabetes and hypertension. Collectively, these findings underscore the auspicious pharmacological panorama of *D. nobile*, offering tantalizing therapeutic prospects across various ailments. However, comprehensive research is imperative to unravel the mechanisms of action, safety parameters and therapeutic efficacy.

3.9.2 Future perspectives and research directions

Exploration of the chemical constituents and pharmacological activities of *D. nobile* has paved the way for novel avenues of drug discovery and development. Nonetheless, several challenges persist, necessitating a thorough examination to unlock the full potential of this botanical resource. One critical avenue for future inquiry lies in pinpointing the active compounds responsible for *D. nobile*'s diverse pharmacological actions. Although numerous compounds have been identified, elucidating their precise mechanisms of action and potential synergies remains elusive. Employing advanced methodologies, such as metabolomics, proteomics and transcriptomics, holds promise in unravelling these complexities. Another imperative research domain is assessment of the safety and toxicity profiles of *D. nobile*. Despite its extensive traditional use, establishing safety parameters in modern clinical settings remains paramount. Animal studies and clinical trials offer invaluable insights into the potential adverse effects and interactions with other medications. Furthermore, there is a pressing need for in-depth investigation of the cultivation, harvesting and processing of *D. nobile*. Implementing sustainable cultivation practices and standardized processing techniques is crucial for ensuring the quality and consistency of plant material, which is pivotal for the development of safe and efficacious pharmaceuticals. *D. nobile* is a promising reservoir of bioactive compounds with diverse pharmacological potential. While existing research forms a robust foundation for drug discovery endeavours, continued investigation is indispensable to harness the complete therapeutic arsenal and guarantee its safe integration into contemporary medical practices. The chemical constituents and pharmacological activities of *D. nobile* have promising prospects and delineate vital research trajectories.

The intricate molecular mechanisms underlying the pharmacological attributes of *D. nobile* require further exploration. This entails delving into the intricate signalling pathways, target receptors and cellular processes implicated in its diverse pharmacological effects, including its antioxidant, anti-inflammatory and anti-cancer activities. Advanced molecular and cellular techniques, along with animal models, offer promising avenues to unravel these mechanisms and provide invaluable insights into their intricacies. Despite the identification of numerous chemical constituents in *D. nobile*, a more exhaustive analysis of its chemical profile remains imperative. Employing sophisticated analytical methodologies such as mass spectrometry and nuclear magnetic resonance spectroscopy can facilitate the identification and characterization of the novel active compounds present in *D. nobile*. Moreover, investigating the synergistic interactions among these compounds and elucidating their mechanisms of action are crucial research frontiers.

Further investigation of the pharmacokinetics and pharmacodynamics of *D. nobile* in human subjects is also warranted. This encompasses studies elucidating its absorption, distribution, metabolism and excretion as well as discerning its pharmacological effects across various doses and treatment durations. Such investigations hold the potential to optimise the dosage regimens and administration protocols of *D. nobile* for diverse medical conditions, thereby ensuring its safety and efficacy in clinical settings.

Further investigations are warranted to comprehensively evaluate the safety profile and potential toxicity of *D. nobile* extract. This entails conducting studies on both acute and chronic toxicity as well as exploring potential interactions with concomitant medications. Safety assessments using animal models and human clinical trials are imperative to provide critical data on the safety parameters of *D. nobile*, which is pivotal for its therapeutic application. Moreover, there is a pressing need for well-designed and robust clinical trials to ascertain the efficacy of *D. nobile* across various medical conditions, including cancer, diabetes and neurodegenerative diseases. Clinical trials with large sample sizes, stringent controls and extended follow-up durations can yield substantial evidence regarding the therapeutic efficacy of *D. nobile* in human subjects. Furthermore, the development of innovative formulations and delivery systems holds promise in enhancing the therapeutic potential of *D. nobile*. Investigations aimed at optimizing extraction techniques, formulation methodologies and delivery systems, such as nanoparticles, liposomes and microemulsions, could augment the bioavailability, stability and efficacy of *D. nobile* for diverse routes of administration.

Ensuring the quality, efficacy and safety of *D. nobile* products through standardization is paramount. Establishing standardized protocols for the extraction, identification and quantification of active compounds in *D. nobile* will guarantee uniform quality and potency across products. Additionally, implementing robust quality control measures, including adherence to good manufacturing practices, is essential to safeguard the safety and effectiveness of *D. nobile* products. Exploration of the chemical constituents and pharmacological activities of *D. nobile* holds significant promise for the development of innovative therapeutics targeting various diseases. Future research endeavours should prioritise elucidating mechanisms of action, characterizing active compounds, understanding pharmacokinetics and pharmacodynamics, assessing safety and toxicity profiles, conducting rigorous clinical trials and devising novel formulations and delivery systems. Furthermore, establishing standardized protocols and quality control measures will be instrumental in advancing our understanding of *D. nobile* and its potential therapeutic applications.

Acknowledgements

The authors express their gratitude for the financial support received through the Distinguished High-Level Talents Research Grant from the Guizhou Science and Technology Corporation Platform Talents Fund (grant nos. [2017]5733-001 and CK-1130-002), the National Natural Science Foundation of China (U1812403, 82373981 and 82060750), 2011 Collaborative Innovation Centre of Traditional Chinese Medicine in Guizhou Province (no. [2022]026) and the support provided by the Zunyi Medical University, China. Special appreciation is extended to all laboratory colleagues and research staff members for their valuable insights, constructive guidance and assistance with the contents of this chapter.

References

Aron-Wisnewsky, J., Warmbrunn, M.V., Nieuwdorp, M. and Clément, K. (2021) Metabolism and metabolic disorders and the microbiome: the intestinal microbiota associated with obesity, lipid metabolism, and metabolic health-pathophysiology and therapeutic strategies. *Gastroenterology* 160(2), 573–599.

Bergasa, N.V. (2022) Approach to the patient with liver disease. In: Bergasa, N.V. (ed.) *Clinical Cases in Hepatology*. Springer, London.

Cakova, V., Bonte, F. and Lobstein, A. (2017) *Dendrobium:* sources of active ingredients to treat age-related pathologies. *Aging and Disease* 8(6), 827–849.

Cervia, D. and Perrotta, C. (2019) Programmed cell death mechanisms and Pheocromocytomas: recent advances in PC12 cells. In: Fernando Martin, J. (ed.) *Pheochromocytoma - A New View of the Old Problem.* InTech, pp. 85–102.

Chen, H., Shi, X., Zhang, L., Yao, L., Cen, L. *et al.* (2022) Ultrasonic extraction process of polysaccharides from *Dendrobium nobile* Lindl.: optimization, physicochemical properties and anti-inflammatory activity. *Foods* 11(19), 1–16.

Ciurea, A.V., Mohan, A.G., Covache-Busuioc, R.A., Costin, H.P., Glavan, L.A. *et al.* (2023) Unraveling molecular and genetic insights into neurodegenerative diseases: advances in understanding Alzheimer's, Parkinson's, and Huntington's diseases and amyotrophic lateral sclerosis. *International Journal of Molecular Sciences* 24(13), 1–41.

Conley, N. and Campisi, J. (2017) Rocket science: an investigation of Magnum ROCKET SCIENCE on memory, athletic performance, endurance, and energy. *Journal of Student Research* 4(1), 53–62.

Crislip, M. (2018) *Flies in the Ointment*. Bitingduck Press LLC, California.

Eva, L., Pleș, H., Covache-Busuioc, R.A., Glavan, L.A., Bratu, B.G. *et al.* (2023) A comprehensive review on neuroimmunology: insights from multiple sclerosis to future therapeutic developments. *Biomedicines* 11(9), 1–27.

Fan, C., Sun, X., Wang, X. and Yu, H. (2023) Therapeutic potential of the chemical composition of *Dendrobium nobile* Lindl. *Frontiers in Pharmacology* 14, 1163830.

Fan, Y., Yu, Q., Wang, G., Tan, J., Liu, S. *et al.* (2020) Effects of non-thermal plasma treatment on the polysaccharide from *Dendrobium nobile* Lindl. and its immune activities in vitro. *International Journal of Biological Macromolecules* 153, 942–950.

Fu, X., Chen, S., Xian, S., Wu, Q., Shi, J. *et al.* (2023) *Dendrobium* and its active ingredients: emerging role in liver protection. *Biomedicine & Pharmacotherapy* 157, 1–13.

Gajendran, M., Loganathan, P., Jimenez, G., Catinella, A.P., Ng, N. *et al.* (2019) A comprehensive review and update on ulcerative colitis. *Disease-a-Month* 65(12), 100851.

Golonka, R.M., Xiao, X., Abokor, A.A., Joe, B. and Vijay-Kumar, M. (2020) Altered nutrient status reprograms host inflammation and metabolic health via gut microbiota. *The Journal of Nutritional Biochemistry* 80, 108360.

Gyawali, C.P. and Fass, R. (2018) Management of gastroesophageal reflux disease. *Gastroenterology* 154(2), 302–318.

Hei, J.W., Hung, W., Yanbo, Z. and Zhang, Z.-J. (2021) *Dendrobium* in diabetes: a comprehensive review of its phytochemistry, pharmacology, and safety. *Journal of Translational Science* 7(6), 1–10.

Huang, M., Bargues-Carot, A., Riaz, Z., Wickham, H., Zenitsky, G. *et al.* (2022) Impact of environmental risk factors on mitochondrial dysfunction, neuroinflammation, protein misfolding, and oxidative stress in the etiopathogenesis of Parkinson's disease. *International Journal of Molecular Sciences* 23(18), 10808.

Jędrejko, K., Lazur, J. and Muszyńska, B. (2021) Risk associated with the use of selected ingredients in food supplements. *Chemistry & Biodiversity* 18(2), e2000686.

Knaggs, H. and Lephart, E.D. (2023) Enhancing skin anti-aging through healthy lifestyle factors. *Cosmetics* 10(142), 1–27.

Kumar, S., Singh, A., Singh, A.P., Ram, S., Gupta, O.P. *et al.* (2024) Gluten-related disorders: current understanding, myths, and facts. In: Gupta, O.P., Kumar, S., Pandey, A., Khan, M.K., Singh, S.K. and Singh, G.P. (eds) *Wheat Science*. CRC Press, Boca Raton, Florida, pp. 321–338.

Lei, H., Zou, S., Lin, J., Zhai, L., Zhang, Y. *et al.* (2022) Antioxidant and anti-inflammatory activity of constituents isolated from *Dendrobium nobile* (Lindl.). *Frontiers in Chemistry* 10, 1–12.

Li, D.D., Zheng, C.Q., Zhang, F. and Shi, J.S. (2022a) Potential neuroprotection by *Dendrobium nobile* Lindl alkaloid in Alzheimer's disease models. *Neural Regeneration Research* 17(5), 972–977.

Li, M., Trapika, I.G.S.C., Tang, S.Y.S., Cho, J.L., Qi, Y. *et al.* (2022b) Mechanisms and active compounds polysaccharides and bibenzyls of medicinal dendrobiums for diabetes management. *Frontiers in Nutrition* 8, 1–13.

Li, P.Y., Li, L. and Wang, Y.Z. (2023) Traditional uses, chemical compositions and pharmacological activities of dendrobium: a review. *Journal of Ethnopharmacology* 310, 116382.

Li, X.W., Huang, M., Lo, K., Chen, W.L., He, Y.Y. *et al.* (2019) Anti-diabetic effect of a Shihunine-rich extract of *Dendrobium loddigesii* on 3T3-L1 cells and db/db mice by up-regulating AMPK-GLUT4-PPARα. *Molecules* 24(14), 1–20.

Liu, W., Cui, X., Zhong, Y., Ma, R., Liu, B. *et al.* (2023a) Phenolic metabolites as therapeutic in inflammation and neoplasms: molecular pathways explaining their efficacy. *Pharmacological Research* 193, 1–10.

Liu, J., Li, Z., Zeng, M., Liang, Y., Liang, B. *et al.* (2023b) Rapid comparison of antitumor chemical constituents and mechanisms between *Dendrobium nobile* and *Dendrobium officinale* by UPLC-IT-TOF, network pharmacology and experimental verification. *Acta Chromatographica* 35(2), 123–138.

Mead, R.J., Shan, N., Reiser, H.J., Marshall, F. and Shaw, P.J. (2023) Amyotrophic lateral sclerosis: a neurodegenerative disorder poised for successful therapeutic translation. *Nature Reviews Drug Discovery* 22(3), 185–212.

Panza, F., Lozupone, M., Seripa, D., Daniele, A. and Watling, M. (2020) Development of disease-modifying drugs for frontotemporal dementia spectrum disorders. *Nature Reviews Neurology* 16(4), 213–228.

Pawar, R.S. and Grundel, E. (2017) Overview of regulation of dietary supplements in the USA and issues of adulteration with phenethylamines (peas). *Drug Testing and Analysis* 9(3), 500–517.

Phillips, M.C. (2022) *Metabolic strategies in healthcare: a new era. Aging and Disease* 13(3), 655–672.

Rabaneda-Bueno, R., Mena-Montes, B., Torres-Castro, S., Torres-Carrillo, N. and Torres-Carrillo, N.M. (2021) Advances in genetics and epigenetic alterations in Alzheimer's disease: a notion for therapeutic treatment. *Genes* 12(12), 1–28.

Ramesh, T., Koperuncholan, M., Praveena, R., Ganeshkumari, K., Vanithamani, J. *et al.* (2019) Medicinal properties of some *Dendrobium* orchids – a review. *Journal of Applied and Advanced Research* 4(4), 119–128.

Rani, V., Deep, G., Singh, R.K., Palle, K. and Yadav, U.C.S. (2016) Oxidative stress and metabolic disorders: Pathogenesis and therapeutic strategies. *Life Sciences* 148, 183–193.

Rahman, S.S.A. and Aziz, Z. (2020) Complementary and alternative medicine: pharmacovigilance in Malaysia and predictors of serious adverse reactions. *Journal of Clinical Pharmacy and Therapeutics* 45(5), 946–958.

Sivakrishnan, S. and Pharm, M. (2019) Liver disease overview. *World Journal of Pharmacy and Pharmaceutical Sciences* 8(1), 1385–1395.

Talley, N.J. (2020) What causes functional gastrointestinal disorders? A proposed disease model. *American Journal of Gastroenterology* 115(1), 41–48.

Teixeira da Silva, J.A. and Ng, T.B. (2017) The medicinal and pharmaceutical importance of *Dendrobium* species. *Applied Microbiology and Biotechnology* 101(6), 2227–2239.

Thellung, S., Corsaro, A. and Florio, T. (2022) Proteostasis unbalance in prion diseases: mechanisms of neurodegeneration and therapeutic targets. *Frontiers in Neuroscience* 16, 1–17.

Wang, G., Wang, J., Deng, Y., Qin, L., He, Y. *et al.* (2022) Chemical constituents and nutritional health functions of *Dendrobium nobile*: A review. *Food Science and Technology* 42, 1–19.

Wang, Y.H. (2021) Traditional uses, chemical constituents, pharmacological activities, and toxicological effects of *Dendrobium* leaves: a review. *Journal of Ethnopharmacology* 270, 113851.

Wu, W., Lin, Y., Farag, M.A., Li, Z. and Shao, P. (2023) *Dendrobium* as a new natural source of bioactive for the prevention and treatment of digestive tract diseases: a comprehensive review with future perspectives. *Phytomedicine* 114, 154784.

Xie, J.H., Jin, M.L., Morris, G.A., Zha, X.Q. and Chen, H.Q. (2016) Advances on bioactive polysaccharides from medicinal plants. *Critical Reviews in Food Science and Nutrition* 56(Suppl 1), S60–S84.

Xie, Z., Jiang, N., Lin, M., He, X., Li, B. *et al.* (2023) The mechanisms of polysaccharides from tonic chinese herbal medicine on the enhancement immune function: a review. *Molecules* 28(21), 1–24.

Zhang, C.C., Kong, Y.L., Zhang, M.S., Wu, Q. and Shi, J.S. (2023a) Two new alkaloids from *Dendrobium nobile* Lindl. exhibited neuroprotective activity, and dendrobine alleviated $A\beta_{1\text{-}42}$-induced apoptosis by inhibiting CDK5 activation in PC12 cells. *Drug Development Research* 84(2), 262–274.

Zhang, J., Xu, H.X., Zhao, Z.L., Xian, Y.F. and Lin, Z.X. (2021a) *Dendrobium nobile* Lindl: a review on its chemical constituents and pharmacological effects. *Chinese Medicine and Culture* 4(4), 235–242.

Zhang, Q., Fu, X., Liu, H., Chen, Y., Chen, S. *et al.* (2023b) A systematical review on ethnobotanical, phytochemical and pharmacological aspects of *Dendrobium nobile* Lindl. *Phytochemistry Reviews* 22(3), 743–780.

Zhang, X., Wang, M., Zhang, C., Liu, Z. and Zhou, S. (2021b) Clinical study of *Dendrobium nobile* Lindl intervention on patients with metabolic syndrome. *Medicine* 100(12), e24574.

Zhang, Y., Wang, J.H., Luo, J.P., Zhang, Q. and Lu, J. (2019) The structure-activity relationship and molecular mechanism of anti-tumor polysaccharide isolated from *Dendrobium nobile* Lindl. *Current Topics in Nutraceutical Research* 17(2), 153–163.

Zhang, Y., Lu, J., Li, H. and Song, H. (2023c) Advances in dietary polysaccharides as hypoglycemic agents: mechanisms, structural characteristics, and innovative applications. *Critical Reviews in Food Science and Nutrition* 14, 1–21.

Zhou, S.S. and Jiang, J.G. (2019) Anti-fatigue effects of active ingredients from traditional Chinese medicine: a review. *Current Medicinal Chemistry* 26(10), 1833–1848.

4 The Role of *Dendrobium nobile* in Enhancing Immunity

Archana Jain, Surendra Sarsaiya, Qihai Gong, Qin Wu and Jingshan Shi*
Key Laboratory of Basic Pharmacology and Joint International Research Laboratory of Ethnomedicine of Ministry of Education, Zunyi Medical University, Zunyi, Guizhou, China

Abstract

This chapter explores the immunomodulatory properties of *Dendrobium nobile*, a traditional Chinese medicinal herb known for its diverse pharmacological activities. *D. nobile*, which is rich in polysaccharides, alkaloids, flavonoids and phenols, exhibits significant potential in enhancing immunity by stimulating immune cell proliferation and cytokine production and by balancing immune responses. This chapter delves into the innate and adaptive immune systems, highlighting the role of inflammation in immunity and detailing *in vitro* and *in vivo* studies on *D. nobile*. It examines the mechanisms of action and clinical applications in immune disorders, cancer and infectious diseases and discusses various formulations and delivery systems. Safety and toxicity considerations, along with future research directions and commercial prospects, are also addressed. Despite these promising findings, further research is necessary to fully understand the immunomodulatory effects of *D. nobile* and to optimize its clinical use.

4.1 Introduction

Dendrobium nobile Lindl., also known as the noble dendrobium, is an orchid species native to Asia, particularly prevalent in various regions of Asia, including China, India, Nepal, Bhutan, Myanmar, Japan, the Korean peninsula, Hong Kong, Taiwan and other South-east Asian regions. Esteemed in traditional Chinese medicine (TCM), *D. nobile* has a rich history of use as a tonic herb, celebrated for its diverse pharmacological properties, including its potential to enhance the immune system. In recent years, significant research efforts have been directed towards uncovering the immunomodulatory capabilities of *D. nobile*, exploring its role in preventing and treating various immune-related conditions (Deva *et al.*, 2023). The immune system acts as a sophisticated defence mechanism, comprising a complex network of cells, tissues and organs that collaborate to protect the body against foreign invaders such as viruses and bacteria, and cancer cells. Immune responses are triggered when immune cells recognize and react to foreign antigens, initiating a cascade of reactions involving the production of antibodies and cytokines. Maintaining a delicate balance between immune activation and suppression is crucial for immune homeostasis, as dysregulation can lead to a spectrum of disorders, including autoimmune diseases, allergies and immunodeficiency syndromes (Shin and Kim, 2019).

*Corresponding author: shijs@zmu.edu.cn

© Jingshan Shi, Archana Jain, Surendra Sarsaiya, Qin Wu and Qihai Gong 2025. Dendrobium nobile: *Research and Therapeutic Applications* (J. Shi, A. Jain, S. Sarsaiya, Q. Wu and Q. Gong)
DOI: 10.1079/9781800627987.0004

D. nobile exhibits a spectrum of immunomodulatory activities, including the stimulation of immune cell proliferation and activation, augmentation of cytokine and antibody production, and regulation of Th1/Th2 and Th17/Treg (regulatory T cells) balance. These effects are largely attributed to its rich array of chemical constituents, encompassing polysaccharides, alkaloids, flavonoids and phenols. Moreover, *D. nobile* has demonstrated additional pharmacological properties such as anti-inflammatory, antioxidant and antitumour effects, which likely contribute to its immunomodulatory actions (Ti, 2020). The promising immunomodulatory profile of *D. nobile* positions it as a potential candidate for natural immunomodulatory agents in managing various immune-related disorders (Fig. 4.1). However, further research is imperative to elucidate its mechanisms of action, optimize dosage and administration, and assess safety and efficacy through clinical trials (Xie *et al.*, 2023). Exploring potential synergies between *D. nobile* and other herbal remedies or conventional treatments for immune-related conditions holds promise for innovative therapeutic strategies and deeper insights into their mechanisms of action. Such endeavours could pave the way for novel treatment modalities and foster a better understanding of the therapeutic potential of *D. nobile* and other herbal remedies in immune-related disorders.

Fig. 4.1. The diverse activity and effects of *D. nobile* (authors' own figure).

4.2 Overview of the Immune System

The immune system is a sophisticated network comprising cells, tissues and organs working together to protect the body against external threats such as bacteria and viruses, and cancer cells. It operates by distinguishing between 'self' and 'non-self' entities, initiating immune responses against foreign antigens (Akhand and Ahsan, 2023). This defence mechanism consists of two primary categories: innate immunity and adaptive immunity. Innate immunity serves as the body's initial defence, mounting non-specific responses immediately upon exposure to pathogens (Smith *et al.*, 2019). Conversely, adaptive immunity provides a specific response triggered by exposure to particular pathogens. This involves the activation of T and B lymphocytes capable of recognizing and responding to specific antigens. T lymphocytes, or T cells, oversee cell-mediated immunity, directly targeting and eliminating infected or aberrant cells. B lymphocytes, or B cells, orchestrate humoral immunity, generating antibodies to bind and neutralize antigens (Abós *et al.*, 2022). Maintaining a delicate balance between immune system activation and suppression is paramount for immune homeostasis. Dysregulation can precipitate a spectrum of immune-related disorders such as autoimmune diseases, allergies and immunodeficiency syndromes.

4.2.1 The innate immune system

The innate immune system serves as the body's primary line of defence against invading pathogens, consisting of physical, chemical and cellular components. Among these, physical barriers like the skin and mucous membranes play a crucial role, serving to physically impede the entry of microorganisms. Chemical barriers encompass antimicrobial peptides and enzymes, possessing properties that directly combat or hinder microbial growth. In the cellular realm, phagocytic cells like macrophages and neutrophils engulf and digest pathogens, while natural killer cells target and eliminate infected or cancerous cells (Yadav, 2022).

The initiation of the innate immune system occurs when pattern recognition receptors on the surfaces of immune cells detect pathogen-associated molecular patterns (PAMPs). PAMPs are conserved molecular structures unique to pathogens, absent in host cells, and include components like chemokines and type I interferons. These signalling molecules recruit and activate immune cells to eradicate invading pathogens (Ezeani, 2021). Recent studies underscore the pivotal role of the innate immune system in shaping adaptive immune responses. Dysregulation of innate immunity has been implicated in various immune-related disorders, including autoimmune diseases, sepsis and cancer (Li and Wu, 2021). Therefore, unravelling the mechanisms governing innate immune system regulation is vital to advance therapies that target these conditions.

4.2.2 The adaptive immune system

The adaptive immune system orchestrates a targeted and highly specific response to pathogens, driven by the recognition of distinct antigens by lymphocytes, encompassing T cells and B cells. B cells are responsible for producing antibodies that bind to specific antigens on pathogen surfaces, facilitating their neutralization and subsequent elimination by phagocytic cells. Conversely, T cells possess the ability to directly identify and eradicate infected or cancerous cells, in addition to aiding in the activation of other immune cells (Sompayrac, 2022). Moreover, the adaptive immune system boasts memory capabilities, whereby initial encounters with pathogens may initially be slow. However, subsequent encounters prompt a swifter and more effective immune response, attributed to the presence of memory B and T cells (Davis, 2019). Activation of the adaptive immune system relies on interactions between antigen-presenting cells (APCs), such as dendritic cells, and lymphocytes. APCs uptake and process antigens from pathogens, presenting them on their surface in conjunction with major histocompatibility complex (MHC) molecules. T cells recognize the antigen–MHC complex via the T cell receptor, instigating T cell activation and differentiation into effector and memory cells (Yadav, 2022).

Disruptions in adaptive immune system functionality can precipitate a spectrum of immune-related disorders, including allergies, immunodeficiencies and autoimmune diseases. Therefore, unravelling the mechanisms governing adaptive immune system regulation holds paramount importance in the quest for novel therapies targeting these conditions.

4.2.3 The role of inflammation in immunity

Inflammation stands as a pivotal element in the immune response, acting as a safeguarding mechanism to eradicate pathogens and damaged cells while fostering tissue repair and regeneration. Initiated by activation of the innate immune system, inflammation entails the release of pro-inflammatory cytokines, chemokines and other mediators, orchestrating the mobilization of immune cells such as neutrophils, macrophages and T cells to the site of infection or tissue injury. It is a crucial aspect of the body's response to pathogens or damage (Gusev and Zhuravleva, 2022). Whereas acute inflammation is a necessary and beneficial process for clearing infections and promoting tissue repair, prolonged inflammation can result in tissue damage and contribute to the onset of diverse diseases, including autoimmune diseases, atherosclerosis and cancer. Chronic inflammation is typified by sustained immune system activation and continuous production of pro-inflammatory mediators, resulting in tissue degradation and ongoing immune cell recruitment (Megha *et al.*, 2021). Thus, meticulous regulation of inflammation is paramount for preserving immune homeostasis and mitigating immune-related disorders. Dysregulation of inflammatory pathways has been implicated in a myriad of diseases, spanning from autoimmune conditions to allergies and cancer. Comprehending the mechanisms governing inflammation regulation is imperative for pioneering novel therapeutic approaches to combat these diseases (Kishen and Muralidharan, 2020).

4.3 Immunomodulatory Effects of *D. nobile*

D. nobile, a traditional Chinese medicinal herb with a history spanning centuries, has been utilized in treating various ailments including fever, inflammation and cancer. Recent investigations have uncovered its immunomodulatory properties, suggesting a potential to bolster the immune response and enhance overall immune function (Wu *et al.*, 2023). These effects are largely attributed to its bioactive compounds, encompassing polysaccharides, alkaloids, flavonoids and phenols. *D. nobile*'s immunomodulatory action centres on activating key components of the innate immune system, including macrophages, natural killer cells and dendritic cells. This activation prompts the production of pro-inflammatory cytokines and chemokines, thus stimulating immune responses. Furthermore, it enhances adaptive immunity by fostering the proliferation and differentiation of T and B lymphocytes, ultimately leading to antibody production and activation of cytotoxic T cells (Deva *et al.*, 2023). Demonstrated through animal models and *in vitro* studies, *D. nobile* administration has shown promising outcomes. In infection models, it has increased survival rates and reduced bacterial loads. Additionally, in elderly mice, it has been shown to improve immune function by boosting immune cell proliferation and activation (Kumar *et al.*, 2022). Clinical trials have further supported these findings, revealing improvements in immune function among human subjects. In a randomized, double-blind, placebo-controlled trial, *D. nobile* administration over 4 weeks resulted in increased white blood cell counts and enhanced immune function in healthy individuals. Collectively, these findings underscore *D. nobile*'s potential as a natural immunostimulant and its prospective application in immune-related disorder management. However, comprehensive studies are warranted to elucidate its mechanisms of immunomodulation and establish its safety and efficacy for human use.

4.3.1 *In vitro* studies

In vitro investigations have illuminated the immunomodulatory potential of *D. nobile*, showcasing its ability to stimulate cytokine and chemokine production by immune cells. For instance, Sun *et al.* (2022) discovered that a polysaccharide extract from *D. nobile* induced the synthesis of interleukin (IL) 6 and tumour necrosis factor alpha (TNF-α) by macrophages *in vitro*. Similarly, Guo *et al.* (2022) demonstrated that a water-soluble polysaccharide from *D. nobile* augmented T lymphocyte proliferation and the secretion of IL-2 and interferon-gamma (IFN-γ) *in vitro*.

Moreover, numerous *in vitro* studies have unveiled *D. nobile*'s capacity to enhance macrophage phagocytic activity. Wu *et al.* (2023) observed that a water-soluble polysaccharide from *D. nobile* amplified macrophage phagocytosis and stimulated reactive oxygen species production *in vitro*. Similarly, Wang *et al.* (2021a) showed that an aqueous extract from *D. nobile* bolstered macrophage phagocytic activity and heightened nitric oxide (NO) production *in vitro*. These findings collectively suggest that *D. nobile* holds promise in bolstering the innate immune response through the stimulation of cytokine and chemokine production, as well as enhancing macrophage phagocytic activity *in vitro*. However, further investigations are warranted to unravel the underlying mechanisms of these immunomodulatory effects and assess their clinical significance.

4.3.2 *In vivo* studies

In mouse models of cyclophosphamide-induced immunosuppression, oral administration of *D. nobile* extract was observed to elevate the counts of various immune cells, including T cells, B cells, and natural killer cells, alongside enhancing the production of cytokines like IL-2 and IFN-γ (Meng *et al.*, 2018). Another study in mice found that *D. nobile* polysaccharides could enhance the immune response to an influenza vaccine, resulting in higher levels of neutralizing antibodies and better protection against influenza virus infection (Wang *et al.*, 2022). In a rat model of ulcerative colitis, oral administration of *D. nobile* extract was found to reduce inflammation and promote tissue repair, possibly through the modulation of immune cell activation and cytokine production (Ti *et al.*, 2021). Similarly, in a rat model of asthma, treatment with *D. nobile* extract was found to reduce airway inflammation and improve lung function, possibly through the role of *D. nobile* in modulating immune cell infiltration and cytokine production (Duan *et al.*, 2022). These findings underscore the potential of *D. nobile* as a potent immunomodulatory agent, with implications for the treatment of immune-related disorders. Nonetheless, comprehensive research is imperative to uncover the underlying mechanisms and assess the safety and efficacy of *D. nobile* in human contexts.

4.4 Mechanisms of Action

While the precise mechanisms underlying the immunomodulatory effects of *D. nobile* remain to be fully understood, various potential pathways have been proposed based on both *in vitro* and *in vivo* investigations. One suggested mechanism involves the activation of the innate immune system. Studies indicate that *D. nobile* can stimulate macrophages to produce pro-inflammatory cytokines, such as TNF-α (tumor necrosis factor-alpha), IL-6 (interleukin-6) and IL-1β (interleukin-1 beta). The immunomodulatory effects of *D. nobile* are multifaceted, involving various mechanisms that interact with both the innate and adaptive immune systems. One proposed mechanism suggests that *D. nobile* engages pattern recognition receptors on macrophages, leading to the activation of the nuclear factor kappa B (NF-κB) signalling pathway and subsequent release of pro-inflammatory cytokines and chemokines, such as TNF-α, IL-1β, IL-6 and monocyte chemoattractant protein-1 (MCP-1) (Wang *et al.*, 2021b). Additionally, research indicates that *D. nobile* enhances T cell proliferation and production of cytokines such as IL-2 and IFN-γ, implying a direct influence on T cells or regulation of APCs like dendritic cells, crucial for T cell activation (Cakova *et al.*, 2017). Moreover, *D. nobile* exhibits antioxidant and anti-inflammatory properties, as evidenced by its ability to scavenge free radicals and inhibit the production of pro-inflammatory cytokines like TNF-α and IL-6, both *in vitro* and *in vivo* (Lam *et al.*, 2015). These properties suggest that *D. nobile* may modulate the immune response by mitigating oxidative stress and inflammation, both of which can impair immune function. Overall, the immunomodulatory effects of *D. nobile* involve a complex interplay between activation and modulation of the immune system, alongside its antioxidant and anti-inflammatory properties. However, further research is essential to fully elucidate these mechanisms and determine the potential therapeutic applications of *D. nobile* in immune-related disorders.

4.5 Clinical Applications of *D. nobile* in Immunity

D. nobile, a staple in TCM for centuries, has garnered increasing attention for its potential clinical applications in immune-related disorders. *D. nobile* exhibits a wide range of potential therapeutic effects across various medical domains. Studies have revealed its anti-cancer properties, including apoptosis induction and inhibition of cell proliferation and migration in cancer cell lines, as well as its ability to enhance the efficacy of chemotherapy drugs and radiation therapy in animal models, suggesting promise as an adjuvant cancer therapy (Wang *et al.*, 2023). Furthermore, it demonstrates anti-allergic effects by suppressing IgE production and alleviating allergic symptoms in animal models, indicating potential as a natural remedy for conditions like asthma and allergic rhinitis (Duan *et al.*, 2022). With its immunomodulatory properties, *D. nobile* shows promise in treating immune system disorders, including autoimmune diseases, and has been found to enhance immune function in older people (Wu *et al.*, 2023). Additionally, recent research highlights its anti-inflammatory and gastroprotective effects, traditionally used for gastrointestinal issues such as diarrhoea and vomiting, suggesting potential benefits in treating conditions like inflammatory bowel disease and gastric ulcers (Zhao *et al.*, 2023). Although the clinical applications of *D. nobile* in immunity are promising, further research is crucial to determine optimal dosages and administration methods for different conditions (Fig. 4.2).

4.5.1 Immune disorders

The immune system constitutes a sophisticated network of cells, tissues and organs collaborating to defend the body against external threats like bacteria and viruses, and harmful substances. However, instances of immune system malfunction can precipitate autoimmune disorders. Immune disorders manifest as either an overactive or underactive immune response,

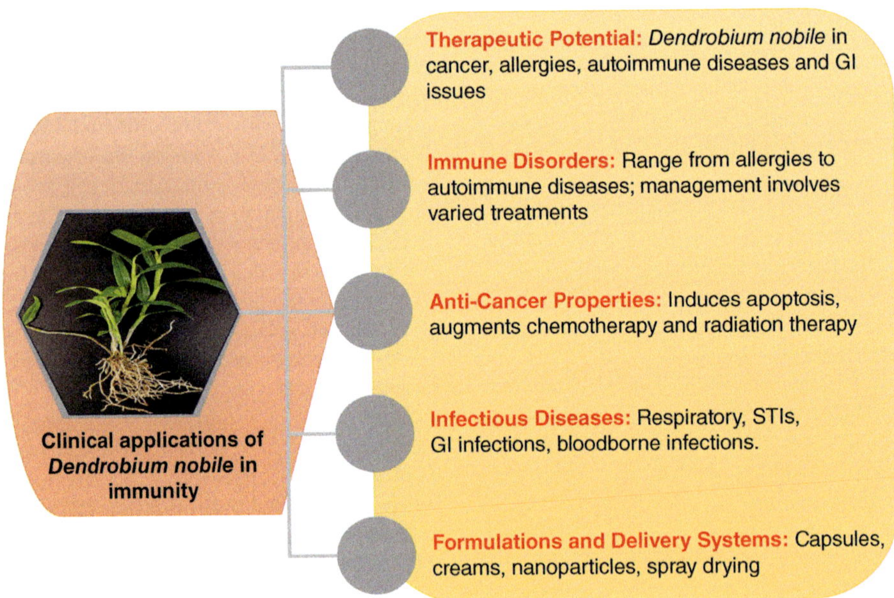

Fig. 4.2. Clinical applications of *D. nobile* in immunity (authors' own figure). GI, gastrointestinal; STIs, sexually transmitted infections.

culminating in a spectrum of conditions encompassing allergies, immunodeficiencies and autoimmune diseases (Smith *et al.*, 2019). Among these, allergies stand as one of the most prevalent immune disorders worldwide, afflicting millions. Allergic reactions occur when the immune system overreacts to benign substances such as pollen, pet dander or certain foods, triggering symptoms like sneezing, coughing, itching and hives. In severe cases, allergies can escalate to life-threatening anaphylaxis, necessitating urgent medical intervention (Duan *et al.*, 2022).

Immunodeficiencies represent another facet of immune disorders, characterized by a compromised immune system unable to combat infections and diseases effectively. These can be classified into primary and secondary immunodeficiencies. Primary immunodeficiencies stem from genetic anomalies affecting immune system development or function, whereas secondary immunodeficiencies arise from external factors like malnutrition, HIV/AIDS or chemotherapy (Bonagura and Rosenthal, 2020).

Autoimmune diseases form a distinct category wherein the immune system erroneously targets the body's own tissues, inciting inflammation and damage. With over 80 identified autoimmune diseases affecting various body parts like joints, skin, thyroid and pancreas, common examples include rheumatoid arthritis, lupus and type 1 diabetes (Frazzei *et al.*, 2022). The aetiology of immune disorders remains incompletely understood, with genetic and environmental factors implicated. Certain genes may predispose individuals to autoimmune diseases, while environmental factors such as infections, toxins and stress can trigger or exacerbate immune disorders (Bieber *et al.*, 2023).

Treatment approaches for immune disorders vary based on type and severity. Management of allergies typically involves antihistamines, corticosteroids and immunotherapy. Immunodeficiencies may necessitate lifelong treatment with immunoglobulin therapy, stem cell transplantation or gene therapy. Autoimmune diseases can be managed with immunosuppressant drugs or biologic drugs targeting specific immune system components (Alexander *et al.*, 2018). In summary, immune disorders encompass a spectrum of conditions stemming from immune system malfunction. These disorders affect diverse aspects of immune function, with various symptoms and health ramifications. While cures remain elusive for most immune disorders, available treatments aim to alleviate symptoms and enhance quality of life.

4.5.2 Cancer

Cancer stands as a prominent global cause of mortality, and conventional treatments like chemotherapy and radiation therapy often entail significant side effects. Consequently, there has been burgeoning interest in exploring alternative cancer therapies, including natural products like *D. nobile*. *D. nobile* exhibits anti-cancer properties, including the capacity to induce apoptosis (programmed cell death) and impede cell proliferation and migration across various cancer cell lines. For instance, Zhang *et al.* (2019) demonstrated that *D. nobile* extract hindered cell proliferation and triggered apoptosis in human liver cancer cells *in vitro*. Similarly, Xu *et al.* (2023) revealed that *D. nobile* extract curtailed the growth and migration of human gastric cancer cells. Beyond its direct anti-cancer effects, *D. nobile* has showcased the potential to augment the efficacy of chemotherapy drugs and radiation therapy in animal studies. Luo *et al.* (2023) showed that *D. nobile* extract bolstered the antitumour effects of cisplatin, a chemotherapy drug, in a mouse model of lung cancer. Similarly, Su *et al.* (2021) illustrated that *D. nobile* extract amplified the antitumour effects of radiation therapy in a mouse model of breast cancer. These findings suggest that *D. nobile* may serve as a promising adjunct therapy for cancer. However, further research is imperative to ascertain optimal dosages and administration methods for different cancer types, as well as to evaluate the safety and efficacy of *D. nobile* through human clinical trials.

4.5.3 Infectious diseases

Infectious diseases stem from pathogens like bacteria, viruses, fungi and parasites infiltrating the body, proliferating and wreaking havoc on healthy cells. While the human immune system has evolved to detect and combat these invaders, at times it may falter, failing to clear the infection adequately or even causing harm to the body itself (Klenerman, 2017). There exists a multitude of infectious diseases, each presenting unique characteristics and requiring tailored treatment approaches. Respiratory infections encompass a spectrum of afflictions targeting the lungs and respiratory tract, driven by various pathogens including viruses (e.g. influenza, coronavirus), bacteria (e.g. *Streptococcus pneumoniae*) and fungi (e.g. *Aspergillus*). Symptoms range from mild, such as with the common cold, to severe manifestations like pneumonia (Unger and Bogaert, 2017). Sexually transmitted infections, transmitted through sexual contact, are caused by bacteria (e.g. *Chlamydia trachomatis*), viruses (e.g. human papillomavirus) and parasites (e.g. *Trichomonas vaginalis*), posing significant long-term health risks like infertility and cancer (Tuddenham *et al.*, 2022). Gastrointestinal infections, affecting the digestive system, are prompted by pathogens including bacteria (e.g. *Salmonella*, *Escherichia coli*), viruses (e.g. norovirus) and parasites (e.g. *Giardia*), leading to symptoms like diarrhoea, vomiting and abdominal pain (Negrut *et al.*, 2020). Bloodborne infections, stemming from pathogens entering the bloodstream, are caused by viruses (e.g. HIV), bacteria (e.g. *Staphylococcus aureus*) and parasites (e.g. *Plasmodium falciparum*), with potential for severe health complications and transmission via contaminated needles or blood transfusions (Fong, 2020).

Treatment of infectious diseases hinges on factors like the pathogen type, infection severity and the patient's overall health. Antibiotics, antiviral drugs or antifungal medications may be administered to eradicate or hinder pathogen growth. In other cases, supportive care, like hydration and rest, constitutes the primary treatment modality. Prevention plays a pivotal role in managing infectious diseases, encompassing practices such as regular handwashing, respiratory etiquette and minimizing contact with infected individuals. Vaccination stands as a crucial preventive measure against various infectious diseases like influenza, measles and human papillomavirus. Despite strides in understanding and managing infectious diseases, they persist as significant global health concerns. Continued research is imperative to develop novel treatments and preventive strategies, alleviating the burden of these diseases on global health.

4.5.4 Formulations and delivery systems for *D. nobile*

D. nobile, commonly utilized in TCM decoctions, is often boiled with other herbs to create medicinal teas (Hong *et al.*, 2022). Additionally, it can be formulated into capsules or tablets for easy oral consumption (De and Singh, 2016). Extracts of *D. nobile*, produced using solvents like ethanol or water, yield concentrated forms suitable for capsules, tablets or liquid preparations (Jia *et al.*, 2023). For topical applications, *D. nobile* can be incorporated into creams, offering benefits in wound healing or dermatitis treatment, as evidenced by Pant (2018). Moreover, the encapsulation of *D. nobile* extracts into nanoparticles, such as liposomes or polymeric nanoparticles, enhances their stability and bioavailability (Zhang *et al.*, 2020). Alternatively, spray drying can be employed to process *D. nobile* extracts into dry powder, suitable for various applications including food supplements or cosmetics, as demonstrated by Qiu *et al.* (2023). The choice of formulation and delivery system for *D. nobile* may vary depending on the intended use, target population and desired pharmacokinetic and pharmacodynamic properties of the active compounds. For instance, oral formulations may be preferred for systemic effects, while topical formulations may be more appropriate for localized effects (Tables 4.1 and 4.2).

4.5.5 Novel formulations and potential benefits for immune enhancement

Recent research has delved into innovative formulations of *D. nobile* and their potential for immune enhancement. Examples include *D. nobile* polysaccharide nanoparticles, which have been shown to boost the immunomodulatory activity of the plant by increasing cytokine production and activating immune cells both *in vitro* and *in vivo* (Li *et al.*, 2023). Additionally, conjugating *D. nobile* polysaccharides with proteins, such as bovine serum albumin, has been found to augment their immune-stimulatory effects, leading to increased antibody production and immune cell activation *in vitro* and *in vivo* (Kim *et al.*, 2015). Fermentation of *D. nobile* with probiotic bacteria enhances its bioactivity and bioavailability, as

Table 4.1. Examples of formulations and delivery systems for *D. nobile* (authors' own table).

Formulation/delivery system	Description	Reference
TCM decoctions	Boiled with other herbs to create medicinal tea	Hong *et al.* (2022)
Capsules and tablets	Formulated for oral consumption	De and Singh (2016)
Extracts	Concentrated extracts for use in capsules, tablets or liquid preparations	Jia *et al.* (2023)
Topical creams	Incorporated into creams for skin applications, such as wound healing or dermatitis treatment	Pant (2018)
Nanoparticles	Encapsulated in nanoparticles, such as liposomes or polymeric nanoparticles, to enhance stability and bioavailability	Zhang *et al.* (2020)
Spray drying	Processed through spray drying to produce a dry powder for various applications, such as food supplements or cosmetics	Qiu *et al.* (2023)

Table 4.2. Comparison of different formulations and delivery systems for *D. nobile* (authors' own table).

Formulation/delivery system	Advantages	Disadvantages
TCM decoctions	Traditional method, combines with other herbs for synergistic effects	Requires more preparation time and not standardized
Capsules and tablets	Convenient for oral consumption and dosing; standardized dosages	May not be as effective as other delivery systems, such as extracts
Extracts	Concentrated and standardized dosages; can be used in various formulations	Extraction process may affect the quality of the active compounds
Topical creams	Localized effects; can be used for skin conditions	Limited to skin applications; may not be as effective for systemic effects
Nanoparticles	Enhanced stability and bioavailability; can be used for various applications	More complex production process; higher cost
Spray drying	Dry powder form for various applications; improved stability and shelf life	May affect the quality of the active compounds; requires additional processing steps

demonstrated by increased immune cell and cytokine production *in vitro* and *in vivo*, along with improved survival rates in mice infected with influenza virus (Lee *et al.*, 2018). Furthermore, the development of *D. nobile*-loaded hydrogels for wound healing applications has shown promise, promoting skin cell migration and proliferation, and enhancing the production of growth factors and cytokines critical for wound healing processes (Tu *et al.*, 2024). These advancements underscore the potential of *D. nobile* in various formulations to bolster immune function and facilitate therapeutic applications. These novel formulations of *D. nobile* have the potential to improve the efficacy and safety of the plant for immune enhancement, as well as expand its applications in various fields, such as drug delivery and tissue engineering.

4.6 Safety and Toxicity Considerations

When developing new drugs for infectious diseases, it is essential to consider safety and toxicity to ensure the health and well-being of patients. Comprehensive toxicology programmes are required to ensure that drugs are safe for patients and that risks are minimized (Samara *et al.*, 2023). Animal studies are commonly used to evaluate the safety and toxicity of new drugs.

For example, studies in the field of translational medicine used animal models to evaluate the safety of a novel antibiotic, finding that it had a favourable safety profile and was well-tolerated by the animals. Animal models were used to evaluate the safety of a new vaccine candidate for tuberculosis, finding that it was safe and well-tolerated (Temrikar *et al.*, 2023). Determining the maximum safe dose of a drug is another important consideration in safety and toxicity testing for all new drugs and biologics. It emphasizes the importance of conducting dose-ranging studies in animal models to identify the maximum safe dose (Lansita *et al.*, 2021). Finally, identifying potential side effects of a drug is essential in ensuring patient safety. This can be accomplished through preclinical testing, as well as through clinical trials in human patients. Grimaldi *et al.* (2020) stated that "assessing the safety of new drugs is an ongoing process that continues throughout the life cycle of a product". In summary, safety and toxicity considerations are critical when developing new drugs for infectious diseases. Through careful evaluation of the safety and toxicity of new therapies, researchers can ensure that they are developing drugs that are effective and safe for patients.

4.6.1 Adverse effects and contraindications

Adverse effects and contraindications are important considerations in the use of any medication, including those used to treat infectious diseases. Adverse effects refer to any negative or unwanted effects that a medication may have on the body, while contraindications refer to any specific circumstances under which a medication should not be used. Adverse effects can range from mild to severe and can include symptoms such as nausea, vomiting, headache and diarrhoea. More severe adverse effects can include liver damage, kidney damage and allergic reactions. Monitoring and reporting any adverse effects of new antimicrobial drugs are important to ensure that the benefits of the drug outweigh the risks (Tsui *et al.*, 2018).

Contraindications refer to situations in which a medication should not be used due to the potential for harm to the patient. For example, some medications may be contraindicated in pregnant women due to the risk of harm to the fetus. Other medications may be contraindicated in patients with certain medical conditions, such as liver or kidney disease. Considering contraindications when prescribing antibiotics, Karasneh *et al.* (2021) stated that "appropriate prescribing is important to minimize the risk of adverse effects and the development of antibiotic resistance". Overall, adverse effects and contraindications are important considerations when using medications to treat infectious diseases. By carefully monitoring for adverse effects and considering contraindications when prescribing medications, healthcare providers can help to ensure that patients receive safe and effective treatments.

4.6.2 Drug interactions

Drug interactions occur when two or more drugs are used together and interact with each other, potentially altering the effectiveness or safety of the medications. This can occur when drugs interact with each other directly, or when they affect the way that the body metabolizes or eliminates one or both of the drugs (Correia, 2018). It is important to consider potential drug interactions when prescribing antibiotics, as certain combinations of antibiotics can increase the risk of adverse events (Tsui *et al.*, 2018). Drug interactions must also be monitored when administering antiretroviral therapy for HIV, as drug interactions can affect both the efficacy and safety of such therapies (Back and Marzolini, 2020). Drug interactions can also occur between medications and other substances, such as food or supplements. For example, grapefruit juice can interact with certain medications, affecting the way that the body metabolizes the drug and potentially increasing the risk of side effects or toxicity. Healthcare providers should be aware of potential interactions between medications and food or supplements to ensure safe and effective treatment (Prescott *et al.*, 2018). Overall, drug interactions are an important consideration in the development and use of medications for infectious diseases. By carefully evaluating potential drug interactions and monitoring patients for adverse events, healthcare providers can ensure that patients receive safe and effective treatment.

4.7 Future Directions and Potential Applications

The potential of *D. nobile* in bolstering immunity is expansive and promising. Yet, numerous avenues of research remain unexplored to comprehend its mechanisms of action fully and unveil its therapeutic applications. A pivotal research focus lies in identifying and isolating the specific bioactive compounds responsible for *D. nobile*'s immunomodulatory effects. While polysaccharides and alkaloids have emerged as potential candidates in some studies, further investigation is warranted to validate these findings and optimize methods for extracting and purifying these compounds. Moreover, assessing the safety and toxicity profile of *D. nobile* is imperative. Despite its extensive traditional use in Chinese medicine, modern scientific scrutiny is essential to ascertain any potential adverse effects or interactions with other medications. Beyond fundamental research inquiries, *D. nobile* holds promise in immunotherapy. Its immunomodulatory effects could prove beneficial in managing autoimmune diseases like rheumatoid arthritis or multiple sclerosis, as well as in cancer immunotherapy, where the aim is to galvanize the immune system to combat cancer cells. In conclusion, while there remains a substantial need for further investigation into the potential applications of *D. nobile* in enhancing immunity, current findings are encouraging. They suggest that the plant harbours significant therapeutic potential for the future.

4.7.1 Research gaps and limitations

Despite the potential advantages of *D. nobile* in augmenting immunity, numerous research gaps and constraints necessitate attention. One constraint is the absence of extensive clinical trials to assess the efficacy and safety of *D. nobile* in human subjects. Predominantly, existing studies have been either preclinical or small-scale clinical trials, restricting the applicability of findings. Another limitation arises from the absence of standardized protocols for preparing and administering *D. nobile*. Diverse studies have employed varied extraction methods and doses, complicating result comparison and reproducibility. Additionally, the chemical makeup of *D. nobile* may fluctuate depending on geographical factors and growth conditions, further complicating standardization efforts. Moreover, a research gap lies in the incomplete understanding of the mechanisms underpinning the immunomodulatory effects of *D. nobile*. Despite evidence demonstrating enhanced immune cell activity, the precise molecular pathways remain elusive. Further investigations are warranted to delineate these pathways and pinpoint the specific bioactive compounds responsible for observed effects. Lastly, it is essential to consider potential adverse effects and drug interactions when utilizing *D. nobile* as an immunomodulatory agent. Although adverse effects have been infrequently reported, comprehensive long-term safety studies are imperative to thoroughly evaluate potential risks.

In summary, while the utilization of *D. nobile* to bolster immunity holds promise, numerous research gaps and limitations persist. Future endeavours should prioritize large-scale clinical trials, standardized protocols for preparation and dosage, mechanistic elucidation and comprehensive long-term safety assessments.

4.7.2 Novel formulations and delivery systems

Novel formulations and delivery systems can play a critical role in improving the effectiveness and safety of *D. nobile*-based immunomodulatory therapies. Nanoparticle-based formulations offer a promising approach to encapsulate *D. nobile* extracts or compounds, safeguarding them from degradation and enhancing their bioavailability. These formulations can also be tailored to target specific immune cells or tissues, potentially magnifying the therapeutic effects of *D. nobile*. Meanwhile, transdermal delivery systems provide an avenue for direct absorption through the skin, circumventing the gastrointestinal tract and potentially improving bioavailability.

Additionally, they can facilitate sustained release of *D. nobile* compounds, reducing the necessity for frequent dosing. Furthermore, combining *D. nobile* with other immunomodulatory compounds or therapies could amplify its efficacy and widen its scope of applications. For instance, using it in synergy with other natural compounds or drugs could target multiple facets of the immune system. Nevertheless, the development of these novel formulations and delivery systems necessitates meticulous consideration of factors like toxicity, stability and compatibility with existing manufacturing processes. Furthermore, conducting clinical studies will be imperative to assess the safety and efficacy of these innovations in humans. Overall, the development of new and improved formulations and delivery systems has the potential to enhance the therapeutic potential of *D. nobile*-based immunomodulatory therapies.

4.7.3 Commercial prospects and industry perspectives

The use of *D. nobile* for immune enhancement shows commercial viability. With its traditional usage in Chinese medicine for immunomodulation, *D. nobile* has garnered attention in both the nutraceutical and pharmaceutical sectors as a potential natural immunostimulant. Market research indicates a significant growth trajectory for the global immune health supplements market in the foreseeable future. This surge is fuelled by escalating awareness of the importance of immune health and the upswing in preventative healthcare practices. The utilization of natural and plant-derived supplements, including *D. nobile*, is anticipated to be a key driver of this growth. Moreover, innovative formulations and delivery systems could further amplify the commercial outlook for *D. nobile*. For instance, nanotechnology-based delivery systems hold promise in augmenting the bioavailability and efficacy of *D. nobile* compounds. Additionally, novel formulations like extracts or isolated compounds present fresh opportunities for both the nutraceutical and pharmaceutical industries. Nevertheless, it is crucial to underscore the significance of regulatory compliance and quality control in ensuring the safety and effectiveness of *D. nobile* products. Employing standardized extracts, adhering to good manufacturing practices and implementing accurate labelling are essential measures to bolster consumer trust and facilitate *D. nobile*'s integration into the commercial market.

4.8 Conclusion

In conclusion, *D. nobile* is a promising natural product with potential immunomodulatory properties. The available literature suggests that *D. nobile* can enhance immunity by stimulating the production of immune cells, cytokines and antibodies, as well as by activating immune signalling pathways. However, one significant deficiency in current research pertains to the absence of standardized methods for extracting and characterizing *D. nobile*. This makes it difficult to compare results across studies and hinders the development of reliable formulations and delivery systems. Additionally, most studies have been conducted *in vitro* or in animal models, and more research is needed to confirm the immunomodulatory effects of *D. nobile* in humans. Despite these limitations, the potential applications of *D. nobile* in enhancing immunity are promising.

D. nobile could be used as a natural supplement to boost the immune system and prevent or treat infectious diseases. It could also be used as an adjuvant therapy to enhance the efficacy of existing immunotherapies, such as cancer immunotherapy. The commercial prospects and industry perspectives of *D. nobile* in enhancing immunity are also promising. With the growing demand for natural and safe products, *D. nobile* could be developed into a marketable product. Additional research and development efforts are required to refine formulations and delivery systems, as well as to guarantee the safety and efficacy of *D. nobile*, thereby unlocking its potential applications in the pharmaceutical and nutraceutical industries. Overall, the role of *D. nobile* in enhancing immunity is a promising area of research that warrants further investigation. The potential applications of *D. nobile* in preventing and treating infectious diseases, as well as in enhancing the efficacy of existing immunotherapies, could have significant impacts on public health.

4.8.1 Summary of key findings

The key findings of the research on the role of *D. nobile* in enhancing immunity are as follows.

- *D. nobile*, a herb entrenched in TCM for its immunomodulatory attributes, has garnered significant attention over centuries.
- This plant species harbours a plethora of bioactive compounds, including polysaccharides, alkaloids and phenols, known for their immunomodulatory effects.
- Numerous studies have showcased the potential of *D. nobile* in bolstering immune responses, evidenced by its ability to augment the production of immune cells and cytokines, and enhance phagocytic activity and antibody production.
- Such findings have propelled exploration into the therapeutic applications of *D. nobile*, spanning from cancer to infectious diseases and autoimmune disorders.
- Despite these promising insights, significant gaps persist in our comprehension of how *D. nobile* orchestrates immune modulation, alongside its potential side effects and toxicity.
- To address these gaps, further research endeavours are imperative, aiming to fully unravel the immunomodulatory properties of *D. nobile* and devise safe and efficacious formulations and delivery systems for clinical applications.
- Despite limitations, existing research underscores the potential of *D. nobile* as a natural immunomodulatory agent, offering promise in bolstering immune function and combating various diseases.
- In summary, investigations into *D. nobile*'s capacity to enhance immunity exemplify the potential of natural products as alternative therapies for immune-related disorders. Nonetheless, a continued exploration is warranted to unlock the full therapeutic potential of *D. nobile* and other natural compounds in immune modulation.

4.8.2 Implications for future research and clinical practice

The findings from studies on the role of *D. nobile* in enhancing immunity suggest that this herb may have potential therapeutic applications for various immune-related disorders. However, there are still several gaps in our understanding of the mechanisms of action and efficacy of *D. nobile*, as well as limitations in the available research. Future research should aim to address these gaps by conducting more rigorous clinical trials with larger sample sizes, standardized dosages and longer treatment durations. Additionally, more studies are needed to explore the potential interactions of *D. nobile* with other medications and to investigate the safety and toxicity of prolonged use of this herb. From a clinical perspective, the findings suggest that *D. nobile* may be a promising alternative or complementary therapy for immune-related disorders such as allergies, infections and autoimmune diseases. However, healthcare professionals should exercise caution and carefully consider the potential risks and benefits before recommending *D. nobile* to their patients. It is also important to note that the use of herbal remedies should not replace conventional medical treatment but rather be used in conjunction with it. The research on the role of *D. nobile* in enhancing immunity has yielded promising results but more research is needed to fully understand its therapeutic potential. As the demand for natural remedies for immune-related disorders continues to grow, it is important that we continue to explore the potential of *D. nobile* and other herbal remedies for the benefit of patients.

Acknowledgements

The authors express their gratitude for the financial support received through the Distinguished High-Level Talents Research Grant from the Guizhou Science and Technology Corporation Platform Talents Fund (grant nos. [2017]5733-001 and CK-1130-002), the National Natural Science Foundation of China (U1812403, 82373981 and 82060750), 2011 Collaborative Innovation Centre of Traditional Chinese Medicine in Guizhou Province (no. [2022]026) and the support provided by the Zunyi Medical University, China. Special appreciation is extended to all laboratory colleagues and research staff members for their valuable insights, constructive guidance and assistance with the contents of this chapter.

References

Abós, B., Bailey, C. and Tafalla, C. (2022) Adaptive immunity. In: Buchmann, K. and Secombes, C.J. (eds) *Principles of Fish Immunology*. Springer, Cham.

Akhand, A.A. and Ahsan, N. (2023) Cells and organs of the immune system. In: Rahman, M.T., Teughels, W. and Lamont, R.J. *Immunology for Dentistry*. Wiley, Oxford, pp. 1–12.

Alexander, T., Farge, D., Badoglio, M., Lindsay, J.O., Muraro, P.A. *et al.* (2018) Hematopoietic stem cell therapy for autoimmune diseases - clinical experience and mechanisms. *Journal of Autoimmunity* 92, 35–46.

Back, D. and Marzolini, C. (2020) The challenge of HIV treatment in an era of polypharmacy. *Journal of the International AIDS Society* 23(2), e25449.

Bieber, K., Hundt, J.E., Yu, X., Ehlers, M., Petersen, F. *et al.* (2023) Autoimmune pre-disease. *Autoimmunity Reviews* 22(2), 103236.

Bonagura, V.R. and Rosenthal, D.W. (2020) Infections that cause secondary immune deficiency. *Stiehm's Immune Deficiencies* 2, 1035–1058.

Cakova, V., Bonte, F. and Lobstein, A. (2017) *Dendrobium:* sources of active ingredients to treat age-related pathologies. *Aging and Disease* 8(6), 827–849.

Correia, M.A. (2018) Drug biotransformation. *Basic and Clinical Pharmacology* 14, 56–73.

Davis, R. (2019) *Cellular and Molecular Immunology*. Scientific e-Resources.

De, L.C. and Singh, D.R. (2016) *Year-Round Production of Orchids*. Pointer Publishers, Rajasthan, pp. 1–44.

Deva, K.D., Bose, B.V. and Basavan, D. (2023) A review on plant-derived immunomodulatory agents: hopes as an alternative medicine in the management of immune-related disorders. *Traditional and Integrative Medicine* 8(2), 1–22.

Duan, F.-P., Li, Y.-S., Hu, T.-Y., Pan, X.-Q., Ma, F. *et al.* (2022) *Dendrobium nobile* protects against ovalbumin-induced allergic rhinitis by regulating intestinal flora and suppressing lung inflammation. *Chinese Journal of Natural Medicines* 20(6), 443–457.

Ezeani, M.C. (2021) Circulating immune complex molecular motifs and association with pathogen recognition receptor ligands and inflammatory stimuli. *Clinicals in Oncology* 1(1), 1–12.

Fong, I.W. (2020) Blood transfusion-associated infections in the twenty-first century: new challenges. In: *Current Trends and Concerns in Infectious Diseases*. pp. 191–215.

Frazzei, G., van Vollenhoven, R.F. and de Jong, B.A. (2022) Preclinical autoimmune disease: a comparison of rheumatoid arthritis, systemic lupus erythematosus, multiple sclerosis and type 1 diabetes. *Frontiers in Immunology* 13, 1–17.

Grimaldi, F., de Leon Izeppi, G.A., Kirschneck, D., Lettieri, P., Escribà-Gelonch, M. *et al.* (2020) Life cycle assessment and cost evaluation of emerging technologies at early stages: the case of continuous flow synthesis of rufinamide. *Journal of Advanced Manufacturing and Processing* 2(2), 1–11.

Guo, R., Chen, M., Ding, Y., Yang, P., Wang, M. *et al.* (2022) Polysaccharides as potential anti-tumor biomacromolecules -a review. *Frontiers in Nutrition* 9, 1–12.

Gusev, E. and Zhuravleva, Y. (2022) Inflammation: a new look at an old problem. *International Journal of Molecular Sciences* 23(9), 1–41.

Hong, H., Wee, K. and Soh, S.B. (2022) *Chinese Medicine for Health: Holistic Healing, Inner Harmony and Herbal Recipes*. World Scientific.

Jia, X., Li, L., Tan, D., Wu, F., He, Y. *et al.* (2023) Effect of superfine-grinding on the physicochemical and antioxidant properties of *Dendrobium nobile* powders. *Food Science and Technology* 43, 1–11.

Karasneh, R.A., Al-Azzam, S.I., Ababneh, M., Al-Azzeh, O., Al-Batayneh, O.B. *et al.* (2021) Prescribers' knowledge, attitudes and behaviors on antibiotics, antibiotic use and antibiotic resistance in Jordan. *Antibiotics* 10(7), 1–13.

Kim, J.H., Oh, S.-Y., Han, S.-B., Uddin, G.M., Kim, C.Y. *et al.* (2015) Anti-inflammatory effects of *Dendrobium nobile* derived phenanthrenes in LPS-stimulated murine macrophages. *Archives of Pharmacal Research* 38(6), 1117–1126.

Kishen, A. and Muralidharan, N.P. (2020) Role of cytokines in inflammation. *International Journal of Pharmaceutical Research* 09752366, 118.

Klenerman, P. (2017) *The Immune System: A Very Short Introduction*. Oxford University Press, Oxford.

Kumar, P., Rai, S., Verma, S.K., Prakash, P.S. and Chitara, D. (2022) Classification, mode of action and uses of various immunomodulators. In: Kesharwani, R.K., Keservani, R.K. and Sharma, A.K. (eds) *Immunomodulators and Human Health*. Springer, Singapore.

Lam, Y., Ng, T.B., Yao, R.M., Shi, J., Xu, K. *et al.* (2015) Evaluation of chemical constituents and important mechanism of pharmacological biology in *Dendrobium* plants. *Evidence-Based Complementary and Alternative Medicine* 2015, 1–25.

Lansita, J., McDougal, A., Sohn, J. and Weis, S.L. (2021) Regulatory toxicology for biopharmaceuticals: preparing for pre-IND interactions and avoiding IND pitfalls for oncology and non-oncology products. *Translational Medicine* 1, 193–217.

Lee, C.T., Kuo, H.C., Chen, Y.H. and Tsai, M.Y. (2018) Current advances in the biological activity of polysaccharides in *Dendrobium* with intriguing therapeutic potential. *Current Medicinal Chemistry* 25(14), 1663–1681.

Li, D. and Wu, M. (2021) Pattern recognition receptors in health and diseases. *Signal Transduction and Targeted Therapy* 6(1), 1–24.

Li, H., Wang, Y., Zhao, P., Guo, L., Huang, L. *et al.* (2023) Naturally and chemically acetylated polysaccharides: structural characteristics, synthesis, activities, and applications in the delivery system. A review. *Carbohydrate Polymers* 313, 120746.

Luo, Y., Liu, G. and Hou, P. (2023) Synergism effect of *Dendrobine* on cisplatin in treatment of H1299 by modulating the balance of Treg/TH17. *Anti-Cancer Agents in Medicinal Chemistry* 23(1), 105–112.

Megha, K.B., Joseph, X., Akhil, V. and Mohanan, P.V. (2021) Cascade of immune mechanism and consequences of inflammatory disorders. *Phytomedicine* 91, 1–17.

Meng, L.-Z., Lv, G.-P., Hu, D.-J., Cheong, K.-L., Xie, J. *et al.* (2018) Effects of polysaccharides from different species of *Dendrobium* (shihu) on macrophage function. *Molecules* 18(5), 5779–5791.

Negrut, N., Khan, S.A., Bungau, S., Zaha, D.C., Corb, A.A.R. *et al.* (2020) Diagnostic challenges in gastrointestinal infections. *Romanian Journal of Military Medicine* 123(2), 83–90.

Pant, B. (2018) Medicinal orchids and their uses: tissue culture a potential alternative for conservation. *African Journal of Plant Science* 7(10), 448–467.

Prescott, J.D., Drake, V.J. and Stevens, J.F. (2018) Medications and micronutrients: identifying clinically relevant interactions and addressing nutritional needs. *Journal of Pharmacy Technology* 34(5), 216–230.

Qiu, Y., Li, R. and Tian, M. (2023) Transcriptome and metabolome analysis reveals the effect of flavonoids on flower color variation in *Dendrobium nobile* Lindl. *Frontiers in Plant Science* 14, 1–11.

Samara, C., Garcia, A., Henry, C., Vallotton, L., Cariolato, L. *et al.* (2023) Safety surveillance during drug development: comparative evaluation of existing regulations. *Advances in Therapy* 40(5), 2147–2185.

Shin, C. and Kim, Y.K. (2019) Autoimmunity in microbiome-mediated diseases and novel therapeutic approaches. *Current Opinion in Pharmacology* 49, 34–42.

Smith, K.G., Kamdar, A.A. and Stark, J.M. (2019) Lung defenses: intrinsic, innate, and adaptive. In: *Kendig's Disorders of the Respiratory Tract in Children*, 9th edn. Elsevier, Oxford, pp. 120–133.

Sompayrac, L.M. (2022) *How the Immune System Works*, 7th edn. Wiley, Oxford.

Su, W., Zeng, L. and Chen, W. (2021) Moscatilin suppresses the breast cancer both in vitro and in vivo by inhibiting HDAC3. *Dose-Response* 19(1), 1–9.

Sun, L., Zhang, Y., Wen, S., Li, Q., Chen, R. *et al.* (2022) Extract of *Jasminum grandiflorum* L. alleviates ccl4-induced liver injury by decreasing inflammation, oxidative stress and hepatic CYP2E1 expression in mice. *Biomedicine & Pharmacotherapy* 152, 1–13.

Temrikar, Z.H., Golden, J.E., Jonsson, C.B. and Meibohm, B. (2023) Clinical and translational pharmacology considerations for anti-infectives approved under the FDA animal rule. *Clinical Pharmacokinetics* 62(7), 943–953.

Ti, H. (2020) Phytochemical profiles and their anti-inflammatory responses against influenza from traditional Chinese medicine or herbs. *Mini Reviews in Medicinal Chemistry* 20(20), 2153–2164.

Ti, H., Zhuang, Z., Yu, Q. and Wang, S. (2021) Progress of plant medicine derived extracts and alkaloids on modulating viral infections and inflammation. *Drug Design, Development and Therapy* 15(15), 1385–1408.

Tsui, V.W., Thomas, D., Tian, S. and Vaida, A.J. (2018) Adverse drug events, medication errors, and drug interactions. *Clinical Pharmacy Education, Practice and Research* 227–245.

Tu, J., Jiang, F., Fang, J., Xu, L., Zeng, Z. *et al.* (2024) Anticipation and verification of *Dendrobium*-derived nanovesicles for skin wound healing targets, predicated upon immune infiltration and senescence. *International Journal of Nanomedicine* 19, 1629–1644.

Tuddenham, S., Hamill, M.M. and Ghanem, K.G. (2022) Diagnosis and treatment of sexually transmitted infections: a review. *Journal of the American Medical Association* 327(2), 161–172.

Unger, S.A. and Bogaert, D. (2017) The respiratory microbiome and respiratory infections. *Journal of Infection* 74(suppl. 1), S84–S88.

Wang, X., Wang, Z., Shen, M., Yi, C., Yu, Q. *et al.* (2022) Acetylated polysaccharides: synthesis, physicochemical properties, bioactivities, and food applications. *Critical Reviews in Food Science and Nutrition* 64(15), 1–16.

Wang, X., Li, J., Chen, R., Li, T. and Chen, M. (2023) Active ingredients from Chinese medicine for combination cancer therapy. *International Journal of Biological Sciences* 19(11), 3499–3525.

Wang, Y., Zhu, H., Wang, X., Yu, Y. and Xie, J. (2021a) Natural food polysaccharides ameliorate inflammatory bowel disease and its mechanisms. *Foods* 10(6), 1–19.

Wang, Y.J., Wan, D.L., Li, Q.M., Zha, X.Q. and Luo, J.P. (2021b) Structural characteristics and immunostimulatory activities of a new polysaccharide from *Dendrobium fimbriatum* hook. *Food & Function* 12(7), 3057–3068.

Wu, W., Lin, Y., Farag, M.A., Li, Z. and Shao, P. (2023) *Dendrobium* as a new natural source of bioactive for the prevention and treatment of digestive tract diseases: a comprehensive review with future perspectives. *Phytomedicine* 114, 154784.

Xie, Z., Jiang, N., Lin, M., He, X., Li, B. *et al.* (2023) The mechanisms of polysaccharides from tonic chinese herbal medicine on the enhancement immune function: a review. *Molecules* 28(21), 1–24.

Xu, X., Yu, Y., Yang, L., Wang, B., Fan, Y. *et al.* (2023) Integrated analysis of *Dendrobium nobile* extract Dendrobin A against pancreatic ductal adenocarcinoma based on network pharmacology, bioinformatics, and validation experiments. *Frontiers in Pharmacology* 14, 1–14.

Yadav, M. (2022) Innate immunity. In: *An Interplay of Cellular and Molecular Components of Immunology*. CRC Press, Boca Raton, pp. 27–59.

Zhang, Y., Zhang, Q., Xin, W., Liu, N. and Zhang, H. (2019) Nudol, a phenanthrene derivative from *Dendrobium nobile*, induces cell cycle arrest and apoptosis and inhibits migration in osteosarcoma cells. *Drug Design, Development and Therapy* 13, 2591–2601.

Zhang, Z., Li, D., Li, X., Guo, Z., Liu, Y. *et al.* (2020) PEI-modified macrophage cell membrane-coated PLGA nanoparticles encapsulating *Dendrobium* polysaccharides as a vaccine delivery system for ovalbumin to improve immune responses. *International Journal of Biological Macromolecules* 165, 239–248.

Zhao, Y., Huang, H., Wang, X., Hu, W., Lu, X. *et al.* (2023) *Dendrobium* officinale polysaccharides inhibit CDCA-induced gastric intestinal metaplasia through activating NRF2/HO-1 and modulating HNF4α/CDX2 signaling pathway. *Journal of Food Biochemistry* 2023, 1–11.

5 Antioxidant Properties of *Dendrobium nobile* and its Potential Health Benefits

Archana Jain, Surendra Sarsaiya, Qihai Gong, Qin Wu and Jingshan Shi*
Key Laboratory of Basic Pharmacology and Joint International Research Laboratory of Ethnomedicine of Ministry of Education, Zunyi Medical University, Zunyi, Guizhou, China

Abstract
This chapter provides an overview of the antioxidant properties of *Dendrobium nobile* and its potential health benefits, including its use for the treatment and prevention of various diseases. This chapter covers the taxonomy, morphology, distribution and traditional uses of *D. nobile* in medicine and cuisine. It highlights bioactive compounds, such as phenolic acids and flavonoids, that contribute to the plant's potent antioxidant properties. The chapter reviews *in vitro* and *in vivo* studies that demonstrate the effectiveness of this plant in scavenging free radicals and protecting against oxidative stress. The potential health benefits, including anti-inflammatory, anti-diabetes, anti-cancer and neuroprotective effects, are discussed. Although promising, further research is needed to validate these findings, understand the mechanisms of action and ensure safety and efficacy for clinical use.

5.1 Introduction

Dendrobium nobile Lindl., a revered species of orchid deeply rooted in traditional Chinese medicine, has attracted scientific interest in recent years owing to its therapeutic potential. This botanical marvel is rich in antioxidants and has been cherished for centuries through holistic healing practices. Antioxidants, pivotal in defending cells against free radicals, hold immense significance in safeguarding against oxidative stress, a precursor to various maladies such as cancer, cardiovascular diseases and neurodegenerative disorders (Fan *et al.*, 2023). Amid the burgeoning enthusiasm for natural remedies, the allure of *D. nobile* has become prominent. The centuries-old lineage of Chinese herbalism offers a testament to its efficacy and safety. Recent studies have shown the remarkable antioxidant potency of this botanical treasure trove. With such a robust antioxidant profile, *D. nobile* has emerged as a promising candidate for therapeutic intervention, opening avenues for exploration of its multifaceted health benefits (Xu *et al.*, 2023).

D. nobile belongs to the Orchidaceae family, which is one of the largest families of flowering plants, with over 25,000 species. *D. nobile* is commonly known as the noble dendrobium, and it is native to various regions of Asia, including China, India, Nepal, Bhutan, Myanmar, Japan, the Korean peninsula, Hong Kong, Taiwan and other South-east Asian regions. *D. nobile* is an epiphytic or lithophytic herb that grows to 40–50 cm in height, with slender stems and lanceolate leaves that are up to 20 cm long and 2 cm

*Corresponding author: shijs@zmu.edu.cn

© Jingshan Shi, Archana Jain, Surendra Sarsaiya, Qin Wu and Qihai Gong 2025. Dendrobium nobile: *Research and Therapeutic Applications* (J. Shi, A. Jain, S. Sarsaiya, Q. Wu and Q. Gong)
DOI: 10.1079/9781800627987.0005

wide. *D. nobile* is also known for its beautiful flowers, which vary from white to pink, lavender and bi-coloured forms, and which bloom in the late winter and early spring (Biswas *et al.*, 2023). *D. nobile* has been used in Chinese medicine for centuries to treat a variety of ailments, including fever, cough, phlegm, thirst and diabetes. *D. nobile* is also used in traditional Chinese cuisine as a tonic and a delicacy, with various forms of preparation, such as soup, tea and wine. The medicinal and culinary uses of *D. nobile* have led to an increased interest in its phytochemical profile and potential health benefits (Ketsa and Warrington, 2023).

Extensive research on the phytochemical composition of *D. nobile* has revealed a diverse array of bioactive compounds including phenolic acids, flavonoids, alkaloids, polysaccharides and glycosides. Numerous studies have highlighted the presence of antioxidants in *D. nobile*, particularly phenolic acids and flavonoids, which are renowned for their robust ability to scavenge free radicals and protect against oxidative damage (Gurjar and Pal, 2021). Numerous *in vitro* studies have explored the antioxidant properties of *D. nobile* using various methods. These studies have consistently revealed the potent antioxidant activity of *D. nobile* extracts, often rivalling or surpassing that of other antioxidant-rich botanicals, such as green tea, grape seed and blueberry (Kumar *et al.*, 2023). *In vivo* investigations into the antioxidant properties of *D. nobile* have utilized animal models of oxidative stress, employing various biomarkers to assess its impact on antioxidant status. The effects of *D. nobile* extract were examined in rats with liver injury induced by carbon tetrachloride (CCl_4). *D. nobile* polysaccharides confer protection against oxidative damage in mice with induced liver injury, as evidenced by enhanced antioxidant enzyme activities and reduced oxidative stress markers (Wang *et al.*, 2023a). Research examining the impact of *D. nobile* on cognitive function and oxidative stress in a rat model of Alzheimer's disease revealed promising outcomes. Treatment with *D. nobile* extract notably enhanced cognitive function while concurrently diminishing oxidative stress markers such as malondialdehyde and protein carbonyl content in the brain. Moreover, the extract boosted antioxidant enzyme activity and reduced the levels of pro-inflammatory cytokines, indicating a potential neuroprotective effect (Ming *et al.*, 2023). Collectively, the findings from *in vivo* studies indicate that *D. nobile* exhibits promising antioxidant effects in diverse animal models of oxidative stress, suggesting potential therapeutic applications for conditions linked to oxidative damage. Nonetheless, additional research is essential to validate these observations and comprehensively understand the underlying mechanisms of action.

5.1.1 Background and rationale

In recent years, the therapeutic properties of *D. nobile*, a plant species traditionally used in Chinese medicine, have been investigated (Zhang *et al.*, 2021). *D. nobile* has been used as a medicinal herb for centuries. Its application spans a range of ailments including fever, diabetes and respiratory conditions. In traditional Chinese medicine, *D. nobile* is revered for its purported ability to invigorate the stomach and lungs, nourish the body and foster longevity. Beyond its medicinal usage, *D. nobile* is also used in Chinese culinary practices, often featured as both a delicacy and functional food item (Hannickel, 2022). Recent studies have revealed that *D. nobile* harbours a diverse array of bioactive compounds including polysaccharides, alkaloids, flavonoids, phenolic acids and other secondary metabolites (Table 5.1). Many of these compounds exhibit robust antioxidant properties. Numerous *in vitro* and *in vivo* investigations have indicated potential health advantages linked to *D. nobile*, including antioxidant, anti-inflammatory, anti-cancer, neuroprotective and metabolic effects (Li *et al.*, 2022a). Despite these encouraging findings, understanding of the underlying mechanisms, as well as studies addressing safety, quality and regulatory concerns regarding the therapeutic use of *D. nobile*, remain limited.

5.2 *D. nobile*: An Overview

D. nobile is a medicinal herb that has been used for centuries in traditional Chinese medicine for its therapeutic properties. *D. nobile* is known as 'shi hu' or 'shi hu gen' in Chinese, which translates to 'stone fungus' (Wang, 2021). The antioxidant attributes of *D. nobile* hold significant

Table 5.1. Bioactive compounds in *D. nobile* (authors' own table).

Bioactive compounds	Sources	Description
Phenolic acids	Extracts from stems, leaves and flowers	Phenolic acids found in *D. nobile* are potent antioxidants, known for their ability to scavenge free radicals and protect against oxidative stress.
Flavonoids	Various parts of the plant	Flavonoids present in *D. nobile* contribute to its antioxidant activity, offering protection against oxidative damage and supporting overall health.
Alkaloids	Extracts derived from the plant	Alkaloids in *D. nobile* possess antioxidant properties and may help reduce oxidative stress in cells, contributing to the plant's overall health benefits.
Polyphenols	Stems, flowers and leaves	Polyphenols abundant in *D. nobile* exhibit strong antioxidant effects, protecting cells from oxidative damage and supporting overall well-being.
Anthocyanins	Extracts from various parts	Anthocyanins extracted from *D. nobile* contribute to its antioxidant activity, offering protection against oxidative stress and promoting overall health.
Terpenoids	Extracts from various plant parts	Terpenoids found in *D. nobile* possess antioxidant properties and contribute to the plant's overall health benefits, protecting cells from oxidative damage.
Carotenoids	Stems, leaves and flowers	Carotenoids in *D. nobile* exhibit antioxidant activity, scavenging free radicals and protecting cells from oxidative stress, thereby promoting health and well-being.
Glucosinolates	Extracts from different plant parts	Glucosinolates found in *D. nobile* contribute to its antioxidant activity, protecting against oxidative stress and supporting overall health.

promise owing to the potential health advantages. Antioxidants serve as agents that counteract free radicals and unstable molecules capable of harming cells and fostering the onset of chronic ailments such as cancer, diabetes and heart disease. *D. nobile* also exhibits anti-inflammatory effects in addition to its antioxidant properties. Inflammation represents a natural immune reaction to injury or infection; however, persistent inflammation can foster the onset of diverse ailments such as arthritis, Alzheimer's disease and cancer (Rabidas *et al.*, 2023). The neuroprotective effects of *D. nobile* have garnered attention, likely because of its antioxidant and anti-inflammatory properties. Neuroprotection involves shielding nerve cells from harm or demise, and holds promise for therapeutic interventions in neurodegenerative conditions such as Alzheimer's and Parkinson's diseases. Its notable attributes include its antioxidant, anti-inflammatory, anti-cancer, neuroprotective and immunomodulatory effects. Despite considerable research, delving deeper into the mechanisms driving these effects and their therapeutic implications remains a priority (Huang *et al.*, 2023). *D. nobile* boasts a rich array of bioactive compounds under rigorous scrutiny for its pharmacological properties (Fig. 5.1).

5.2.1 Morphology and taxonomy

D. nobile is a herbaceous perennial plant that can grow up to 40–50 cm in height. The stem of *D. nobile* is cylindrical and has many nodes, with each node giving rise to a single leaf and a cluster of flowers (De and Pathak, 2020). Several revisions have been made to *D. nobile* taxonomy. It was first described by Carl Linnaeus in 1753 as *Epidendrum nobile*, and was later transferred to the genus *Dendrobium* by Olof Swartz in 1799. However, owing to the high morphological variation within the genus, the taxonomy of *Dendrobium* has been the subject of debate and revision (LaBouff, 2021). Currently, *D. nobile* is classified under the section *Dendrobium* of the subgenus *Dendrobium* in the genus *Dendrobium*. This section is characterized by the presence of pseudobulbs,

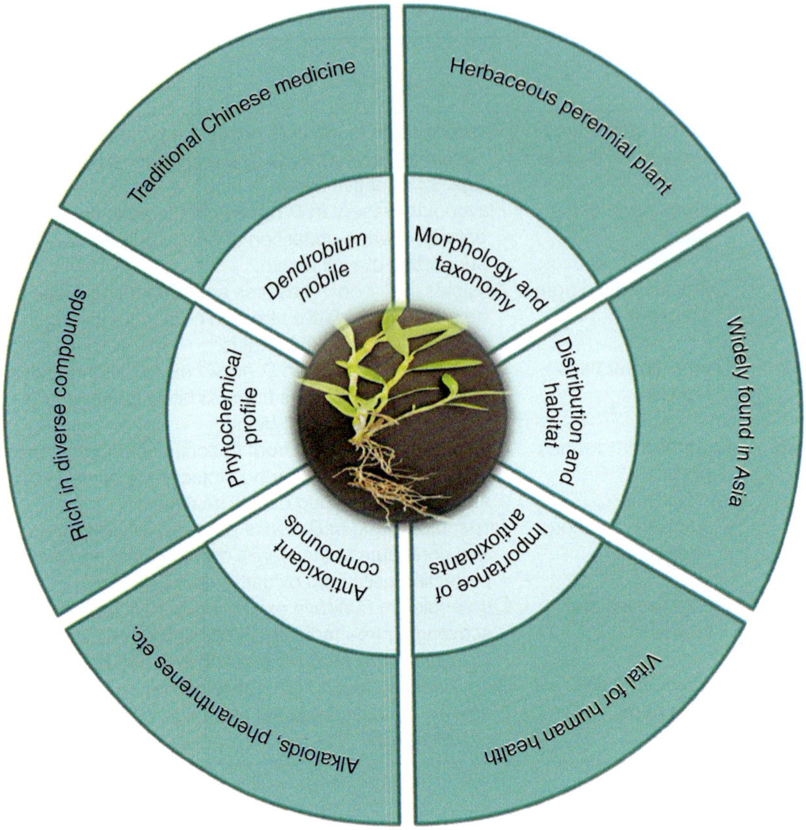

Fig. 5.1. Morphology, taxonomy, distribution, habitat, importance and antioxidant compounds of *D. nobile* (authors' own figure).

which are specialized storage organs that enable plants to survive periods of drought. Recent molecular studies have shed light on the phylogenetic relationships in the genus *Dendrobium*. These studies have shown that the genus is highly diverse and complex, and can be divided into several major clades. However, taxonomic classification of many species within the genus remains unresolved.

5.2.2 Distribution and habitat

D. nobile is a widely distributed species found in many parts of Asia. It is commonly found in China, India, Nepal, Bhutan, Myanmar, Japan, the Korean peninsula, Hong Kong, Taiwan and other South-east Asian regions (Zhang *et al.*, 2021). In China, *D. nobile* is widely distributed and is found in provinces such as Sichuan, Yunnan, Guizhou, Hubei, Hunan, Jiangxi, Zhejiang, Fujian and Guangdong (Pujari and Babu, 2022). *D. nobile* typically grows in rocky areas, trees and bamboo forests at elevations of 500–2500 m above sea level (De, 2023). The *D. nobile* plant prefers a temperate climate, with moderate humidity and bright sunlight. It is often found growing on the branches of trees, but can also grow on rocks or other surfaces with some amount of moisture (Rai and Moktan, 2023).

5.2.3 Importance of antioxidants in maintaining human health

Antioxidants are indispensable guardians of human health and they shield the body from free radicals. These highly reactive molecules pose a threat by damaging cells, proteins and DNA, culminating in various diseases and accelerating the ageing process. Antioxidants neutralize free radicals and mitigate the risk of afflictions such as cancer, cardiovascular ailments and neurodegenerative conditions such as Alzheimer's disease (Das *et al.*, 2021). Beyond their protective roles, research has indicated that antioxidants have multifaceted benefits, including fortifying the immune system, curbing inflammation and enhancing cognitive function (Mesa *et al.*, 2023). Although an abundance of antioxidants can be found in fruits, vegetables, nuts and whole grains, supplements are also available for those seeking additional support. Nevertheless, it is vital to exercise caution, because excessive antioxidant supplementation may cause adverse effects. Striking balance is key, with experts advising on a preference for obtaining antioxidants from a well-rounded diet rather than solely relying on supplements (Shaw *et al.*, 2022). Antioxidants are important in the quest for robust health, emphasizing the importance of incorporating antioxidant-rich foods into daily dietary habits to unlock their full spectrum of benefits.

5.2.4 Antioxidant compounds in *D. nobile*

D. nobile contains various bioactive compounds that exhibit antioxidant properties. These compounds include alkaloids, phenanthrenes, bibenzyls, polysaccharides and flavonoids. Alkaloids are a class of nitrogen-containing organic compounds that have been identified in *D. nobile*. They exhibit antioxidant properties, thereby enhancing the overall antioxidant capacity of the plants. One study found that alkaloids isolated from *D. nobile* exhibit strong antioxidant activity *in vitro* and can effectively scavenge free radicals (Wang, 2021). Phenanthrenes are another class of bioactive compounds found in *D. nobile* that exhibit antioxidant properties. One study found that phenanthrenes isolated from *D. nobile* exhibited strong antioxidant activity *in vitro* and protected cells against oxidative stress (Wang *et al.*, 2022).

Bibenzyls are a unique class of compounds found in *D. nobile* that also possess antioxidant properties. One study found that bibenzyls isolated from *D. nobile* exhibited strong antioxidant activity *in vitro* and were able to scavenge free radicals effectively (Rao *et al.*, 2023). Polysaccharides are another important class of bioactive compounds found in *D. nobile* that exhibit antioxidant properties. Polysaccharides isolated from *D. nobile* effectively scavenge free radicals and protect the cells against oxidative stress (Li *et al.*, 2020). Flavonoids have also been widely studied for their antioxidant properties. Flavonoids isolated from *D. nobile* have been shown to have strong antioxidant activity *in vitro* and can effectively scavenge free radicals. Flavonoids isolated from *D. nobile* can protect cells against oxidative stress by upregulating the expression of antioxidant enzymes (Rao *et al.*, 2022). Overall, the antioxidant compounds found in *D. nobile* have potential health benefits that warrant further investigation (Table 5.2).

5.2.5 Phytochemical profile

D. nobile, also known as noble dendrobium, presents a diverse phytochemical composition pivotal to its esteemed position in traditional Chinese medicine. *D. nobile* contains a diverse range of phytochemicals, including alkaloids, phenanthrenes, bibenzyls, polysaccharides and flavonoids (Wang *et al.*, 2022). Its array of compounds includes alkaloids such as dendrobine and nobilonine, which are renowned for their analgesic and anti-inflammatory properties, along with phenanthrenes, including dendrocandins A and B and moscatilin, with potential anti-cancer and neuroprotective effects (Wang, 2021). The flavonoids found in the plant, such as kaempferol and quercetin glycosides, contribute to its antioxidant capacity, whereas phenolic acids, such as caffeic and ferulic acids, further bolster its therapeutic potential for combating oxidative stress and inflammation (Rao *et al.*, 2023). Moreover, its polysaccharides exhibit promising immunomodulatory traits, complemented by an abundance of amino acids such as proline and alanine. Although these constituents collectively underscore their multifaceted medicinal properties, further exploration is warranted to elucidate the intricate mechanisms underlying their pharmacological actions and therapeutic applications (Fan *et al.*, 2023). Overall,

Table 5.2. Antioxidant properties of *D. nobile* (authors' own table).

Antioxidant property	Description
Scavenging of free radicals	Neutralizes harmful free radicals, reducing oxidative stress
Inhibition of lipid peroxidation	Hinders lipid peroxidation processes, protecting cell membranes from oxidative damage
Enhancement of antioxidant enzymes	Boosts activity of key antioxidant enzymes like superoxide dismutase, catalase and glutathione peroxidase
Modulation of immune responses	Modulates immune responses, indirectly aiding antioxidant effects
Reduction of oxidative stress	Overall reduction in oxidative stress in cells
Protection against DNA damage	Guards against DNA damage caused by reactive oxygen species, preserving genetic integrity
Anti-inflammatory activity	Reduces inflammation by inhibiting pro-inflammatory cytokines and signalling pathways
Protection against neurodegenerative diseases	Helps protect against neurodegeneration by reducing oxidative stress and inflammation in the brain

the diverse phytochemical profiles of *D. nobile* suggest that it has a significant potential for the development of new natural antioxidants and other bioactive compounds with health benefits.

D. nobile has been compared to other antioxidant-rich plants, including fruits, vegetables and herbs. One study compared the antioxidant activity of *D. nobile* with that of several other fruits, such as blueberries, raspberries and strawberries, and found that *D. nobile* had higher antioxidant capacity than all the fruits tested (Kumar *et al.*, 2023). The antioxidant activity of *D. nobile* was compared to that of two other *Dendrobium* species, *Dendrobium chrysotoxum* and *Dendrobium moniliforme*, and it was found that *D. nobile* had the highest antioxidant activity of the three species (He *et al.*, 2020). *D. nobile* has been shown to have comparable or higher antioxidant activity than other antioxidant-rich plants, suggesting that it may have potential as a dietary antioxidant supplement. However, further research is required to fully understand the comparative antioxidant capacities of *D. nobile* and other plants (Natta *et al.*, 2022).

5.3 *In Vitro* Studies on Antioxidant Properties of *D. nobile*

In vitro studies have explored the antioxidant properties of *D. nobile* by employing diverse methodologies to scrutinize various plant components, such as stems, leaves and flowers. A comparative study gauged the antioxidant potential of different parts of *D. nobile* using a 2,2-diphenyl-1-picrylhydrazyl (DPPH) assay. The stems exhibited the highest antioxidant activity, followed by the leaves and flowers. Similarly, the polysaccharide extract showed robust antioxidant efficacy and proficient free radical scavenging activity. Mechanistic insights into the antioxidant properties of *D. nobile* have emerged from *in vitro* studies. A previous study observed that a polysaccharide extract from *D. nobile* could augment the activity of antioxidant enzymes such as superoxide dismutase and catalase in cells, suggesting a potential mechanism by which the extract bolsters the cellular antioxidant defence system. *In vitro* studies assessing the antioxidant properties of *D. nobile* typically employed assays to gauge the ability of plant extracts or isolated compounds to counteract free radicals and mitigate oxidative stress.

The antioxidant potential of *D. nobile* serves as a foundation for the further exploration of its health benefits. A greater decrease in fluorescence indicates higher antioxidant activity (Güçlü *et al.*, 2021). In addition to the above assays, other methods can also be used to assess antioxidant activity, such as total phenolic and flavonoid content assays. The specific protocols for conducting these assays may vary depending on the study but generally involve preparing plant extracts or isolating compounds, diluting samples to the appropriate concentration, and measuring absorbance or fluorescence using a spectrophotometer or fluorometer.

5.3.1 Results and interpretation

The results and interpretation of *in vitro* studies on the antioxidant properties of *D. nobile* can vary depending on the specific study, extraction method and assay employed. However, these studies have demonstrated that *D. nobile* contains significant amounts of antioxidants and can effectively scavenge free radicals. For instance, one study observed that an ethanol extract of *D. nobile* displayed notable antioxidant activity in the DPPH assay as well as in the ferric reducing antioxidant power (FRAP) and 2,2′-azino-bis (3-ethylbenzothiazoline-6-sulfonic acid) (ABTS) assays. Moreover, the study revealed substantial levels of phenolic and flavonoid compounds in the extract, indicating their potential contribution to the antioxidant efficacy shown (Swainson *et al.*, 2023). Another study investigated the antioxidant potential of a polysaccharide extract derived from *D. nobile* using DPPH and hydroxyl radical scavenging assays (Bao *et al.*, 2020). These findings revealed robust antioxidant activity in the extract, demonstrating its efficacy in scavenging free radicals. Researchers have suggested that the polysaccharides present in *D. nobile* may be responsible for its antioxidant properties (Sunanda *et al.*, 2023). Furthermore, one study explored the impact of various extraction techniques on the antioxidant activity of *D. nobile*. The investigation revealed that different extraction methods yielded extracts with varying degrees of antioxidant activity, with the highest activity observed in a hot water extract. The authors proposed that the antioxidant potential of *D. nobile* can be attributed to its polysaccharides, flavonoids and phenolic compounds (Fan *et al.*, 2023). Collectively, these and other studies underscore the substantial antioxidant activity of *D. nobile*, suggesting its potential utility in mitigating oxidative stress and associated conditions. Nevertheless, further research is warranted to comprehensively elucidate the mechanisms underlying its antioxidant properties and ascertain the optimal extraction methods and dosages for therapeutic applications.

5.3.2 Limitations and future directions

In vitro studies on the antioxidant properties of *D. nobile* have provided valuable insights into the potential health benefits of this plant. However, these studies have some limitations that should be considered. First, *in vitro* assays may not fully reflect the complexity of the human body and its responses to oxidative stress. Although these assays can provide information on the ability of plant extracts or compounds to scavenge free radicals, it is important to validate these findings in animal models and clinical trials. Second, the antioxidant activity of *D. nobile* may vary depending on various factors such as the stage of growth, location of cultivation and extraction method used. Therefore, it is important to standardize these factors in future studies to ensure consistent results. Lastly, most *in vitro* studies have focused on isolated compounds or extracts of *D. nobile* rather than evaluating the antioxidant properties of the whole plant. Further studies are needed to investigate the potential synergistic effects of multiple compounds and their interactions with other plant components. In summary, although *in vitro* studies have provided important insights into the antioxidant properties of *D. nobile*, further research is needed to validate these findings in animal models and clinical trials, using standardized testing methods, and to investigate the potential synergistic effects of multiple compounds in the plant.

5.4 *In Vivo* Studies on Antioxidant Properties of *D. nobile*

In vivo investigations of the antioxidant properties of *D. nobile* have typically used animal models to assess the effects of its extract or isolated compounds on oxidative stress and related ailments. These studies sought to substantiate the therapeutic potential of *D. nobile* in combating the conditions associated with oxidative stress and inflammation. Zhou *et al.* (2020) explored the impact of *D. nobile* extract on oxidative stress in rats by exposing them to high oxygen levels for a brief period. Subsequently, they administered *D. nobile* extract and evaluated oxidative stress markers in the blood and tissues. The results indicated a significant reduction in oxidative stress levels in rats treated with *D. nobile* extract, accompanied by enhanced antioxidant capacity. This study suggests promising therapeutic applications of *D. nobile* extract under conditions linked to oxidative stress.

In another study, Huang *et al.* (2022) explored the effects of a polysaccharide derived from *D. nobile* on oxidative stress and inflammation in mice with alcohol-induced liver injury. Researchers observed notable mitigation of oxidative stress and inflammation in mice following polysaccharide administration, leading to improved liver function. These findings underscore the potential therapeutic value of *D. nobile* polysaccharides in addressing liver-related disorders. Zhang *et al.* (2018) explored the impact of *D. nobile* extract on oxidative stress and inflammation in rats with sepsis. This study revealed a significant reduction in oxidative stress and inflammation upon administration of the extract, resulting in improved survival rate in rats. This study highlights the potential therapeutic efficacy of *D. nobile* extract in managing sepsis, a condition characterized by systemic inflammation and oxidative stress.

Jiang and Wang (2022) explored the impact of *D. nobile* extract on cognitive function and oxidative stress in ageing mice. Administration of the extract notably enhanced cognitive function and mitigated oxidative stress in the mice. These findings indicate the potential therapeutic advantages of *D. nobile* extract in addressing age-related cognitive decline. Collectively, these and additional *in vivo* investigations suggest that *D. nobile* extract and its components could offer therapeutic advantages for conditions associated with oxidative stress and inflammation. Nonetheless, further research is imperative to comprehensively elucidate the mechanisms underlying these effects and to ascertain the optimal dosages and administration routes tailored to specific conditions (Fig. 5.2).

5.4.1 Animal models and experimental design

Animal models and experimental designs are crucial components of preclinical research to investigate the effects of drugs, compounds or interventions on biological processes, diseases or conditions. In studies on the antioxidant properties of natural products like *D. nobile*, animal models are often used to mimic human diseases or conditions characterized by oxidative stress and inflammation (Peng *et al.*, 2021). The selection of an appropriate animal model depends on several factors such as the specific research question, availability of resources and ethical considerations. Commonly used animal models in studies of the antioxidant properties of *D. nobile* include rats, mice and rabbits. The selection of an animal model depends on factors such as the specific disease or condition under investigation and the genetic, physiological and immunological similarities to humans (Wang *et al.*, 2023b).

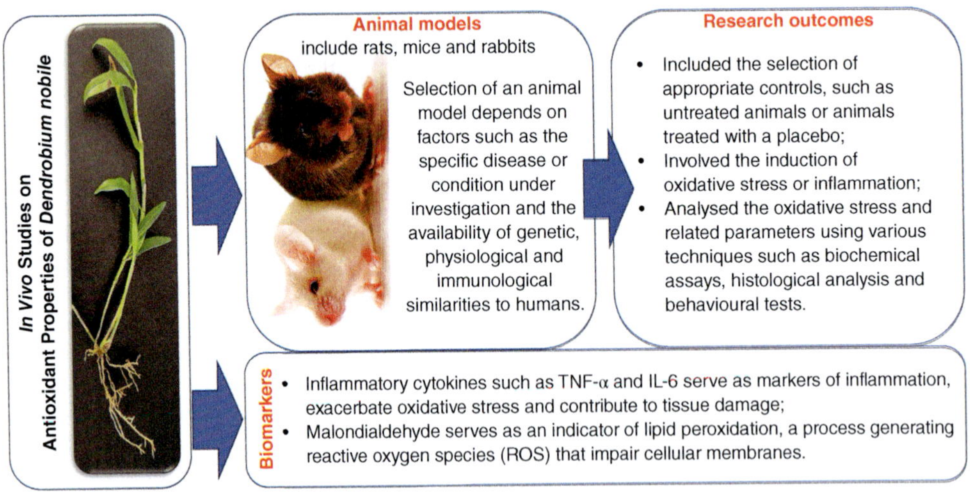

Fig. 5.2. Antioxidant properties of *D. nobile in vivo*, biomarkers and animal models (authors' own figure). IL-6, interleukin 6; TNF-α, tumour necrosis factor alpha.

For example, rats are commonly used in studies of liver injury, whereas mice are commonly used in studies of cognitive decline or neurodegenerative diseases (Huang *et al.*, 2022).

The experimental design of *in vivo* studies on the antioxidant properties of *D. nobile* typically involves the induction of oxidative stress or inflammation in an animal model and the administration of *D. nobile* extract or isolated compounds. The effects of *D. nobile* on oxidative stress and related parameters have been evaluated using various techniques such as biochemical assays, histological analysis and behavioural tests (Ri *et al.*, 2023). Experimental design also includes the selection of appropriate controls, such as untreated animals or animals treated with a placebo, to compare the effects of *D. nobile* treatment. In conclusion, the choice of animal model and experimental design in studies of the antioxidant properties of *D. nobile* are essential to ensure the relevance, reliability and reproducibility of the results and to minimize the ethical concerns associated with animal research. By selecting an appropriate animal model and experimental design, researchers can obtain valuable insights into the potential benefits of *D. nobile* and other natural products for oxidative stress-related diseases and conditions.

5.4.2 Biomarkers and outcomes

In vivo studies on the antioxidant properties of *D. nobile* often measure various biomarkers and outcomes to evaluate the effects of *D. nobile* treatment on oxidative stress and related parameters. Biomarkers are measurable indicators of biological processes or responses to interventions, whereas outcomes refer to changes or improvements in health- or disease-related parameters after intervention (Xu *et al.*, 2018). Malondialdehyde serves as an indicator of lipid peroxidation, a process generating reactive oxygen species that impair cellular membranes. Conversely, glutathione, superoxide dismutase and catalase act as antioxidants that protect cells from oxidative damage by scavenging reactive oxygen species or curtailing their production. Inflammatory cytokines such as tumour necrosis factor alpha (TNF-α) and interleukin 6 (IL-6) serve as markers of inflammation, exacerbate oxidative stress and contribute to tissue damage (Park *et al.*, 2023). Furthermore, alongside biomarkers, outcomes assessed in *in vivo* studies on the antioxidant properties of *D. nobile* encompass enhancements in cognitive function, liver function and other disease-specific parameters. For instance, research has demonstrated the ability of *D. nobile* to ameliorate cognitive function and mitigate oxidative stress in animal models of Alzheimer's disease (Shah *et al.*, 2020) as well as to enhance liver function and attenuate oxidative stress in animal models of liver injury (Huang *et al.*, 2022).

5.4.3 Implications for human health

In vivo studies on the antioxidant properties of *D. nobile* have shown promising results in improving oxidative stress and related parameters in animal models, indicating its potential therapeutic application in human diseases or conditions characterized by oxidative stress and inflammation. These conditions include cardiovascular diseases, neurodegenerative diseases and cancers. Antioxidant therapies, including natural products such as *D. nobile*, have been suggested as potential strategies for the prevention and treatment of these diseases (Mou *et al.*, 2021). Nevertheless, additional research is required to ascertain the safety, effectiveness and ideal dosing protocols of *D. nobile* in humans. Clinical trials are essential to assess the impact of *D. nobile* on biomarkers associated with oxidative stress and inflammation, as well as to analyse clinical outcomes in individuals with diverse diseases or health conditions (Nie *et al.*, 2020). The selection of appropriate biomarkers and outcome measures is crucial to ensure the relevance and reliability of the results and guide the translation of preclinical findings into clinical practice (Davis *et al.*, 2020). Moreover, the potential therapeutic benefits of *D. nobile* may vary depending on the type and severity of the disease, the patient population and other individual factors. Therefore, personalized medicine approaches may be necessary to optimize the use of *D. nobile* and other natural products as antioxidant therapies (Zhang *et al.*, 2023).

5.5 Potential Health Benefits of *D. nobile*

D. nobile has garnered attention because of its diverse health-promoting properties. Recent studies have highlighted its potential therapeutic use across a range of health conditions, which is attributed to its antioxidative, anti-inflammatory and immunomodulatory properties (Table 5.3).

5.5.1 Antioxidant properties

D. nobile has been observed to modulate immune responses and suppress inflammation, further contributing to its antioxidant effects. These properties highlight the potential of *D. nobile* as a natural antioxidant and its possible therapeutic application in combating oxidative stress-related conditions (Rao *et al.*, 2023).

5.5.2 Anti-inflammatory properties

D. nobile extracts and isolated compounds can alleviate inflammation in various experimental models, including those of liver injury, sepsis and cardiovascular diseases (Liu *et al.*, 2020). Moreover, *D. nobile* has been shown to modulate immune cell activity, further contributing to its anti-inflammatory effects. The anti-inflammatory properties of *D. nobile* make it a promising natural remedy for mitigating inflammation-related disorders and warrant further investigation for its therapeutic potential in clinical settings (Wu *et al.*, 2023).

5.5.3 Immune-modulatory properties

The immune system is vital in protecting the body against infections and diseases. However, an overactive or weakened immune system can lead to various health problems. Studies have demonstrated that extracts and compounds isolated from *D. nobile* can regulate the activity of immune cells, including T, B and natural killer (NK) cells (Xie *et al.*, 2023). *D. nobile* has potential therapeutic applications in diseases or conditions characterized by immune dysfunction, such as autoimmune diseases, allergies and infections (Yang *et al.*, 2022). Overall, the potential health benefits of *D. nobile* are promising; however, further research is required to determine its safety, efficacy and optimal dosing regimens in humans. Clinical trials are necessary to evaluate the effects of *D. nobile* on biomarkers of oxidative stress, inflammation and immune function, as well as clinical outcomes in patients with various diseases or conditions. The selection of appropriate biomarkers and outcome measures is crucial to ensure the relevance and reliability of the results and guide the translation of preclinical findings into clinical practice.

5.5.4 Anti-cancer and chemopreventive properties

Numerous *in vitro* and *in vivo* investigations have indicated that *D. nobile* may possess promising anti-cancer and chemopreventive properties. Studies on various types of cancer have illustrated the potential therapeutic value of *D. nobile* in oncology. For instance, in a study using human hepatocellular carcinoma (HCC) cells, researchers observed that a polysaccharide extracted from

Table 5.3. Potential health benefits of *D. nobile* (authors' own table).

Health condition/area of action	Potential benefit
Diabetes	Blood sugar regulation, improvement in insulin sensitivity
Cancer	Antitumour activity, inhibition of cancer cell proliferation
Inflammation	Reduction of inflammation and related symptoms
Cardiovascular health	Improvement in blood lipid profile, prevention of atherosclerosis
Cognitive function	Improvement in learning and memory, neuroprotection
Gastrointestinal health	Protection against gastric ulcers, promotion of gut health
Immune function	Enhancement of immune function, protection against infections

D. nobile exerted notable inhibitory effects on cancer cell proliferation and induced programmed cell death (apoptosis) (Wang *et al.*, 2021). Similarly, in a mouse model of colon cancer, an extract derived from *D. nobile* demonstrated significant reductions in tumour growth and metastasis, potentially mediated through the inhibition of angiogenesis (Song *et al.*, 2022). Moreover, emerging evidence suggests that *D. nobile* is promising for the chemoprevention of breast cancer. Investigations involving human breast cancer cells revealed that treatment with a polysaccharide extracted from *D. nobile* suppressed cell growth and stimulated apoptosis by upregulating tumour suppressor genes (Da Silva *et al.*, 2017).

Although these findings are compelling, further comprehensive research is warranted to elucidate the underlying mechanisms and therapeutic potential of *D. nobile* in cancer management. Rigorous clinical trials are imperative for evaluating the safety, efficacy and optimal dosing regimens of *D. nobile* for cancer prevention and treatment. Such endeavours hold immense promise for advancing our understanding of the role of *D. nobile* in combating cancer and may pave the way for novel therapeutic interventions.

5.5.5 Neuroprotective and cognition-enhancing effects

In addition to its antioxidant and anti-inflammatory properties, *D. nobile* has also been found to have neuroprotective and cognition-enhancing effects. In preclinical studies, *D. nobile* has been demonstrated to improve cognitive function, memory and learning ability as well as protect against neurodegenerative diseases. One study found that *D. nobile* extract improved spatial learning and memory in rats, potentially through the upregulation of neurotrophic factors and the enhancement of synaptic plasticity (Li *et al.*, 2022b). A separate study demonstrated that *D. nobile* extract provided protection against cognitive decline in a mouse model of Alzheimer's disease, potentially by suppressing neuroinflammation and oxidative stress (Pi *et al.*, 2022). These results imply that *D. nobile* holds promise as a therapeutic intervention for cognitive impairment and neurodegenerative disease. Nevertheless, additional research is essential to establish optimal dosing regimens and potential side effects of *D. nobile* in humans.

5.5.6 Metabolic and cardiovascular benefits

Numerous studies have investigated the metabolic and cardiovascular benefits of *D. nobile* extracts. In a study involving diabetic rats, oral administration of *D. nobile* extract over a period of 6 weeks notably enhanced glucose metabolism and insulin sensitivity, while concurrently reducing oxidative stress and inflammation markers compared to the control group (Luo *et al.*, 2024). Additionally, in another study focusing on rats with high-fat diet-induced obesity, *D. nobile* extract mitigated body weight gain, and alleviated liver inflammation (Liu *et al.*, 2021). Moreover, evidence suggests that *D. nobile* has cardiovascular benefits. In a study conducted on rats with Nω-nitro-l-arginine methyl ester (l-NAME)-induced hypertension, oral administration of *D. nobile* extract for 6 weeks resulted in notable decreases in blood pressure levels, enhancement of endothelial function and reduced oxidative stress compared to the control group (Oskouei *et al.*, 2023). Similarly, another study involving rats with myocardial infarction demonstrated that *D. nobile* extract improved cardiac function, decreased the size of infarcted tissue and alleviated oxidative stress and inflammation in the heart (Ding and Zhang, 2023). Overall, these findings suggest that *D. nobile* may have potential metabolic and cardiovascular benefits. However, further research is needed to determine the optimal dosing regimens and mechanisms of action in humans.

5.6 Safety, Quality and Regulatory Issues

D. nobile is a natural product with potential health benefits; however, its safety, quality and regulatory issues should be carefully evaluated. Some guidelines have aimed to ensure the safety and efficacy of natural products. One concern with *D. nobile* is the potential for contamination with heavy metals and pesticides owing to its cultivation in polluted environments or improper handling during processing. Therefore, strict quality control measures should be implemented to ensure the purity and safety of *D. nobile* products. A study conducted in China found that some *D. nobile* products contained high levels of lead and arsenic, indicating a need for stricter quality control measures (Liao, 2023).

Another concern is the potential for adverse interactions between medications and supplements. *D. nobile* contains alkaloids that may interact with drugs that affect the nervous system, blood pressure or heart function. Therefore, individuals taking medications should consult a healthcare professional before using *D. nobile* or any other natural product. One study reported a case of a patient who experienced an adverse reaction when using a supplement containing *D. nobile* along with a medication for hypertension (Zhang *et al.*, 2021).

Additionally, the lack of standardized dosing guidelines and the variability in the composition of *D. nobile* preparations can make it difficult to assess their safety and efficacy. Therefore, more research is needed to establish standardized protocols for the production and dosing of *D. nobile* products. In conclusion, the safety, quality and regulatory issues associated with *D. nobile* should be carefully evaluated to ensure its safe and effective use in humans. Strict quality control measures and standardized dosing guidelines are necessary to ensure the purity and safety of *D. nobile* products. Healthcare professionals should also be aware of the potential for adverse interactions between *D. nobile* and medications or other supplements.

5.6.1 Toxicity and adverse effects

There is limited scientific understanding of the safety profile of *D. nobile*. Although traditionally regarded as safe when used appropriately, concerns include potential allergic reactions, drug interactions, contaminants, gastrointestinal discomfort and the absence of long-term safety data. Variability in product quality and regulations in the herbal supplement industry underscore the importance of caution. Consulting a healthcare professional before supplementation is crucial, especially for individuals with underlying health conditions or those taking medications, to mitigate the potential risks associated with their consumption. Despite its historical use, further research is required to elucidate the safety profile of *D. nobile* in contemporary settings.

Some studies have reported the potential adverse effects associated with its use. In one study, participants who consumed *D. nobile* extract experienced mild gastrointestinal symptoms, such as abdominal pain, diarrhoea and nausea (Gao *et al.*, 2022). Another study reported adverse effects, such as dizziness and headaches, in participants who consumed a dendrobium-containing supplement (Jędrejko *et al.*, 2021). Additionally, concerns have been raised about the potential of *D. nobile* to interact with medications that affect the nervous system, blood pressure and heart function, due to its alkaloid content (Zhang *et al.*, 2021). It is also important to note that the safety and purity of *D. nobile* products can be affected by contamination with heavy metals and pesticides, which can occur due to improper cultivation or processing.

5.6.2 Quality control and standardization

Ensuring the quality and standardization of *D. nobile* preparations is important to ensure their efficacy and safety for human use. The variability in the composition of *D. nobile* preparations and the lack of standardized dosing guidelines can make it difficult to assess their safety and efficacy. One approach to ensuring quality control is to develop standardized protocols for the production and dosing of *D. nobile* products. This could involve establishing guidelines for the cultivation, harvesting, processing and packaging of the plant, as well as standardized methods for extracting and measuring the active compounds in the plant (Kaushik, 2019). Another important aspect of quality control is testing for contaminants, such as heavy metals and pesticides, that can accumulate in *D. nobile* due to cultivation in polluted environments or improper handling during processing. Testing for microbial contamination, such as by bacteria or fungi, is also important to ensure the purity and safety of *D. nobile* products (Pujari and Babu, 2022).

5.6.3 Interactions with drugs and other supplements

Certain supplements, including *D. nobile*, may interact with drugs and other supplements, potentially leading to adverse effects. For example, *D. nobile* may interact with medications used to lower blood sugar levels, such as insulin and metformin, increasing the risk of hypoglycaemia. In addition, *D. nobile* may interact with other supplements or herbs with hypoglycaemic effects, such as bitter melon and ginseng, further increasing the risk of low blood sugar levels (Luo *et al.*, 2024). Other potential interactions include those with anticoagulant medications, such as warfarin, as *D. nobile* may increase the risk of bleeding. It may also interact with medications metabolized in the liver as some compounds in *D. nobile* may inhibit certain liver enzymes involved in drug metabolism (Zhang *et al.*, 2021). As with any supplement or medication, it is important to consult a healthcare professional before taking *D. nobile*, especially if a patient is taking any prescription medications or having a medical condition. Healthcare professionals can help determine the potential risks and benefits of consuming *D. nobile* and monitor any potential interactions or adverse effects (Wang *et al.*, 2023a,b).

D. nobile exhibits significant anti-inflammatory properties, attributed to its diverse array of bioactive compounds. Studies have demonstrated that *D. nobile* extracts and isolated compounds can alleviate inflammation in various experimental models, including those of liver injury, sepsis and cardiovascular diseases. Moreover, *D. nobile* has been shown to modulate immune cell activity, further contributing to its anti-inflammatory effects (Li *et al.*, 2023). Overall, the anti-inflammatory properties of *D. nobile* make it a promising natural remedy for mitigating inflammation-related disorders, and warrant further investigation for its therapeutic potential in clinical settings.

5.7 Conclusion and Future Directions

D. nobile, a traditional Chinese medicinal plant, has potent antioxidant properties. Its bioactive compounds, including phenolic acids, flavonoids and alkaloids, have demonstrated the ability to scavenge free radicals and mitigate oxidative stress, both *in vitro* and *in vivo*. Consequently, researchers have explored *D. nobile* and its extracts for their potential health benefits, including anti-inflammatory, anti-diabetes, anti-cancer and neuroprotective effects. Although there is limited clinical evidence to support these benefits, preclinical studies and traditional use suggest that *D. nobile* may have therapeutic potential for numerous health conditions. Nonetheless, additional research is imperative to deepen our understanding of the mechanism of action, ascertain its safety and efficacy in humans, and formulate standardized preparations suitable for clinical application.

In light of its potential health benefits, it is important to consider the interactions between *D. nobile* and other drugs or supplements that a person may take. While there is currently limited information on potential interactions, it is recommended to consult a healthcare provider before using *D. nobile* or its extracts as a dietary supplement or alternative treatment. *D. nobile* presents promising prospects as a natural source of antioxidants with potential health benefits. However, further research is required to comprehensively understand their properties and applications. Similar to any natural product, its use should be approached with caution and under the guidance of a healthcare professional.

Future investigations into the antioxidant properties of *D. nobile* and its potential health benefits should focus on several key areas. First, extensive studies are required to elucidate the molecular mechanisms underlying the antioxidant effects of *D. nobile*. This knowledge will aid in identifying the specific bioactive compounds responsible for these effects and provide valuable insights into their functions. Second, further investigation is required to determine the optimal dosage and duration of *D. nobile* supplementation to achieve antioxidant benefits. This would help to establish appropriate recommendations for the use of *D. nobile* as a dietary supplement to promote health and prevent disease. Third, further studies on the effects of *D. nobile* in combination with other antioxidants or supplements are required. This would help to determine whether *D. nobile* has synergistic effects when used in combination with other compounds, and whether it may be useful as part of a broader antioxidant supplement regimen. Fourth, further research is needed to explore the potential therapeutic applications of *D. nobile*, beyond its antioxidant properties. This would involve investigating its effects on other aspects of health such as immune function, inflammation and metabolism. Finally, more studies are needed to assess the safety of *D. nobile* supplementation, particularly with regard to its long-term use and potential interactions with medications or other supplements. In summary, further research on the antioxidant properties of *D. nobile* and its potential health benefits has the potential to shed light on the underlying mechanisms of action, optimize dosing regimens, explore synergistic effects with other compounds, investigate potential therapeutic applications and assess safety concerns.

5.7.1 Summary of key findings and implications

The key findings from research on the antioxidant properties of *D. nobile* and its potential health benefits are as follows.

- *D. nobile* contains various phenolic compounds and flavonoids that possess antioxidant activity.
- Research conducted both in laboratory settings (*in vitro*) and in living organisms (*in vivo*) indicates that *D. nobile* extract has protective properties against damage caused by oxidative stress, inflammation and cell death.
- Promising results from previous research indicate that *D. nobile* extract has the potential for both the prevention and treatment of various diseases, including diabetes, cardiovascular disease, cancer and neurodegenerative diseases.
- *D. nobile* extract may exert its health benefits through various mechanisms, including free radical scavenging, modulation of antioxidant enzyme activity and regulation of signalling pathways involved in inflammation and cell survival.
- *D. nobile* extract has been shown to be safe and well tolerated in animal and human clinical trials.

These findings imply that *the D. nobile* extract holds significant promise as a natural antioxidant, potentially conferring health benefits for the prevention and treatment of diverse diseases. Further research is necessary to deepen our understanding of the mechanisms driving its antioxidant properties, and to ascertain the optimal therapeutic dosage and administration method. Therefore, the antioxidant properties and potential health benefits of *D. nobile* are significant. These findings suggest that *D. nobile* is a valuable source of natural antioxidants that can protect the body from oxidative stress, and potentially prevent or manage various chronic diseases. The potential health benefits of *D. nobile* include improved cognitive function, reduced inflammation, regulation of blood sugar levels, enhanced immunity, and cancer prevention and treatment. These benefits stem from the presence of antioxidant compounds, including phenolic compounds and flavonoids, in plants. Additionally, the incorporation of *D. nobile* in traditional medicinal practices, including Chinese medicine, suggests its suitability for human consumption. Nevertheless, further investigations are necessary to confirm the safety and effectiveness of *D. nobile* as a dietary supplement or a medicinal treatment. In conclusion, the antioxidant properties of *D. nobile* and its potential health benefits indicate its potential role in promoting wellness and in preventing chronic diseases. Further studies are required to explore its therapeutic potential and develop safe and effective products for human use.

5.7.2 Knowledge gaps and research needs

Despite a growing body of evidence suggesting that *D. nobile* possesses potent antioxidant properties and potential health benefits, there are still several knowledge gaps and research needs that must be addressed to fully understand its therapeutic potential. First, there is a lack of standardized methods for the extraction and analysis of bioactive compounds from *D. nobile*. This has led to inconsistencies in the reported antioxidant activity and bioavailability of the constituents. The development of validated and standardized methods for the extraction and analysis of bioactive compounds in *D. nobile* is needed to improve the reliability and comparability of research findings. Second, most studies on the antioxidant properties and potential health benefits of *D. nobile* have been conducted *in vitro* or in animal models. Clinical trials involving human participants are needed to establish its safety and efficacy in humans, and to determine the optimal dose and duration of treatment. Third, although *D. nobile* has been traditionally used in Chinese medicine for centuries, there is limited understanding of its mechanism of action at the molecular level. Further research is required to identify the specific bioactive compounds and pathways involved in the antioxidant activity and other potential health benefits. Fourth, the potential interactions between *D. nobile* and other drugs or supplements need to be explored. While *D. nobile* is generally considered safe, its use in combination with other medications or supplements may lead to adverse effects. Overall, further research is needed to fully understand the potential health benefits of *D. nobile* and establish its safety and efficacy in humans. Addressing these knowledge gaps is crucial to unlocking the therapeutic potential of this promising natural product.

5.7.3 Potential applications and challenges for commercialization

The potential applications of *D. nobile* and its antioxidant properties are vast, ranging from nutraceuticals and functional foods to pharmaceuticals and cosmetic products. However, several challenges must be addressed before commercialization can be achieved (Table 5.4). One major challenge is the lack of standardized manufacturing and quality control processes. There is currently no uniform method for extracting and purifying *D. nobile* compounds, and efforts towards standardization are crucial to mitigate variations in product quality and efficacy, thereby facilitating consistent product quality and enabling reliable clinical studies. Another challenge is the limited clinical data available on the safety and efficacy of *D. nobile* extracts in humans. Although preclinical studies have shown promising results, human studies are needed to confirm these findings and to establish appropriate dosing and safety profiles. In addition, potential interactions between drugs and other supplements must be fully understood to avoid adverse effects. Lastly, the cost of producing high-quality *D. nobile* extracts can be prohibitive, which may limit its commercial viability. Innovative approaches to reduce production costs, such as the use of alternative cultivation methods or genetic engineering techniques, may be needed

Table 5.4. Challenges for commercialization of *D. nobile* products (authors' own table).

Challenge	Description
Standardization and quality control	Ensuring consistent quality and potency of *D. nobile* extracts, as well as verifying the presence and concentration of specific bioactive compounds
Safety and toxicity	Evaluating the safety and potential toxicity of *D. nobile* extracts, particularly with long-term use or at high doses
Regulatory approval	Obtaining regulatory approval for *D. nobile* products as food supplements or medicines in different countries and regions, which can involve complex and lengthy processes
Extraction and formulation	Developing effective and efficient methods for extracting and formulating *D. nobile* extracts into products with desirable properties, such as stability, bioavailability and palatability
Market competition and consumer demand	Facing competition from other natural products and supplements with similar claims, as well as addressing consumer skepticism and education about the benefits of *D. nobile*
Sustainability and ethical considerations	Ensuring sustainable harvesting and cultivation practices for *D. nobile*, as well as addressing potential ethical concerns about the impact of commercialization on the environment and local communities

to make *D. nobile* products more affordable. Despite these challenges, the potential applications of *D. nobile* and its antioxidant properties make it an attractive target for further research and commercialization efforts. With continued research and development, it may be possible to overcome these challenges and bring safe and effective *D. nobile* products to the market.

Acknowledgements

The authors express their gratitude for the financial support received through the Distinguished High-Level Talents Research Grant from the Guizhou Science and Technology Corporation Platform Talents Fund (grant nos. [2017]5733-001 and CK-1130-002), the National Natural Science Foundation of China (U1812403, 82373981 and 82060750), 2011 Collaborative Innovation Centre of Traditional Chinese Medicine in Guizhou Province (no. [2022]026) and the support provided by the Zunyi Medical University, China. Special appreciation is extended to all laboratory colleagues and research staff members for their valuable insights, constructive guidance and assistance with the contents of this chapter.

References

Bao, Y., Ren, X., Zhu, Y., Zhang, Y., Peng, Z. *et al.* (2020) Comparison of lipid radical scavenging capacity of spice extract in situ in roast beef with DPPH and peroxy radical scavenging capacities in vitro models. *LWT - Food Science and Technology* 130, 109626.

Biswas, S.S., Pal, R., Kalaivanan, N.S., Ngangom, N.M. and Natta, S. (2023) Production and postharvest management for *Dendrobium* orchids. *Indian Horticulture* 68(4), 30–35.

Das, N., Raymick, J. and Sarkar, S. (2021) Role of metals in Alzheimer's disease. *Metabolic Brain Disease* 36(7), 1627–1639.

Davis, K.D., Aghaeepour, N., Ahn, A.H., Angst, M.S., Borsook, D. *et al.* (2020) Discovery and validation of biomarkers to aid the development of safe and effective pain therapeutics: challenges and opportunities. *Nature Reviews Neurology* 16(7), 381–400.

De, L.C. (2023) Technological advancements, innovation and entrepreneurship development in orchids. *Journal of Agriculture and Technology* 10(2), 109–120.

De, L.C. and Pathak, P. (2020) Good agricultural practices of *Dendrobium* orchids. *Journal of The Orchid Society of India* 34(1), 35–43.

Ding, C. and Zhang, C. (2023) Reducing myocardial infarction by combination of irisin and *Dendrobium nobile* Lindl through inhibiting nod-like receptor protein-3-related pyroptosis and activating PINK1/Parkin-mitophagy during aging. *Chinese Journal of Physiology* 66(5), 351–358.

Fan, C., Sun, X., Wang, X. and Yu, H. (2023) Therapeutic potential of the chemical composition of *Dendrobium nobile* Lindl. *Frontiers in Pharmacology* 14, 1163830.

Gao, X., Liu, J., Luo, Y., Lei, Y., Long, W. *et al.* (2022) Various fractions of alcoholic extracts from *Dendrobium nobile* functionalized antioxidation and antiaging in D-galactose-induced aging mice. *Frontiers in Bioscience-Landmark* 27(11), 315.

Güçlü, K., Kıbrıslıoğlu, G., Özyürek, M. and Apak, R. (2021) Development of a fluorescent probe for measurement of peroxyl radical scavenging activity in biological samples. *Journal of Agricultural and Food Chemistry* 62(8), 1839–1845.

Gurjar, V.K. and Pal, D. (2021) Natural compounds extracted from medicinal plants and their immunomodulatory activities. *Bioactive Natural Products for Pharmaceutical Applications* 140, 197–261.

Hannickel, E. (2022) *Orchid Muse: A History of Obsession in Fifteen Flowers*. WW Norton and Company, New York.

He, L., Su, Q., Bai, L., Li, M., Liu, J. *et al.* (2020) Recent research progress on natural small molecule bibenzyls and its derivatives in *Dendrobium* species. *European Journal of Medicinal Chemistry* 204, 112530.

Huang, J., Huang, N., Mao, Q., Shi, J. and Qiu, Y. (2023) Natural bioactive compounds in Alzheimer's disease: From the perspective of type 3 diabetes mellitus. *Frontiers in Aging Neuroscience* 15, 1–8.

Huang, X., Yang, S., Sun, J., Li, X., Zhou, S.Y. *et al.* (2022) Transcriptome analysis of protection by *Dendrobium nobile* alkaloids (DNLA) against chronic alcoholic liver injury in mice. *Biomedicines* 10(11), 1–16.

Jędrejko, K., Lazur, J. and Muszyńska, B. (2021) Risk associated with the use of selected ingredients in food supplements. *Chemistry & Biodiversity* 18(2), e2000686.

Jiang, N., Li, Y.-J., Wang, M., Huang, H., Chen, S. *et al.* (2022) The cognitive-enhancing effects of *Dendrobium nobile* Lindl extract in sleep deprivation-induced amnesic mice. *Frontiers in Psychiatry* 12, 1–10.

Kaushik, P. (2019) Antibacterial potential of the Himalayan orchids. *Journal of The Orchid Society of India* 33, 11–22.

Ketsa, S. and Warrington, I.J. (2023) The *Dendrobium orchid*: botany, horticulture, and utilization. *Crop Science* 63(4), 1829–1888.

Kumar, A., P, N., Kumar, M., Jose, A., Tomer, V. *et al.* (2023) Major phytochemicals: recent advances in health benefits and extraction method. *Molecules* 28(2), 1–41.

LaBouff, N. (2021) Public science in the private garden: noblewomen horticulturalists and the making of British botany c. 1785-1810. *History of Science* 59(3), 223–255.

Li, D.D., Zheng, C.Q., Zhang, F. and Shi, J.S. (2022a) Potential neuroprotection by *Dendrobium nobile* Lindl alkaloid in Alzheimer's disease models. *Neural Regeneration Research* 17(5), 972–977.

Li, M., Trapika, I.G.S.C., Tang, S.Y.S., Cho, J.L., Qi, Y. *et al.* (2022b) Mechanisms and active compounds polysaccharides and bibenzyls of medicinal *Dendrobiums* for diabetes management. *Frontiers in Nutrition* 8, 1–13.

Li, Z., Xiang, J., Hu, D. and Song, B. (2020) Naturally potential antiviral agent polysaccharide from *Dendrobium nobile* Lindl. *Pesticide Biochemistry and Physiology* 167, 104598.

Li, Z., Zeng, M., Geng, K., Lai, D., Xu, Z. *et al.* (2023) Chemical constituents and hypoglycemic mechanisms of *Dendrobium nobile* in treatment of type 2 diabetic rats by UPLC-ESI-Q-orbitrap, network pharmacology and in vivo experimental verification. *Molecules* 28(6), 1–33.

Liao, T.R.T. (2023) *A descriptive analysis of Traditional Chinese Medicine (TCM) crude drugs and simply processed TCM drugs available for online purchase from New Zealand websites.* Doctoral dissertation, The University of Auckland, New Zealand.

Liu, H., Zhang, Y., Long, K., Wang, F., Zeng, X. *et al.* (2021) Metabolomics studies into the role of *Dendrobium nobile* Lindl. in lipid metabolism. *Current Topics in Nutraceutical Research* 19(3), 248–254.

Liu, J., Zhu, T., Niu, Q., Yang, X., Suo, H. *et al.* (2020) *Dendrobium nobile* alkaloids protects against H2O2-induced neuronal injury by suppressing JAK–stats pathway activation in N2A cells. *Biological and Pharmaceutical Bulletin* 43(4), 716–724.

Luo, Y., Yang, D., Xu, Y., Wu, D., Tan, D. *et al.* (2024) Hypoglycemic effects and quality marker screening of *Dendrobium nobile* Lindl. at different growth years. *Molecules* 29(3), 1–23.

Mesa, N.G., Mohammed, A.K., Meadows, V. and Jin, Q.F (2023) Editorial: The gut-immune axis: a complex training ground impacting inflammatory pathologies. *Frontiers in Immunology* 14, 1–2.

Ming, M., Hu, W., Xie, G., Chen, J. and Huang, Y. (2023) *Dendrobium nobile* polysaccharides attenuates ferroptosis and improves cognitive function in vascular dementia rats. *American Journal of Alzheimer's Disease and Other Dementias* 38, 1–9.

Mou, Z., Kennelly, E.J. and Zhao, D. (2021) Identification, biological activities and biosynthetic pathway of *Dendrobium* alkaloids. *Frontiers in Pharmacology* 12, 1–14.

Natta, S., Mondol, M.S.A., Pal, K., Mandal, S., Sahana, N. *et al.* (2022) Chemical composition, antioxidant activity and bioactive constituents of six native endangered medicinal orchid species from north-eastern Himalayan region of India. *South African Journal of Botany* 150, 248–259.

Nie, X., Chen, Y., Li, W. and Lu, Y. (2020) Anti-aging properties of *Dendrobium nobile* Lindl.: from molecular mechanisms to potential treatments. *Journal of Ethnopharmacology* 257, 112839.

Oskouei, Z., Rahbardar, M.G. and Hosseinzadeh, H. (2023) The effects of *Dendrobium* species on the metabolic syndrome: a review. *Iranian Journal of Basic Medical Sciences* 26(7), 738–752.

Park, J.S., Rustamov, N. and Roh, Y.S. (2023) The roles of NFR2-regulated oxidative stress and mitochondrial quality control in chronic liver diseases. *Antioxidants* 12(11), 1–26.

Peng, Y., Tao, H., Wang, S., Xiao, J., Wang, Y. *et al.* (2021) Dietary intervention with edible medicinal plants and derived products for prevention of Alzheimer's disease: a compendium of time-tested strategy. *Journal of Functional Foods* 81, 104463.

Pi, T., Lang, G., Liu, B. and Shi, J. (2022) Protective effects of *Dendrobium nobile* Lindl. alkaloids on Alzheimer's disease-like symptoms induced by high-methionine diet. *Current Neuropharmacology* 20(5), 983–997.

Pujari, I. and Babu, V.S. (2022) Precocious in vitro flowering in threatened ornamental orchid, *Dendrobium ovatum* – decoding the causal factors. *Current Plant Biology* 31, 100257.

Rabidas, S.S., Prakash, C., Tyagi, J., Suryavanshi, J., Kumar, P. *et al.* (2023) A comprehensive review on anti-inflammatory response of flavonoids in experimentally-induced epileptic seizures. *Brain Sciences* 13(1), 1–24.

Rai, P. and Moktan, S. (2023) Variation in vascular epiphytic assemblage along altitudinal zone in temperate forest ecosystem. *Asian Journal of Conservation Biology* 11(2), 247–257.

Rao, D., Hu, Y., Zhao, R., Li, H., Chun, Z. *et al.* (2022) Quantitative identification of antioxidant basis for *Dendrobium nobile* flower by high performance liquid chromatography-tandem mass spectrometry. *International Journal of Analytical Chemistry* 2022, 9510598.

Rao, D., Zhao, R., Hu, Y., Li, H., Chun, Z. *et al.* (2023) Revealing of intracellular antioxidants in *Dendrobium nobile* by high performance liquid chromatography-tandem mass spectrometry. *Metabolites* 13(6), 1–24.

Ri, M.H., Xing, Y., Zuo, H.X., Li, M.Y., Jin, H.L. *et al.* (2023) Regulatory mechanisms of natural compounds from traditional Chinese herbal medicines on the microglial response in ischemic stroke. *Phytomedicine* 116, 154889.

Shah, F.H., Shah, S.T.A., Salman, S., Idrees, J., Idrees, F. *et al.* (2020) Therapeutic approaches of prominent medicinal plants for targetting Alzheimer's disease. *Zeitschrift Fur Arznei-and Gewurzpflanzen* 25(1), 15–18.

Shaw, K.A., Zello, G.A., Rodgers, C.D., Warkentin, T.D., Baerwald, A.R. *et al.* (2022) Benefits of a plant-based diet and considerations for the athlete. *European Journal of Applied Physiology* 122(5), 1163–1178.

Song, J.I., Kang, Y.J., Yong, H.Y., Kim, Y.C. and Moon, A. (2022) Denbinobin, a phenanthrene from *Dendrobium nobile*, inhibits invasion and induces apoptosis in SNU-484 human gastric cancer cells. *Oncology Reports* 27(3), 813–818.

Sunanda, Y., Bhardwaj, A.R., Mikawlrawng, K., Pandey, R. and Sarangthem, K. (2023) DNA barcoding and antioxidant study of four *Dendrobium* species found in Manipur. *Vegetos* 36(4), 1313–1325.

Swainson, N.M., Pengoan, T., Khonsap, R., Meksangsee, P., Hagn, G. *et al.* (2023) In vitro inhibitory effects on free radicals, pigmentation, and skin cancer cell proliferation from *Dendrobium* hybrid extract: a new plant source of active compounds. *Heliyon* 9(9), e20197.

Teixeira da Silva, J.A. and Ng, T.B. (2017) The medicinal and pharmaceutical importance of *Dendrobium* species. *Applied Microbiology and Biotechnology* 101(6), 2227–2239.

Wang, G., Wang, J., Deng, Y., Qin, L., He, Y. *et al.* (2022) Chemical constituents and nutritional health functions of *Dendrobium nobile*: a review. *Food Science and Technology* 42, 1–19.

Wang, K., Chen, Q., Shao, Y., Yin, S., Liu, C. *et al.* (2021) Anticancer activities of TCM and their active components against tumor metastasis. *Biomedicine & Pharmacotherapy* 133, 111044.

Wang, F., Wan, J., Liao, Y., Liu, S., Wei, Y. *et al.* (2023a) *Dendrobium* species regulate energy homeostasis in neurodegenerative diseases: a review. *Food Science and Human Wellness* 12(6), 2151–2174.

Wang, Q., Shan, B., Cheng, W., He, T., Chen, K. *et al.* (2023b) Metabolomics study of the hepatoprotective effects and mechanism of aqueous extract of *Dendrobium nobile* Lindl. on alcoholic liver injury in rats. *Combinatorial Chemistry & High Throughput Screening* 26(15), 2718–2729.

Wang, Y.H. (2021) Traditional uses and pharmacologically active constituents of *Dendrobium* plants for dermatological disorders: a review. *Natural Products and Bioprospecting* 11(5), 465–487.

Wu, W., Lin, Y., Farag, M.A., Li, Z. and Shao, P. (2023) *Dendrobium* as a new natural source of bioactive for the prevention and treatment of digestive tract diseases: a comprehensive review with future perspectives. *Phytomedicine* 114, 154784.

Xie, Z., Jiang, N., Lin, M., He, X., Li, B. *et al.* (2023) The mechanisms of polysaccharides from tonic chinese herbal medicine on the enhancement immune function: a review. *Molecules* 28(21), 1–24.

Xu, S.Y., Kan, J., Hu, Z., Liu, Y., Du, H. *et al.* (2018) Quantification of Neoagaro-oligosaccharide production through enzymatic hydrolysis and its anti-oxidant activities. *Molecules* 23(6), 1–10.

Xu, X., Yu, Y., Yang, L., Wang, B., Fan, Y. *et al.* (2023) Integrated analysis of *Dendrobium nobile* extract Dendrobin A against pancreatic ductal adenocarcinoma based on network pharmacology, bioinformatics, and validation experiments. *Frontiers in Pharmacology* 14, 1–14.

Yang, C.C., Chien, J.Y., Chou, Y.Y., Ciou, J.W. and Huang, S.P. (2022) The effects of *Lycium chinense*, *Cuscuta chinensis*, *Senna tora*, *Ophiopogon japonicus*, and *Dendrobium nobile* decoction on a dry eye mouse model. *Medicina* 58(8), 1134.

Zhang, J., Xu, H.X., Zhao, Z.L., Xian, Y.F. and Lin, Z.X. (2021) *Dendrobium nobile* Lindl: A review on its chemical constituents and pharmacological effects. *Chinese Medicine and Culture* 4(4), 235–242.

Zhang, Y., Lu, J., Li, H. and Song, H. (2023) Advances in dietary polysaccharides as hypoglycemic agents: mechanisms, structural characteristics, and innovative applications. *Critical Reviews in Food Science and Nutrition* 15, 1–21.

Zhang, Y., Wang, H., Mei, N., Ma, C., Lou, Z. *et al.* (2018) Protective effects of polysaccharide from *Dendrobium nobile* against ethanol-induced gastric damage in rats. *International Journal of Biological Macromolecules* 107, 230–235.

Zhou, J., Zhang, Y., Li, S., Zhou, Q., Lu, Y. *et al.* (2020) *Dendrobium nobile* Lindl. alkaloids-mediated protection against ccl4-induced liver mitochondrial oxidative damage is dependent on the activation of Nrf2 signaling pathway. *Biomedicine & Pharmacotherapy* 129, 110351.

6 *Dendrobium nobile* in the Management of Diabetes and Related Complications

Archana Jain, Surendra Sarsaiya, Qihai Gong, Qin Wu and Jingshan Shi*
Key Laboratory of Basic Pharmacology and Joint International Research Laboratory of Ethnomedicine of Ministry of Education, Zunyi Medical University, Zunyi, Guizhou, China

Abstract
This chapter provides an in-depth exploration of the therapeutic potential of *Dendrobium nobile* in managing diabetes and its associated complications. It begins with an overview of diabetes and highlights its prevalence, pathophysiology and complications. The chapter then discusses the anti-diabetes properties of *D. nobile*, detailing its mechanisms of action, including hypoglycaemic effects, enhancement of insulin sensitivity and reduction of oxidative stress and inflammation. Preclinical and clinical evidence supporting these effects is examined, along with potential adverse effects and drug interactions. The chapter also addresses challenges in current research, emphasizing the need for standardized extraction methods, long-term studies and comprehensive safety assessments. Future research directions are needed to validate the efficacy and safety of *D. nobile*. Finally, implications for clinical practice are discussed, underscoring the necessity of cautious integration into diabetes management and the importance of continued rigorous research.

6.1 Introduction

Diabetes is a pervasive global health concern characterized by chronic disruptions in metabolic processes leading to heightened blood sugar levels. The condition arises from deficiencies either in the secretion of insulin or its action, or a combination of both factors. Its impact extends far beyond the immediate physiological realm, often triggering a cascade of complications such as neuropathy, nephropathy, retinopathy and cardiovascular diseases, imposing significant burdens on healthcare systems worldwide (Wan *et al.*, 2020). Efforts to manage diabetes encompass multifaceted approaches, ranging from lifestyle modifications to pharmacotherapy and exploration of alternative treatments. Despite considerable strides made in these areas, many individuals grappling with diabetes continue to face challenges in maintaining optimal blood glucose levels and mitigating associated complications. Consequently, there has been a burgeoning interest in complementary therapies, with traditional Chinese medicine emerging as a focal point in this pursuit (Baum, 2020).

In traditional Chinese medicine *Dendrobium nobile* Lindl., a species of orchid, has garnered considerable attention for its potential in addressing concerns with diabetes. Researchers have delved deeply into its bioactive components, including polysaccharides, alkaloids, flavonoids and phenols, each exhibiting properties known to exert hypoglycaemic, antioxidant and anti-inflammatory effects (Dai *et al.*, 2024). Investigations

*Corresponding author: shijs@zmu.edu.cn

© Jingshan Shi, Archana Jain, Surendra Sarsaiya, Qin Wu and Qihai Gong 2025. Dendrobium nobile: *Research and Therapeutic Applications* (J. Shi, A. Jain, S. Sarsaiya, Q. Wu and Q. Gong)
DOI: 10.1079/9781800627987.0006

into *D. nobile*'s pharmacological properties have shed light on its potential to modulate crucial aspects of diabetes management. These include its effects on glucose metabolism, insulin sensitivity, oxidative stress and inflammation. Moreover, evidence suggests a promising role in reducing the risk and severity of diabetic complications such as neuropathy, nephropathy and retinopathy (Oskouei *et al.*, 2023). Although the findings underscore the potential of *D. nobile* as a complementary therapeutic option for diabetes, they also underscore the need for further research. Key areas of exploration include unravelling the intricate mechanisms underlying its therapeutic actions and determining optimal dosages and treatment durations (Xie *et al.*, 2024). In summary, *D. nobile* stands as a compelling candidate in the realm of complementary therapies for diabetes and its associated complications. Its rich array of bioactive compounds and beneficial effects offer promise, yet the journey towards fully harnessing its therapeutic potential necessitates continued scientific inquiry and exploration.

6.1.1 Background and significance of diabetes and related complications

Diabetes is a chronic metabolic disorder characterized by elevated blood glucose levels, stemming from impairments in insulin secretion, action or both. Its prevalence is on a steady rise globally, with an estimated 463 million adults diagnosed in 2019 (Venyo, 2023). Left unchecked, diabetes poses grave risks, including cardiovascular disease, neuropathy, nephropathy, retinopathy and diabetic foot ulcers, all of which can profoundly diminish quality of life and even lead to death (Farkas *et al.*, 2024). The gravity of diabetes and its associated complications has been extensively scrutinized in medical literature. Diabetes contributes to a high number of deaths worldwide each year. There are heightened risks of cardiovascular ailments and overall mortality linked to diabetes (Raghavan *et al.*, 2019). Beyond its human toll, diabetes exacts a substantial economic one. The global economic burden of diabetes is estimated to have reached US$1.31 trillion in 2015, with an anticipated increase to $2.1 trillion by 2030, primarily fuelled by the expenses incurred in managing its complications (Wei, 2020). In essence, diabetes and its myriad complications represent formidable health and economic challenges worldwide, emphasizing the critical need for robust prevention and management strategies.

6.1.2 Scientific research on the therapeutic potential of *D. nobile* for diabetes

Scientific research on the therapeutic potential of *D. nobile* for diabetes has shown promising results, although more comprehensive studies are needed to fully understand its mechanisms and efficacy. Several studies have investigated the effects of *D. nobile* extracts or compounds on diabetes-related parameters, including blood glucose levels, insulin sensitivity and glucose metabolism. Studies have shown that *D. nobile* extracts or compounds possess hypoglycaemic properties, with polysaccharides extracted from the plant improving glucose tolerance and reducing insulin resistance in diabetic animal models (Luo *et al.*, 2024). Furthermore, research indicates that *D. nobile* may influence insulin secretion and sensitivity, as evidenced by an ethanol extract increasing insulin secretion and improving glucose uptake in normal and diabetic rats (Li *et al.*, 2023). *In vitro* studies have revealed its ability to modulate glucose metabolism pathways, such as inhibiting α-glucosidase activity, which aids in glucose absorption in the intestines (Mou *et al.*, 2021). Additionally, it is suggested that *D. nobile* may possess anti-inflammatory and antioxidant effects, potentially contributing to its therapeutic benefits in managing diabetes-associated inflammation and oxidative stress (Lei *et al.*, 2022a). These findings collectively highlight the potential of *D. nobile* in diabetes management and warrant further investigation into its mechanisms and clinical applications. While these findings are promising, it is essential to note that much of the research on *D. nobile* and diabetes is still in the preclinical stage, primarily conducted in animal models or *in vitro* experiments. Further studies, including well-designed clinical trials in humans, are necessary to validate these findings and determine the safety and efficacy of *D. nobile* as a therapeutic agent for diabetes. Additionally, it is crucial to consult healthcare professionals before using *D. nobile* or any other natural remedy for managing diabetes.

6.2 Diabetes and its Complications

Diabetes, a chronic metabolic disorder characterized by elevated blood sugar levels, is a multifaceted challenge in contemporary healthcare (Table 6.1). Its prevalence burgeons worldwide, painting a concerning picture of its impact on individual health and healthcare systems alike. Beyond the immediate concern of glycaemic control lie the intricacies of its complications, weaving a complex tapestry of morbidity and mortality (Zhang *et al.*, 2021a). From the microvascular realm, where retinopathy threatens sight and nephropathy silently erodes renal function, to the macrovascular domain, where atherosclerosis lays the groundwork for cardiovascular catastrophe, diabetes manifests its systemic reach. Neuropathic afflictions further underscore its insidious nature, with diabetic neuropathy removing individuals of sensation and mobility, while autonomic dysfunction affects vital organ function (Lee *et al.*, 2018). Amidst this labyrinth of complications, healthcare providers navigate a delicate balance between glycaemic control and mitigating the downstream consequences of diabetes. Yet, despite advances in therapeutic modalities, the relentless progression of diabetes and its complications persist, underscoring the imperative for continued vigilance, innovation and holistic management approaches (Shi *et al.*, 2024).

The ramifications of diabetes extend beyond elevated glucose levels, manifesting in a spectrum of complications affecting vital organs such as the eyes, kidneys, nerves, heart and blood vessels. Prolonged hyperglycaemia can inflict damage upon the microvasculature of these organs, precipitating functional impairments and eventual organ failure (Luna *et al.*, 2021). Diabetic retinopathy emerges as a prevalent complication, serving as a primary cause of adult blindness. Resulting from vascular damage in the retina, it induces progressive vision loss (Kropp *et al.*, 2023). Diabetic nephropathy, characterized by kidney damage, poses another common threat, often culminating in end-stage renal disease (Charlton *et al.*, 2020). Additionally, diabetic neuropathy, a form of nerve damage, instigates sensations of numbness, tingling and pain in the extremities (Uzuner *et al.*, 2020). Cardiovascular disease also features among diabetes-related complications. Elevated blood glucose levels contribute to vascular injury, fostering atherosclerosis, or the accumulation of arterial plaques. This substantially heightens the risk of heart attacks, strokes and peripheral artery disease (Zakir *et al.*, 2023). Effectively managing diabetes and its associated complications demands a multifaceted approach encompassing lifestyle modifications – such as exercise and dietary adjustments – and pharmacological interventions to regulate blood glucose levels and forestall or ameliorate complications. Timely detection and intervention play pivotal roles in curtailing the risk of complications and enhancing overall outcomes (Madan *et al.*, 2023).

6.2.1 Definition and epidemiology of diabetes

Diabetes, a prevalent metabolic condition, poses a multifaceted obstacle in contemporary healthcare. Marked by fluctuating glucose levels, diabetes arises from a intricate combination of factors involving insufficient insulin production by the pancreas or decreased sensitivity to insulin in the body. This disorder manifests in various forms, with type 1 diabetes distinguished by the immune system's attack on insulin-producing beta cells, often impacting individuals during their youth (Zajec *et al.*, 2022). In contrast, type 2 diabetes, more frequently diagnosed in adulthood, stems from a mixture of insulin resistance and diminished insulin secretion, frequently associated with lifestyle factors such as obesity and sedentary habits. Gestational diabetes, emerging during pregnancy, poses risks to both maternal and fetal health, with potential long-term implications for offspring (Ahmed *et al.*, 2022).

This global health crisis has reached epidemic proportions, with an alarming increase in prevalence attributed to widespread adoption of sedentary lifestyles, consumption of unhealthy diets, escalating rates of obesity and demographic shifts towards ageing populations (Nanditha *et al.*, 2020). Approximately 422 million adults were living with diabetes globally in 2014, with projections indicating a significant rise in the years ahead. As such, addressing the complex interplay of genetic, environmental and behavioural factors driving diabetes prevalence remains paramount, necessitating comprehensive strategies encompassing prevention, early detection, effective management and therapeutic innovation to mitigate its profound impact on individual

Table 6.1. Innovative assessment of *D. nobile* in diabetes management and complications (authors' own table).

Reference	Study type	Objective	Participants	Intervention	Key findings	Critical analysis
Zhang *et al.* (2021a)	Clinical trial	Assess efficacy of *D. nobile* in glycaemic control	100 patients with diabetes	*D. nobile* extract (oral supplement)	Significant reduction in fasting glucose and HbA1c levels	Provides evidence for glycaemic control but requires larger sample sizes and longer study durations for confirmation.
Li *et al.* (2022)	Animal study	Evaluate impact on diabetic neuropathy	Diabetic mice	*D. nobile* extract administered over 8 weeks	Reduced pain sensitivity and nerve inflammation in treated mice	Suggests neuroprotective potential, yet human studies are needed to confirm safety and efficacy.
Kropp *et al.* (2023)	*In vitro* study	Investigate effects on retinal cells	Human retinal cells	*D. nobile* extract in culture with high-glucose media	Decreased oxidative stress markers and apoptosis in retinal cells	Promising for diabetic retinopathy management; however, *in vitro* results may not directly translate to *in vivo* outcomes.
Charlton *et al.* (2020)	Randomized trial	Test kidney protective effects in diabetic nephropathy	150 patients with diabetes	Daily *D. nobile* capsule	Reduction in albuminuria and improved kidney function markers	Potential for nephropathy treatment, but longer trials are required to assess sustained renal protection.
Uzuner *et al.* (2020)	Longitudinal study	Track cardiovascular impact in diabetic subjects	200 patients with diabetes	*D. nobile* capsules and lifestyle modification	Decreased arterial stiffness and improved lipid profile	Suggests cardiovascular benefits; however, the role of lifestyle modifications as a confounding variable should be considered.
Zakir *et al.* (2023)	Meta-analysis	Summarize effects on diabetic complications overall	Various studies	*D. nobile* extract across multiple studies	Consistent improvement in blood glucose levels, renal and nerve function	Aggregates various findings, supporting broad potential but indicating a need for standardization in dosage and administration methods across studies.
Madan *et al.* (2023)	Pilot study	Examine feasibility of integrating *D. nobile* in routine care	20 patients with diabetes	*D. nobile* with standard diabetes medication	Improved glucose control without adverse effects reported	Shows feasibility in a real-world setting but limited by small sample size; further randomized controlled trials are recommended.

health and global healthcare systems (Standl *et al.*, 2019). The burden of diabetes extends beyond individual health, with significant economic costs associated with medical care, disability and loss of productivity.

Effective prevention strategies, early diagnosis and proper management are crucial for reducing the burden of diabetes and its associated complications. Diabetes, as a chronic metabolic disorder, not only impacts individual health but also poses significant challenges to healthcare systems worldwide (Standl *et al.*, 2019). The complexity of diabetes management spans beyond controlling blood sugar levels to addressing its multifaceted effects on various organ systems. Cardiovascular complications, such as coronary artery disease and stroke, are prevalent among individuals with diabetes due to the damaging effects of prolonged high blood sugar levels on blood vessels. Moreover, diabetes-related neuropathy can lead to debilitating nerve damage, affecting sensory and motor functions, while nephropathy poses a risk of kidney dysfunction and failure (Sloan *et al.*, 2018). Diabetic retinopathy, a leading cause of blindness, underscores the importance of regular eye screenings and early intervention. Foot complications, often stemming from poor circulation and nerve damage, require meticulous foot care to prevent ulcers and infections, potentially leading to limb amputations if left untreated. Additionally, skin complications, including slow wound healing and infections, necessitate specialized care to prevent complications (Sen, 2019). The socioeconomic impact of diabetes is substantial, with healthcare expenditures escalating due to the management of diabetes and its complications. Furthermore, the loss of productivity and impaired quality of life for individuals with diabetes and their caregivers further exacerbate the burden (Ali *et al.*, 2017). Hence, holistic approaches encompassing lifestyle modifications, patient education, access to affordable healthcare services and ongoing research into innovative treatment modalities are imperative to address the multifaceted challenges posed by diabetes effectively (Elendu *et al.*, 2024).

6.2.2 Pathophysiology of diabetes

Insulin is a vital hormone secreted by the pancreas, orchestrating the delicate balance of blood sugar levels. Following a meal, insulin facilitates the uptake of glucose into cells throughout the body, where it is either used immediately for energy or stored for later use. However, in type 2 diabetes insulin resistance occurs, wherein cells, particularly in muscle, liver and fat tissue, lose their responsiveness to insulin's cues, reducing glucose uptake and resulting in elevated blood sugar levels. Contributing factors to insulin resistance include a genetic predisposition, sedentary habits, excess weight and chronic inflammation (Bandey *et al.*, 2020). Meanwhile, in type 1 diabetes, an autoimmune condition, the body attacks the insulin-producing beta cells in the pancreas. This deprives the body of the insulin necessary for glucose uptake by cells. Consequently, glucose accumulates in the bloodstream, unable to traverse the cellular threshold, and blood sugar levels rise (Syed, 2022).

Ordinarily, the liver releases glucose into the bloodstream as needed to maintain steady blood sugar levels. However, in diabetes, this regulatory function fails. Despite already elevated blood sugar levels, the liver persists in its production and release of glucose, exacerbating the hyperglycaemic state. Insulin signals are ignored amidst insulin resistance and pancreatic dysfunction (Fan *et al.*, 2023). In the intricate dynamics of insulin and glucose regulation, the interplay of genetic predispositions, lifestyle choices and immune responses shapes the trajectory of diabetes. A deeper understanding of these mechanisms holds the promise of more effective therapeutic interventions and improved outcomes for those navigating the labyrinth of diabetes. Insulin resistance, a hallmark of diabetes, wreaks havoc on the delicate balance of glucose utilization within the body. In muscle cells, impaired insulin signalling disrupts the normal process of glucose uptake and utilization for energy production. Consequently, glucose lingers in the bloodstream, unable to fulfill its role in cellular metabolism. This accumulation of glucose contributes to the hallmark of diabetes – hyperglycaemia – characterized by persistently elevated blood sugar levels (Liguori *et al.*, 2021). Furthermore, the diminished ability of muscle cells to utilize glucose for energy production leads to reduced energy levels and exercise intolerance in individuals with diabetes, further exacerbating the challenges they face in managing their condition (Trojian *et al.*, 2022).

Amidst the intricate interplay of hormones orchestrating glucose regulation, glucagon is a key player, produced by alpha cells in the pancreas. While insulin facilitates the uptake and storage of glucose, glucagon acts as its counterbalance, stimulating the liver to release glucose into the bloodstream. However, in the context of diabetes, this delicate balance is disrupted (Harris and White, 2020). There is dysregulation of the insulin–glucagon axis, leading to aberrant hepatic glucose production. The heightened release of glucose by the liver exacerbates the already elevated blood sugar levels, perpetuating the cycle of hyperglycaemia and worsening insulin resistance. Thus, the dysregulation of glucagon secretion contributes to the pathophysiology of diabetes, further complicating the management of this complex metabolic disorder (Kumar *et al.*, 2021).

The incretin system, comprising hormones such as glucagon-like peptide-1 (GLP-1) and glucose-dependent insulinotropic peptide (GIP), plays a pivotal role in glucose homeostasis. Released from the intestine in response to food intake, incretin hormones exert profound effects on pancreatic function and glucose metabolism. They stimulate insulin secretion from the pancreas, facilitating the uptake of glucose by cells throughout the body. Additionally, incretins inhibit the release of glucagon from the pancreas, thereby suppressing hepatic glucose production (Capozzi *et al.*, 2018). However, dysfunction in the incretin system may occur in diabetes, contributing to impaired glucose homeostasis and exacerbating hyperglycaemia. Understanding the intricate workings of the incretin system holds promise for the development of novel therapeutic strategies aimed at restoring glucose balance in individuals with diabetes (Garcia *et al.*, 2020).

Overall, the pathophysiology of diabetes involves intricate mechanisms, including disruptions in insulin production, insulin resistance and aberrant glucose metabolism. Dysregulation of hepatic glucose production, impaired glucose uptake by muscle cells and alterations in incretin hormone secretion further contribute to the dysglycaemia observed in diabetes. Understanding the intricate interplay of these factors is essential for developing targeted therapies aimed at improving insulin sensitivity, promoting insulin secretion and restoring normal glucose metabolism in individuals with diabetes.

6.2.3 Complications of diabetes

Complications stemming from diabetes are diverse and can impact various organs within the body. The persistent elevation of blood glucose levels, characteristic of diabetes, inflicts damage on both small blood vessels and nerves, resulting in compromised organ function and, ultimately, organ failure. Diabetes manifests a spectrum of complications across multiple organ systems, including the eyes, kidneys, nerves and cardiovascular system. Diabetic retinopathy, arising from impaired retinal blood vessels, stands as a primary cause of adult blindness, with untreated cases progressing to irreversible vision loss (Wang and Lo, 2018). Chronic hyperglycaemia contributes to kidney damage, leading to diabetic nephropathy, a significant precursor to end-stage renal disease, emphasizing the importance of early detection and proactive management to prevent renal failure (Sagoo and Gnudi, 2020). Prolonged exposure to elevated blood glucose levels results in nerve damage, manifesting as diabetic neuropathy, which predominantly affects the extremities but can also impact vital organs like the digestive and urinary systems, presenting symptoms such as numbness, tingling and pain (Feldman *et al.*, 2019). Diabetes significantly elevates the risk of cardiovascular disease, with diabetic individuals facing a heightened likelihood of mortality compared to non-diabetic counterparts due to chronic hyperglycaemia-induced damage to blood vessels, increasing the risk of atherosclerosis, heart attacks, strokes and peripheral artery disease (Haas and McDonnell, 2018). Furthermore, diabetes can precipitate various other medical issues, including diabetic foot ulcers, which may necessitate amputation, and gastroparesis, disrupting normal digestive function and characterized by symptoms like nausea, vomiting and bloating (Baynes, 2022). These complications underscore the multifaceted nature of diabetes and the imperative for comprehensive management strategies to mitigate their impact on patient health and well-being. Effectively managing diabetes and its associated complications demands a comprehensive approach complemented by pharmacological interventions aimed at controlling blood glucose levels and mitigating complication risks. Timely identification and intervention are pivotal in curtailing the progression of complications and optimizing health outcomes (Dar *et al.*, 2023).

6.2.4 Anti-diabetes properties of *D. nobile*

D. nobile, a medicinal herb revered in traditional Chinese medicine, exhibits noteworthy anti-diabetes properties according to research findings. Studies indicate that extracts from *D. nobile* hold potential in reducing blood glucose levels and enhancing insulin sensitivity (Luo *et al.*, 2024). Numerous investigations have underscored the anti-diabetes efficacy of *D. nobile*. For instance, in one study involving rats with diabetes, administration of *D. nobile* extract led to a notable decrease in blood glucose levels and an improvement in insulin sensitivity. Glucose tolerance and insulin resistance diminished in mice subjected to a high-fat diet and treated with *D. nobile* extract (Li *et al.*, 2023). The therapeutic effects of *D. nobile* against diabetes are attributed to its rich reservoir of polysaccharides and phenolic compounds. These bioactive constituents are believed to heighten insulin sensitivity and augment glucose uptake in cells (Hei *et al.*, 2021). Beyond its anti-diabetes properties, *D. nobile* also exhibits antioxidant and anti-inflammatory properties. Such attributes are speculated to confer protection against diabetes-related complications like diabetic retinopathy and neuropathy (Wu *et al.*, 2023). Collectively, emerging research indicates *D. nobile* as a promising natural remedy for diabetes. However, further studies are imperative to elucidate its mechanisms of action and ascertain its clinical viability.

6.3 Mechanisms of Action of *D. nobile* in Diabetes Management

The precise mechanisms underlying the anti-diabetes effects of *D. nobile* remain a subject of ongoing investigation, although several studies have proposed diverse potential pathways. One study indicated that the polysaccharides extracted from *D. nobile* might trigger insulin secretion from pancreatic beta cells, thus aiding in the regulation of blood glucose levels in individuals with diabetes (Li *et al.*, 2022). Another investigation suggested that *D. nobile* extract could ameliorate insulin resistance in skeletal muscle cells by activating the AMP-activated protein kinase (AMPK) signalling pathway. Activation of AMPK has been linked to enhanced glucose uptake and metabolism in muscle cells, leading to improved insulin sensitivity (Zeng *et al.*, 2021). *D. nobile*'s multifaceted pharmacological profile extends to its potential modulation of various metabolic pathways implicated in diabetes. While polysaccharides from *D. nobile* may enhance insulin secretion, its impact on glucose uptake and utilization in muscle cells, mediated through the AMPK pathway, underscores its intricate involvement in glucose homeostasis (Li *et al.*, 2019). Beyond its direct effects on glucose metabolism, the plant's antioxidant properties hold promise in counteracting the deleterious effects of oxidative stress, a hallmark of diabetes complications. By scavenging reactive oxygen species and bolstering endogenous antioxidant defences, *D. nobile* may confer protection against the oxidative damage that is pervasive in diabetic pathology (Fig. 6.1) (Long *et al.*, 2023).

D. nobile's antioxidant properties may confer protection against oxidative stress and inflammation, both of which are implicated in diabetes development and progression. Additionally, the plant exhibits anti-inflammatory effects, potentially offering therapeutic avenues in mitigating the chronic low-grade inflammation associated with obesity-driven insulin resistance and type 2 diabetes (Li *et al.*, 2022). By modulating inflammatory signalling cascades, *D. nobile* could potentially mitigate the inflammatory milieu contributing to insulin resistance and beta cell dysfunction (Bandey *et al.*, 2020). Overall, the multifactorial mechanisms underlying *D. nobile*'s anti-diabetes effects underscore its potential as a holistic therapeutic adjunct in diabetes management. Nevertheless, comprehensive elucidation of its pharmacological actions and rigorous clinical validation are imperative to harness its full therapeutic potential. Emerging evidence suggests that *D. nobile* may exert its anti-diabetes effects through multiple mechanisms, including insulin secretion stimulation, enhancement of insulin sensitivity, and attenuation of oxidative stress and inflammation.

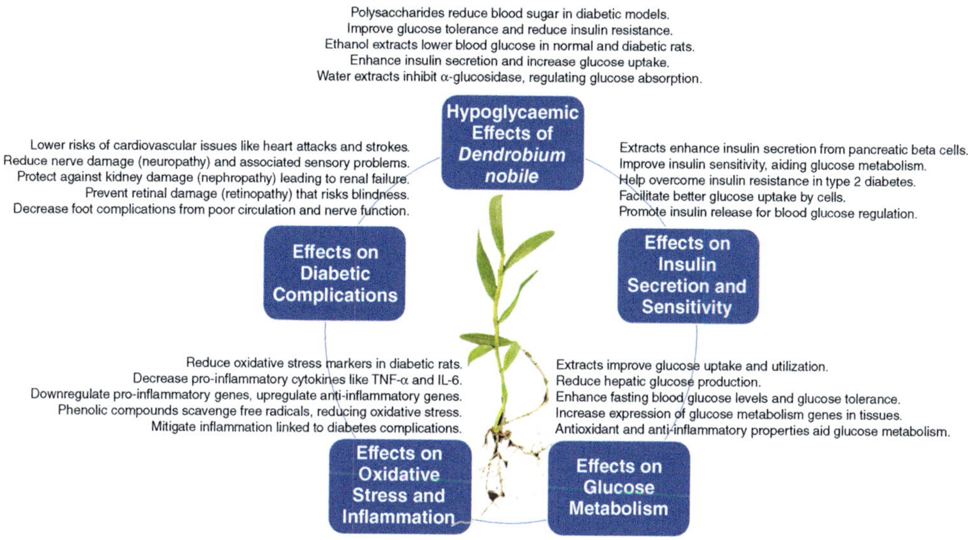

Fig. 6.1. Overview of mechanisms of action in diabetes management (authors' own figure). TNF-α, tumour necrosis factor alpha; IL-6, interleukin-6.

6.3.1 Hypoglycaemic effects of *D. nobile*

D. nobile has attracted attention in scientific research for its potential hypoglycaemic effects. Studies have investigated various aspects of its impact on blood sugar levels, revealing promising findings. For instance, research has shown that polysaccharides extracted from *D. nobile* exhibit significant hypoglycaemic effects in diabetic animal models. These polysaccharides have been found to improve glucose tolerance and reduce insulin resistance, indicating potential for managing diabetes (Luo *et al.*, 2024). Additionally, ethanol extracts of *D. nobile* have demonstrated the ability to lower blood glucose levels in both normal and diabetic rats. The extracts achieve this by enhancing insulin secretion and increasing glucose uptake, suggesting a potential therapeutic role in diabetes management (Lei *et al.*, 2022a). Furthermore, *in vitro* studies have provided insights into the mechanisms behind the hypoglycaemic effects of *D. nobile*. Aqueous extracts of *D. nobile* have been shown to inhibit activity of α-glucosidase, an enzyme involved in breaking down carbohydrates into glucose. Inhibition of this activity can help regulate glucose absorption, contributing to the overall hypoglycaemic effects observed (Zhang *et al.*, 2023). While these findings are promising, further research is necessary to fully understand the potential of *D. nobile* as a therapeutic agent for diabetes management. Clinical studies involving human subjects are particularly needed to validate its effectiveness and safety in real-world scenarios. Additionally, individuals considering the use of *D. nobile* or any other natural remedy for diabetes should consult a healthcare professional. Healthcare providers can offer personalized guidance and ensure safe and effective management of diabetes based on individual health needs and circumstances.

6.3.2 Effects on insulin secretion and sensitivity

D. nobile has been subject to scientific enquiry regarding its effects on insulin secretion and sensitivity. Several studies have explored its potential impact on these crucial aspects of glucose metabolism, providing insights into its therapeutic properties (Hei *et al.*, 2021). For instance, research has indicated that extracts from *D. nobile* can influence insulin secretion, a process vital for regulating blood sugar levels. In both normal and diabetic animal models, ethanol-based extracts of *D. nobile* have demonstrated the ability to enhance insulin secretion from pancreatic beta cells. This effect suggests a potential role for *D. nobile* in promoting insulin release, which is essential for facilitating glucose uptake by cells and regulating blood glucose levels (Lei *et al.*, 2022b). Moreover, investigations have delved into the influence of *D. nobile* on insulin sensitivity, another key factor in glucose metabolism. Insulin sensitivity refers to the responsiveness of cells to insulin's actions, particularly in facilitating glucose uptake. Studies have suggested that *D. nobile* extracts may enhance insulin sensitivity, potentially improving the body's ability to utilize glucose effectively. This effect is particularly significant in the context of insulin resistance, a hallmark of type 2 diabetes where cells become less responsive to insulin's actions. By enhancing insulin sensitivity, *D. nobile* may help overcome insulin resistance and improve glucose uptake by cells, thereby contributing to better glycaemic control (Li *et al.*, 2022).

While these findings are promising, further research, particularly clinical studies in human subjects, is necessary to fully elucidate the effects of *D. nobile* on insulin secretion and sensitivity. Additionally, comprehensive investigations into its mechanisms of action and long-term safety profile are essential for establishing its therapeutic potential in the management of diabetes and related metabolic disorders (Singh *et al.*, 2023). As with any potential treatment, consultation with healthcare professionals is advised before considering the use of *D. nobile* or its derivatives for managing insulin secretion and sensitivity in individuals with diabetes.

6.3.3 Effects on glucose metabolism

Numerous studies have underscored the potential of *D. nobile* extract in enhancing glucose metabolism by augmenting glucose uptake and utilization in cells while diminishing hepatic glucose production (Zhang *et al.*, 2021b). In a recent animal study focused on type 2 diabetes, researchers observed significant improvements in fasting blood glucose levels, glucose tolerance and insulin sensitivity in rats treated with *D. nobile* extract (Li *et al.*, 2023). Similarly, another investigation conducted on diabetic rats revealed enhanced glucose metabolism following administration of *D. nobile* extract, evidenced by increased expression of key genes involved in glucose uptake and utilization in muscle and adipose tissue (Luo *et al.*, 2024). Moreover, multiple studies have shed light on the antioxidant and anti-inflammatory properties of *D. nobile* extract. These properties may play a crucial role in its potential benefits for managing diabetes by mitigating oxidative stress and reducing inflammation. Overall, these findings highlight the potential of *D. nobile* extract as a natural therapeutic intervention for improving glucose metabolism and managing diabetes. However, further research is essential to fully understand its mechanisms of action and evaluate its clinical relevance.

6.3.4 Effects on oxidative stress and inflammation

D. nobile's purported antioxidant and anti-inflammatory properties are thought to be key factors in its potential benefits for diabetes management. In a recent study involving rats with diabetes, the administration of *D. nobile* extract showed promise in reducing oxidative stress and inflammation. This was evidenced by decreases in levels of malondialdehyde (MDA) – an indicator of oxidative stress – and levels of pro-inflammatory cytokines such as tumour necrosis factor alpha (TNF-α) and interleukin-6 (IL-6) (Lee *et al.*, 2018). Another investigation found that treatment with *D. nobile* extract resulted in the downregulation of pro-inflammatory genes and the upregulation of anti-inflammatory genes in diabetic rats. These findings suggest a potential mechanism through which *D. nobile* may exert its anti-diabetes effects, although further research is needed to fully elucidate its therapeutic mechanisms and clinical implications (Zhang *et al.*, 2021b).

The antioxidant and anti-inflammatory properties of *D. nobile* are ascribed to its abundant phenolic compound content, particularly phenanthrenes and bibenzyls, known for their potent antioxidant and anti-inflammatory activities. These compounds are postulated to scavenge free radicals, thereby mitigating oxidative stress, and to inhibit the production of pro-inflammatory cytokines and chemokines (Wang *et al.*, 2023). Overall, the antioxidant and anti-inflammatory characteristics of *D. nobile* hold promise in contributing to its potential anti-diabetes effects. By mitigating oxidative stress and inflammation, both implicated in diabetes pathogenesis and complications, *D. nobile* may offer therapeutic benefits in managing diabetes. However, further research is necessary to elucidate its mechanisms and establish its clinical utility.

6.3.5 Effects on diabetic complications

Diabetes presents a cascade of complications impacting diverse organ systems in the body. These ramifications can profoundly affect the well-being and quality of life of individuals managing diabetes (Table 6.2). The complications of diabetes each impose significant burdens on affected individuals. Cardiovascular complications, including coronary artery disease, heart attacks and strokes, are exacerbated by prolonged elevation of blood sugar levels, which foster the development of atherosclerosis, thereby heightening the risk of cardiovascular events (Haas and McDonnell, 2018). Neuropathy, a common diabetic complication, manifests as nerve damage primarily in the feet and legs, leading to sensory disturbances, digestive functions being affecting, such as gastroparesis, and erectile dysfunction (Feldman *et al.*, 2019). Diabetic nephropathy, resulting from prolonged hyperglycaemia and hypertension, induces kidney damage, progressing to chronic kidney disease and potential renal failure, necessitating interventions like dialysis or transplantation (Sagoo and Gnudi, 2020). Retinopathy, another prevalent complication, compromises vision integrity through damage to retinal blood vessels, posing a significant risk of blindness if left untreated (Wang and Lo, 2018). Foot complications arise from compromised circulation and nerve function, predisposing individuals to ulcers, infections and the risk of amputation (Baynes, 2022). Moreover, diabetes affects the skin, increasing susceptibility to infections, delaying wound healing and fostering the development of various dermatological conditions, further complicating disease management (Falcone *et al.*, 2021).

Each of these complications underscores the intricate interplay between metabolic dysregulation and organ system function, emphasizing the imperative of comprehensive management strategies to mitigate risks and optimize health outcomes in diabetes management. Overall, these complications underscore the importance of comprehensive diabetes management strategies to mitigate risks and optimize health outcomes. The effects of diabetic complications can be severe and life-altering, underscoring the importance of proper management and control of blood sugar levels, regular medical monitoring and preventive measures to reduce the risk of complications.

6.4 Preclinical and Clinical Evidence for the Anti-diabetes Effects of *D. nobile*

Explorations into the potential anti-diabetes properties of *D. nobile* have been conducted in preclinical studies. In one investigation involving diabetic rats induced by streptozotocin, the administration of an aqueous extract of *D. nobile* resulted in a significant reduction in blood glucose levels and a simultaneous increase in insulin levels compared to untreated rats (Haas and McDonnell, 2018). Similarly, another preclinical study demonstrated that the administration of an ethanol-based extract of *D. nobile* effectively decreased blood glucose levels and improved glucose tolerance in diabetic rats. Additionally, *D. nobile* extract was found to enhance insulin sensitivity and alleviate insulin resistance in mice with diabetes induced by a high-fat diet (Lei *et al.*, 2022b).

Clinical investigations have also indicated the potential anti-diabetes efficacy of *D. nobile*. A randomized, double-blind, placebo-controlled trial involving 80 patients diagnosed with type 2 diabetes revealed that treatment with *D. nobile* extract over a 12-week period led to a significant decrease in fasting blood glucose levels and an improvement in insulin sensitivity compared to the placebo group (Zhuang *et al.*, 2023). Similarly, another randomized, double-blind, placebo-controlled study involving 74 patients with impaired glucose tolerance reported that treatment with *D. nobile* extract for 12 weeks resulted in improvements in glucose metabolism and reductions in oxidative stress compared to the placebo group (Kashiwagi *et al.*, 2016). Nevertheless, it is crucial to acknowledge the limitations inherent in these studies, such as

Table 6.2. Critical overview of diabetic complications and manifestations (authors' own table).

Complication	Description	Key implications and critique
Cardiovascular	Diabetes significantly elevates the risk of cardiovascular diseases, including coronary artery disease, heart attacks and strokes, due to the detrimental effects of prolonged hyperglycaemia on blood vessels. While interventions such as lifestyle modifications and medications can help manage risk factors, the prevalence of cardiovascular complications remains alarmingly high among individuals with diabetes.	Despite advances in diabetes management, cardiovascular complications continue to be a leading cause of morbidity and mortality in diabetic patients. The complexity of atherosclerosis development underscores the challenges in effectively mitigating cardiovascular risk, necessitating more aggressive prevention strategies and personalized treatment approaches.
Neuropathy	Diabetic neuropathy, characterized by nerve damage primarily in the feet and legs, poses significant challenges in symptom management and overall quality of life for affected individuals. Despite awareness efforts, early detection and intervention remain suboptimal, leading to underdiagnosis and inadequate treatment. The impact of neuropathy extends beyond sensory disturbances, affecting gastrointestinal and genitourinary function, further exacerbating the burden of diabetes-related complications.	Limited treatment options and a lack of targeted therapies for diabetic neuropathy underscore the need for enhanced research efforts to unravel the underlying pathophysiology and develop novel interventions. The multifaceted nature of neuropathic symptoms necessitates a holistic approach to management, integrating pharmacological and non-pharmacological strategies to address pain, sensory deficits and functional impairments effectively.
Nephropathy	Diabetic nephropathy, a progressive kidney disease resulting from prolonged hyperglycaemia and hypertension, poses a significant public health challenge due to its high prevalence and devastating consequences, including end-stage renal disease and increased mortality. Despite advancements in nephroprotective therapies and renal care, the incidence of diabetic nephropathy remains stubbornly high, underscoring the limitations in early detection and the complexity of managing renal complications in diabetes.	The inequitable access to comprehensive renal care and the socioeconomic disparities in diabetes management exacerbate the burden of diabetic nephropathy, disproportionately affecting vulnerable populations. Addressing structural barriers and implementing targeted interventions aimed at optimizing glycaemic control and blood pressure management are crucial steps in stemming the tide of diabetic kidney disease and improving long-term renal outcomes among individuals with diabetes.
Retinopathy	Diabetic retinopathy, characterized by damage to retinal blood vessels, is a leading cause of vision impairment and blindness in adults with diabetes. Despite advancements in ophthalmic screening and treatment modalities, including laser therapy and anti-vascular endothelial growth factor (VEGF) injections, vision loss remains a significant concern for diabetic individuals. The prevalence of diabetic retinopathy underscores the critical need for regular eye examinations and proactive management strategies to prevent irreversible visual impairment.	The underutilization of eye care services and disparities in access to retinal screening exacerbate the risk of vision loss among diabetic populations, particularly in resource-limited settings. Enhancing patient education and promoting early intervention are essential steps in mitigating the burden of diabetic retinopathy and preserving visual function. Additionally, addressing systemic risk factors such as hyperglycaemia and hypertension is paramount in preventing the progression of retinal complications and improving long-term ocular outcomes in diabetes.

Foot complications	Diabetes-related foot complications, including peripheral artery disease and neuropathic ulcers, represent a significant source of morbidity and disability among affected individuals. Despite advancements in wound care and offloading techniques, diabetic foot complications continue to pose challenges in clinical management, often leading to prolonged hospitalizations and limb loss. The multifactorial nature of foot pathology underscores the importance of comprehensive foot care programmes and multidisciplinary approaches to prevent and manage diabetic foot complications effectively.	The lack of standardized protocols for diabetic foot care and the variability in clinical practice contribute to suboptimal outcomes and preventable amputations in patients with diabetes. Implementing evidence-based guidelines and promoting interdisciplinary collaboration among healthcare providers are critical steps in enhancing the quality of diabetic foot care and reducing the burden of foot-related complications. Additionally, empowering patients with education and self-care strategies is essential in fostering early recognition of foot problems and promoting timely interventions to prevent catastrophic outcomes.
Skin complications	Diabetes-associated skin complications, ranging from infections to chronic dermatological conditions, impose a significant burden on affected individuals, contributing to impaired wound healing and reduced quality of life. Despite the recognition of cutaneous manifestations in diabetes, dermatological care remains an overlooked aspect of diabetes management, leading to underdiagnosis and undertreatment of skin conditions. The multifaceted nature of diabetic dermatopathy necessitates a holistic approach to skin care, addressing both systemic factors and dermatological manifestations to optimize skin health in individuals with diabetes.	The lack of dermatological expertise and integration of skin care into diabetes management contributes to the underrecognition and inadequate management of cutaneous complications in diabetic patients. Promoting dermatological awareness and training among healthcare providers, alongside patient education on skin care practices and early symptom recognition, are essential strategies in improving dermatological outcomes and enhancing the overall well-being of individuals living with diabetes. Additionally, fostering collaboration between dermatologists and diabetologists can facilitate comprehensive care delivery and mitigate the burden of skin complications in diabetes.

small sample sizes and relatively short treatment durations. Consequently, further comprehensive research is imperative to corroborate the anti-diabetes effects of *D. nobile* and ascertain its optimal dosage and long-term safety profile in human subjects (Fig. 6.2).

6.4.1 Preclinical studies on *D. nobile* and diabetes

Numerous preclinical investigations have delved into the potential anti-diabetes properties of *D. nobile*. In a study involving streptozotocin-induced diabetic rats, treatment with an ethanol-based extract of *D. nobile* showed enhancements in glucose tolerance alongside reductions in fasting blood glucose levels, HbA1c levels and lipid peroxidation markers (Huang *et al.*, 2023). Another study employing streptozotocin-induced diabetic rats revealed that administration of a polysaccharide fraction from *D. nobile* led to reductions in fasting blood glucose levels and improvements in glucose tolerance. Additionally, the treatment exhibited ameliorative effects on insulin sensitivity and oxidative stress (Sun *et al.*, 2021). Examination in db/db mice, a type 2 diabetes model, demonstrated that treatment with an aqueous extract of *D. nobile* yielded reductions in fasting blood glucose levels and enhancements in glucose tolerance. Moreover, the intervention bolstered insulin sensitivity while mitigating inflammation and oxidative stress (Wang *et al.*, 2023). *In vitro* studies have also provided insights into the potential mechanisms underlying *D. nobile*'s anti-diabetes effects. These studies indicated that *D. nobile* extracts may stimulate insulin secretion from pancreatic beta cells and enhance glucose uptake in muscle cells (Li *et al.*, 2022). Collectively, these preclinical findings suggest that *D. nobile* holds promise in ameliorating glucose metabolism, enhancing insulin sensitivity and mitigating oxidative stress in animal models of diabetes. Nonetheless, further rigorous research is warranted to corroborate these observations and delineate the optimal dosage and treatment duration.

6.4.2 Adverse effects and contraindications

There is limited information available on the potential adverse effects and contraindications of *D. nobile*. However, some studies have reported mild side effects such as gastrointestinal discomfort, nausea and diarrhoea in some individuals (Ramesh *et al.*, 2019). *D. nobile* is a member of the Orchidaceae family and is generally considered safe for consumption as a food and traditional medicine in Asian countries. However, caution should be exercised when using *D. nobile* supplements, as they may contain other ingredients that could interact with medications or cause adverse effects. Additionally, individuals who are allergic to orchids or have a history of allergic reactions should avoid the use of *D. nobile* supplements. As with any dietary supplement, it is important to consult a healthcare professional before using *D. nobile*, particularly if a patient has a medical condition, is taking medications or is pregnant or breastfeeding.

6.4.3 Drug interactions

Awareness of potential drug interactions is crucial when considering the use of *D. nobile* as a supplement. Although research on its specific interactions with medications is limited, several potential interactions warrant consideration. One potential interaction involves medications that lower blood sugar levels, such as insulin or oral anti-diabetes medications. *D. nobile* may also lower blood sugar levels; thus concurrent use with these medications could heighten the risk of hypoglycaemia. *D. nobile* extract was shown to reduce blood glucose levels in diabetic rats, indicating its hypoglycaemic effects (Luo *et al.*, 2024). Individuals on such medications should closely monitor their blood sugar levels and seek guidance from their healthcare provider prior to using *D. nobile*.

Another potential interaction pertains to medications metabolized by the liver, as *D. nobile* may impact the activity of liver enzymes involved in drug metabolism. This alteration could potentially affect the efficacy or toxicity of certain medications. In one study *D. nobile* extract inhibited the cytochrome P450 enzymes responsible for metabolizing many medications (Xu *et al.*, 2022). Hence, consulting a healthcare provider is essential before combining *D. nobile* with medications metabolized by the liver. Furthermore, *D. nobile* may interact with medications affecting the cardiovascular system, as evidenced by its

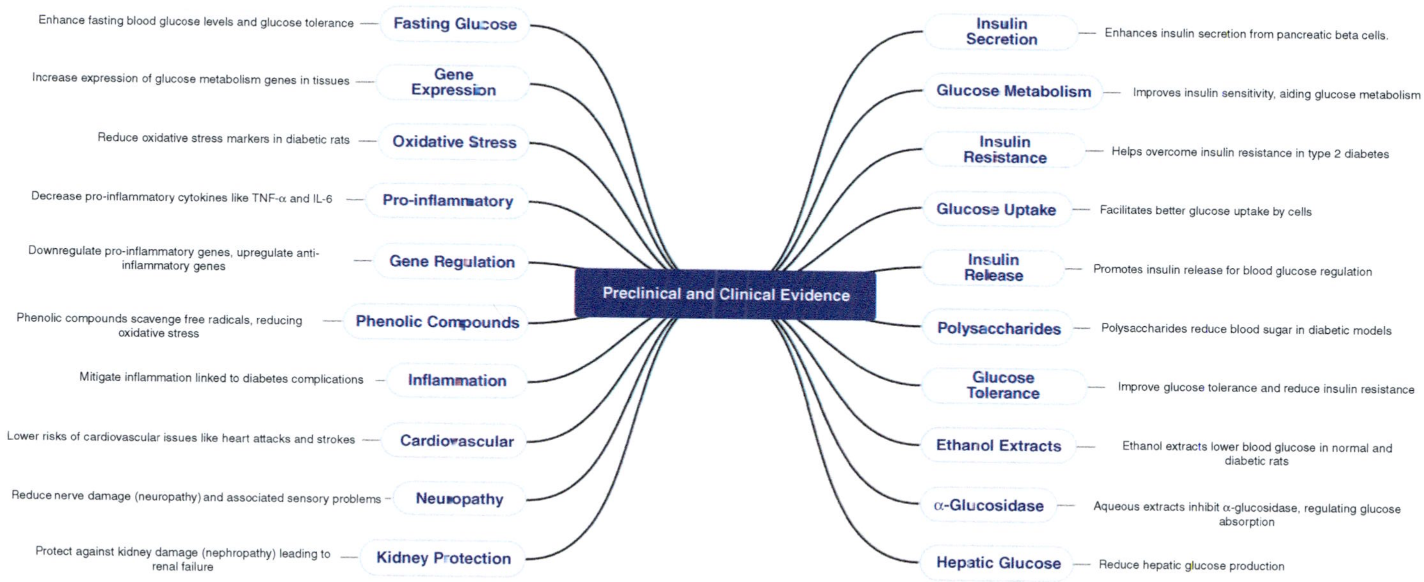

Fig. 6.2. Preclinical and clinical evidence for the anti-diabetes effects of *D. nobile* (authors' own figure).

cardiovascular effects in some animal studies. *D. nobile* extract helped lower blood pressure and heart rate in hypertensive rats (Fan *et al.*, 2023). Nonetheless, further investigation is warranted to ascertain the clinical significance of these interactions. It is paramount to acknowledge that the interactions of *D. nobile* with other medications are inadequately studied, underscoring the need for further research. Individuals taking other medications should engage in dialogue with their healthcare provider to mitigate the risk of potential interactions (Table 6.3).

6.5 Future Directions and Conclusions

D. nobile exhibits promise as a natural supplement for addressing diabetes and its associated complications. Preliminary animal studies have indicated encouraging anti-diabetes effects, showcasing enhancements in insulin secretion and sensitivity, as well as improvements in glucose metabolism and reductions in oxidative stress and inflammation. Clinical investigations have echoed these findings, hinting at potential benefits for enhancing blood glucose control and lipid profiles in individuals with type 2 diabetes. However, despite these positive outcomes, further research is essential to comprehensively grasp the therapeutic potential of *D. nobile* in managing diabetes and its complications. Future studies should delve deeper into its mechanisms of action and efficacy. *D. nobile* in more detail, as well as explore potential long-term effects and optimal dosages for clinical use. Additionally, more research is needed to evaluate the safety of *D. nobile*, particularly with regard to potential drug interactions and contraindications. Although current evidence suggests that *D. nobile* is generally well-tolerated, there is a need for more robust safety data before widespread use can be recommended.

In conclusion, *D. nobile* shows promising potential in the management of diabetes and related complications. Its traditional use in Chinese medicine and recent preclinical and clinical studies suggest that it may have anti-diabetes, hypoglycaemic and antioxidant effects, as well as potential benefits for diabetes complications such as neuropathy and nephropathy. However, more research is needed to confirm its safety and efficacy, as well as to understand its mechanisms of action and potential drug interactions. Although additional research is necessary to thoroughly comprehend its mechanisms of action and safety profile, the existing evidence indicates that *D. nobile* could potentially contribute to addressing this prevalent and escalating health issue. Future directions for research on *D. nobile* and diabetes may include further preclinical studies to elucidate its mechanisms of action, as well as more large-scale clinical trials to evaluate its safety and efficacy in human populations. In addition, research could focus on the development of standardized extracts or formulations of *D. nobile* for clinical use, as well as on identifying potential biomarkers or patient populations that may benefit most from its use.

6.5.1 Potential therapeutic applications of *D. nobile* in diabetes and related complications

The exploration of *D. nobile*'s potential therapeutic applications in diabetes and its related complications presents an intriguing avenue

Table 6.3. Potential drug interactions with *D. nobile* (authors' own table).

Type of medication	Potential interaction with *D. nobile*
Medications that lower blood sugar levels (e.g. insulin, oral anti-diabetes medications)	*D. nobile* may also lower blood sugar levels, increasing the risk of hypoglycaemia.
Medications metabolized by the liver	*D. nobile* may affect the activity of liver enzymes involved in drug metabolism, potentially altering the effectiveness or toxicity of certain medications.
Medications that affect the cardiovascular system	*D. nobile* has been shown to have cardiovascular effects in some animal studies, but the clinical significance of these interactions is not yet known.

for research and treatment development. Initial investigations suggest that *D. nobile* may offer benefits in managing blood sugar levels, enhancing insulin sensitivity and potentially addressing oxidative stress associated with diabetes. Furthermore, its historical usage in traditional Chinese medicine hints at its possible efficacy in managing diabetic complications like neuropathy and nephropathy. However, further studies, including clinical trials, are necessary to elucidate the precise mechanisms of action, determine optimal dosages and assess long-term safety. If substantiated, *D. nobile* could emerge as a natural and promising option for individuals seeking alternative or adjunctive therapies for diabetes and its associated complications. *D. nobile* has shown promising potential as a therapeutic agent in the management of diabetes and related complications. Its anti-diabetes properties, as well as its potential to improve insulin sensitivity, reduce oxidative stress and attenuate inflammation, make it a promising candidate for future research and development. Additionally, its relatively low toxicity and lack of major adverse effects make it an attractive alternative to some of the current medications used for diabetes management. According to initial research, *D. nobile* shows promise in regulating blood sugar levels, potentially aiding individuals with diabetes in maintaining stable glucose levels. Early studies also indicate its potential to enhance insulin sensitivity, crucial for efficient glucose uptake and utilization by cells, and potentially contributing to improved glycaemic control. Additionally, emerging evidence suggests that *D. nobile* possesses antioxidative properties, which may help reduce oxidative stress associated with diabetes, offering protective effects against cellular and tissue damage. Furthermore, traditional usage in Chinese medicine suggests that *D. nobile* may manage diabetes complications like neuropathy and nephropathy. Although further research is warranted to confirm these effects, the historical use of *D. nobile* highlights its potential therapeutic value in diabetes management.

These preliminary findings are promising, but additional research, including well-designed clinical trials, is necessary to validate the efficacy, safety profile and optimal dosage of *D. nobile* for diabetes management and its related complications. If substantiated, *D. nobile* could emerge as a valuable natural option for individuals seeking alternative or adjunctive therapies for diabetes. The potential therapeutic applications of *D. nobile* in diabetes management are promising, and more extensive investigation is required to completely grasp how *D. nobile* functions and its potential clinical uses. Further clinical trials are essential to assess its safety, effectiveness and long-term impacts in human subjects. Additionally, more exploration is necessary to establish the best dosing and formulation for therapeutic purposes. *D. nobile* displays encouraging potential as a treatment option for managing diabetes and its associated complications. Its anti-diabetes properties, low toxicity and lack of major adverse effects make it an attractive candidate for further research and development. As the prevalence of diabetes continues to rise worldwide, the development of safe and effective new therapies is crucial in the fight against this chronic disease.

6.5.2 Limitations and challenges of current research

Although preclinical and clinical studies have yielded promising results, the current research on the potential use of *D. nobile* in managing diabetes and its complications is still limited by several factors. The current body of research on *D. nobile* and its potential role in diabetes management is predominantly preclinical, with limited clinical studies conducted in humans, which hampers our understanding of its efficacy and safety in real-world settings. Moreover, many clinical studies have involved small sample sizes and short durations, diminishing statistical power and hindering the generalizability of findings. Additionally, variability in the composition and quality of *D. nobile* extracts used in research further complicates comparison across studies, emphasizing the need for standardized extraction methods and product formulations. While preliminary evidence suggests potential mechanisms of action, such as improved insulin sensitivity and antioxidative effects, the exact pathways remain poorly understood. Furthermore, while *D. nobile* is generally considered safe, limited information on its long-term safety profile necessitates comprehensive safety assessments, particularly with prolonged use or at higher doses. Addressing these limitations through well-designed, large-scale clinical trials with standardized protocols and longer follow-up periods is crucial to ascertain the true therapeutic potential and safety of *D. nobile* in diabetes management and related complications.

6.6 Challenges

The current evidence on *D. nobile*'s potential effects on diabetes primarily consists of preclinical research or small-scale clinical trials, limiting the validation of its efficacy and safety in human subjects. Moreover, the heterogeneity of *D. nobile* products used in studies introduces variability in composition and quality, necessitating standardized extraction methods and product formulations to ensure consistency and reproducibility. Many clinical trials have been of short duration, underscoring the need for longer-term studies to evaluate sustained effects and safety. While potential mechanisms of action, such as improved insulin sensitivity and antioxidative effects, have been proposed, the exact pathways remain poorly understood, requiring further research. Despite its traditional use and perceived safety, limited information exists on the long-term safety profile of *D. nobile*, warranting comprehensive safety assessments. Even if clinical efficacy is established, challenges persist in translating research findings into clinical practice, including patient adherence, product standardization and healthcare provider acceptance, impacting its adoption as a therapeutic option for diabetes management. Addressing these challenges through rigorous research efforts, including well-designed clinical trials with standardized protocols and longer follow-up periods, will be crucial to unlocking the full potential of *D. nobile* in the management of diabetes and related complications.

6.6.1 Future research directions

Future research directions for *D. nobile* in the management of diabetes and related complications include the following. To advance our understanding and utilization of *D. nobile* in diabetes management, comprehensive strategies are imperative. Mechanistic studies should delve into the biological pathways underpinning its effects on diabetes, including insulin sensitivity, glucose metabolism, oxidative stress, inflammation and diabetes complications, at the cellular and molecular levels. Large-scale clinical trials with diverse populations and long-term follow-up are essential to further evaluate its efficacy and safety, assessing outcomes like glycaemic control, insulin sensitivity, lipid profiles and complications. Standardization of extraction protocols and formulations is crucial to ensure consistency and reproducibility across studies, facilitating reliable comparisons. Safety assessments, including acute and chronic toxicity studies, are necessary to evaluate its safety profile, particularly with prolonged or higher-dose use, while identifying potential adverse effects and drug interactions. Determining optimal dosages and formulations through dose–response studies and investigating combination therapies with existing anti-diabetes medications or natural compounds can enhance glycaemic control and minimize side effects. Long-term studies are needed to assess the sustained effects on disease progression, quality of life and mortality. Additionally, patient-centred research is essential to gauge acceptability, adherence and satisfaction with *D. nobile* supplementation among individuals with diabetes, incorporating patient preferences and attitudes towards natural therapies for diabetes management.

6.6.2 Implications for clinical practice

Although *D. nobile* holds potential as a complementary therapy for managing diabetes and its associated complications, the clinical evidence supporting its use remains limited. In navigating the potential of *D. nobile* as a complementary therapy for diabetes management, healthcare professionals are presented with a landscape of promise intertwined with caution. While glimpses of its efficacy are visible in preliminary research, the current clinical evidence remains veiled, urging a tempered approach in its integration into clinical practice. Patients embarking on the exploration of *D. nobile* should embark with their eyes open, guided by their healthcare provider. A consultation, akin to a compass guiding through uncharted waters, is paramount, particularly for those using diabetes medications or other health regimens. Through this dialogue, the potential risks and benefits of *D. nobile* can be weighed with clarity and insight. Yet, amidst the allure of novelty, the foundation of health remains rooted in time-honoured principles. Encouragement to improve lifestyle habits – such as regular exercise and a balanced diet – is important. *D. nobile*, a delicate bloom in the tapestry of therapeutic options, should be nurtured as a bud rather than a full blossom, complementing the overarching canopy of conventional medical treatment.

In this care, orchestrated by the collaboration of patient and healthcare provider, *D. nobile* may find its place, enriching diabetes management. As the journey unfolds, with each note resonating through the corridors of health, may it be guided by the steady hand of knowledge, tempered by caution and illuminated by the promise of possibility. In summary, while *D. nobile* may offer potential therapeutic benefits for the management of diabetes and related complications, its use should be approached with caution and its effectiveness and safety should be further investigated through rigorous clinical research.

Acknowledgements

The authors express their gratitude for the financial support received through the Distinguished High-Level Talents Research Grant from the Guizhou Science and Technology Corporation Platform Talents Fund (grant nos. [2017]5733-001 and CK-1130-002), the National Natural Science Foundation of China (U1812403, 82373981 and 82060750), 2011 Collaborative Innovation Centre of Traditional Chinese Medicine in Guizhou Province (no. [2022]026) and the support provided by the Zunyi Medical University, China. Special appreciation is extended to all laboratory colleagues and research staff members for their valuable insights, constructive guidance and assistance with the contents of this chapter.

References

Ahmad, E., Lim, S., Lamptey, R., Webb, D.R. and Davies, M.J. (2022) Type 2 diabetes. *The Lancet* 400(10365), 1803–1820.

Ali, M.K., Galaviz, K.I., Weber, M.B. and Narayan, K.V. (2017) The global burden of diabetes. *Textbook of Diabetes* 5, 65–83.

Banday, M.Z., Sameer, A.S. and Nissar, S. (2020) Pathophysiology of diabetes: an overview. *Avicenna Journal of Medicine* 10(4), 174–188.

Baum, E. (2020) Medicine and public health in twentieth-century China: histories of modernization and change. *History Compass* 18(7), e12616.

Baynes, K. (2022) Diabetes mellitus, obesity, lipoprotein disorders and other metabolic diseases. *In Medicine for Finals and Beyond* 1, 429–470.

Capozzi, M.E., DiMarchi, R.D., Tschöp, M.H., Finan, B. and Campbell, J.E. (2018) Targeting the incretin/glucagon system with triagonists to treat diabetes. *Endocrine Reviews* 39(5), 719–738.

Charlton, A., Garzarella, J., Jandeleit-Dahm, K.A.M. and Jha, J.C. (2020) Oxidative stress and inflammation in renal and cardiovascular complications of diabetes. *Biology* 10(1), 1–18.

Dai, Y.P., Duan, Y., Lu, Y.T., Ni, X.T., Zhang, Y.K. *et al.* (2024) Nourishing yin traditional Chinese medicine: potential role in the prevention and treatment of type 2 diabetes. *American Journal of Translational Research* 16(1), 234–254.

Dar, M.A., Maqbool, M., Ara, I. and Qadrie, Z. (2023) Preserving sight: managing and preventing diabetic retinopathy. *Open Health* 4(1), 1–10.

Elendu, C., Amaechi, D.C., Elendu, T.C., Fiemotonghan, B.E., Okoye, O.K. *et al.* (2024) A comprehensive review of heart failure: unraveling the etiology, decoding pathophysiological mechanisms, navigating diagnostic modalities, exploring pharmacological interventions, advocating lifestyle modifications, and charting the horizon of emerging therapies in the complex landscape of chronic cardiac dysfunction. *Medicine* 103(3), e36895.

Falcone, M., Meier, J.J., Marini, M.G., Caccialanza, R., Aguado, J.M. *et al.* (2021) Diabetes and acute bacterial skin and skin structure infections. *Diabetes Research and Clinical Practice* 174, 1–9.

Fan, C., Sun, X., Wang, X. and Yu, H. (2023) Therapeutic potential of the chemical composition of *Dendrobium nobile* Lindl. *Frontiers in Pharmacology* 14, 1–15.

Farkas, K., Stanek, A., Zbinden, S., Borea, B., Ciurica, S. *et al.* (2024) Vascular diseases in women: do women suffer from them differently? *Journal of Clinical Medicine* 13(4), 1–30.

Feldman, E.L., Callaghan, B.C., Pop-Busui, R., Zochodne, D.W., Wright, D.E. *et al.* (2019) Diabetic neuropathy. *Nature Reviews Disease Primers* 5(1), 41.

Galicia-Garcia, U., Benito-Vicente, A., Jebari, S., Larrea-Sebal, A., Siddiqi, H. *et al.* (2020) Pathophysiology of type 2 diabetes mellitus. *International Journal of Molecular Sciences* 21(17), 1–34.

Haas, A.V. and McDonnell, M.E. (2018) Pathogenesis of cardiovascular disease in diabetes. *Endocrinology and Metabolism Clinics of North America* 47(1), 51–63.

Harris, G.D. and White, R.D. (2012) Diabetes in the competitive athlete. *Current Sports Medicine Reports* 11(6), 309–315.

Hei, J.W., Hung, W., Yanbo, Z. and Zhang, Z.J. (2021) *Dendrobium* in diabetes: a comprehensive review of its phytochemistry, pharmacology, and safety. *Journal of Translational Science* 7(6), 1–10.

Huang, W., Wang, J., Kuang, M., Xiao, Z., Fan, B. *et al.* (2023) Exploring global research status and trends in anti-obesity effects of traditional Chinese medicine through intestinal microbiota: a bibliometric study. *Frontiers in Cellular and Infection Microbiology* 13, 1271473.

Kashiwagi, A., Takahashi, H., Ishikawa, H., Yoshida, S., Kazuta, K. *et al.* (2015) A randomized, double-blind, placebo-controlled study on long-term efficacy and safety of ipragliflozin treatment in patients with type 2 diabetes mellitus and renal impairment: results of the long-term ASP1941 safety evaluation in patients with type 2 diabetes with renal impairment (LANTERN) study. *Diabetes, Obesity and Metabolism* 17(2), 152–160.

Kropp, M., Golubnitschaja, O., Mazurakova, A., Koklesova, L., Sargheini, N. *et al.* (2023) Diabetic retinopathy as the leading cause of blindness and early predictor of cascading complications—risks and mitigation. *EPMA Journal* 14(1), 21–42.

Kumar, V., Xin, X., Ma, J., Tan, C., Osna, N. *et al.* (2021) Therapeutic targets, novel drugs, and delivery systems for diabetes associated NAFLD and liver fibrosis. *Advanced Drug Delivery Reviews* 176, 113888.

Lee, C.T., Kuo, H.C., Chen, Y.H. and Tsai, M.Y. (2018) Current advances in the biological activity of polysaccharides in *Dendrobium* with intriguing therapeutic potential. *Current Medicinal Chemistry* 25(14), 1663–1681.

Lei, H., Zou, S., Lin, J., Zhai, L., Zhang, Y. *et al.* (2022a) Antioxidant and anti-inflammatory activity of constituents isolated from *Dendrobium nobile* (Lindl.). *Frontiers in Chemistry* 10, 988459.

Lei, X., Huo, P., Xie, Y.J., Wang, Y., Liu, G. *et al.* (2022b) *Dendrobium nobile* Lindl. polysaccharides improve testicular spermatogenic function in streptozotocin-induced diabetic rats. *Molecular Reproduction and Development* 89(4), 202–213.

Li, M., Trapika, I.G.S.C., Tang, S.Y.S., Cho, J.L., Qi, Y. *et al.* (2022) Mechanisms and active compounds polysaccharides and bibenzyls of medicinal dendrobiums for diabetes management. *Frontiers in Nutrition* 8, 1–13.

Li, Z., Zeng, M., Geng, K., Lai, D., Xu, Z. *et al.* (2023) Chemical constituents and hypoglycemic mechanisms of *Dendrobium nobile* in treatment of type 2 diabetic rats by UPLC-ESI-Q-orbitrap, network pharmacology and in vivo experimental verification. *Molecules* 28(6), 1–33.

Liguori, F., Mascolo, E. and Vernì, F. (2021) The genetics of diabetes: what we can learn from *drosophila International Journal of Molecular Sciences* 22(20), 1–21.

Long, Y., Wang, W., Zhang, Y., Du, F., Zhang, S. *et al.* (2023) Photoprotective effects of *Dendrobium nobile* Lindl. polysaccharides against UVB-induced oxidative stress and apoptosis in HaCaT cells. *International Journal of Molecular Sciences* 24(7), 1–17.

Luna, R., Talanki Manjunatha, R., Bollu, B., Jhaveri, S., Avanthika, C. *et al.* (2021) A comprehensive review of neuronal changes in diabetics. *Cureus* 13(10), e19142.

Luo, Y., Yang, D., Xu, Y., Wu, D., Tan, D. *et al.* (2024) Hypoglycemic effects and quality marker screening of *Dendrobium nobile* Lindl. at different growth years. *Molecules* 29(3), 1–23.

Madan, K., Paliwal, S., Sharma, S., Kesar, S., Chauhan, N. *et al.* (2023) Metabolic syndrome: the constellation of co-morbidities, a global threat. *Endocrine, Metabolic & Immune Disorders Drug Targets* 23(12), 1491–1504.

Mou, Z., Zhao, Y., Ye, F., Shi, Y., Kennelly, E.J. *et al.* (2021) Identification, biological activities and biosynthetic pathway of *Dendrobium* alkaloids. *Frontiers in Pharmacology* 12, 605994.

Nanditha, A., Chamukuttan, S., Raghavan, A. and Ramachandran, A. (2020) Global epidemic of type 2 diabetes mellitus: an epidemiologist's perspective. *Current Trends in Diabetes* 2, 36–51.

Oskouei, Z., Rahbardar, M.G and Hosseinzadeh, H. (2023) The effects of *Dendrobium* species on the metabolic syndrome: a review. *Iranian Journal of Basic Medical Sciences* 26(7), 738–752.

Raghavan, S., Vassy, J.L., Ho, Y.L., Song, R.J., Gagnon, D.R. *et al.* (2019) Diabetes mellitus-related all-cause and cardiovascular mortality in a national cohort of adults. *Journal of the American Heart Association* 8(4), e011295.

Ramesh, T., Koperuncholan, M., Praveena, R., Ganeshkumari, K., Vanithamani, J. *et al.* (2019) Medicinal properties of some dendrobium orchids – a review. *Journal of Applied and Advanced Research* 4(4), 119–128.

Sagoo, M.K. and Gnudi, L. (2020) Diabetic nephropathy: an overview. *Methods and Protocols* 3–7.

Sen, C.K. (2019) Human wounds and its burden: an updated compendium of estimates. *Advances in Wound Care* 8(2), 39–48.

Shi, B., Li, H. and He, X. (2024) Advancing lifelong precision medicine for cardiovascular diseases through gut microbiota modulation. *Gut Microbes* 16(1), 2323237.

Singh, S., Sarma, D.K., Verma, V., Nagpal, R. and Kumar, M. (2023) Unveiling the future of metabolic medicine: omics technologies driving personalized solutions for precision treatment of metabolic disorders. *Biochemical and Biophysical Research Communications* 682, 1–20.

Sloan, G., Shillo, P., Selvarajah, D., Wu, J., Wilkinson, I.D. *et al.* (2018) A new look at painful diabetic neuropathy. *Diabetes Research and Clinical Practice* 144, 177–191.

Standl, E., Khunti, K., Hansen, T.B. and Schnell, O. (2019) The global epidemics of diabetes in the 21st century: current situation and perspectives. *European Journal of Preventive Cardiology* 26(2), 7–14.

Sun, J., Ren, J., Hu, X., Hou, Y. and Yang, Y. (2021) Therapeutic effects of Chinese herbal medicines and their extracts on diabetes. *Biomedicine & Pharmacotherapy* 142, 111977.

Syed, F.Z. (2022) Type 1 diabetes mellitus. *Annals of Internal Medicine* 175(3), 33–48.

Trojian, T., Colberg, S., Harris, G., Oh, R., Dixit, S. *et al.* (2022) American medical society for sports medicine position statement on the care of the athlete and athletic person with diabetes. *Clinical Journal of Sport Medicine* 32(1), 8–20.

Uzuner, B., Ketenci, S. and Salbas, E. (2020) General approach to diabetic neuropathy. *Acta Medica Alanya* 4(3), 296–308.

Venyo, A.K.G. (2023) Diabetes mellitus: a review and update. *Journal of Ophthalmology Research Reviews and Reports* 4(144), 2–24.

Wan, J.H., Lam, K.H., Song, T.H., Lin, X., Leung, L.L. *et al.* (2020) Review of the therapeutic effects of the traditional Chinese medicine yuye decoction on diabetes mellitus and its complications. *Journal of Translational Science* 7, 1–16.

Wang, F., Wan, J., Liao, Y., Liu, S., Wei, Y. *et al.* (2023) *Dendrobium* species regulate energy homeostasis in neurodegenerative diseases: a review. *Food Science and Human Wellness* 12(6), 2151–2174.

Wang, W. and Lo, A.C.Y. (2018) Diabetic retinopathy: pathophysiology and treatments. *International Journal of Molecular Sciences* 19(6), 1816.

Wei, T.K. (2020) Modelling to assess the impact of interventions for preventing chronic diseases and their complications. Doctoral dissertation, National University of Singapore, Singapore.

Wu, C., Duan, X., Wang, X. and Wang, L. (2023) Advances in the role of epigenetics in homocysteine-related diseases. *Epigenomics* 15(15), 769–795.

Xie, J., Luo, M., Chen, Q., Zhang, Q., Qin, L. *et al.* (2024) Hypolipidemic effect and gut microbiota regulation of Gypenoside aglycones in rats fed a high-fat diet. *Journal of Ethnopharmacology* 328, 118066.

Xu, Q., Niu, S.C., Li, K.L., Zheng, P.J., Zhang, X.J. *et al.* (2022) Chromosome-scale assembly of the *Dendrobium nobile* genome provides insights into the molecular mechanism of the biosynthesis of the medicinal active ingredient of dendrobium. *Frontiers in Genetics* 13, 1–15.

Zajec, A., Trebušak Podkrajšek, K., Tesovnik, T., Šket, R., Čugalj Kern, B. *et al.* (2022) Pathogenesis of type 1 diabetes: established facts and new insights. *Genes* 13(4), 1–19.

Zakir, M., Ahuja, N., Surksha, M.A., Sachdev, R., Kalariya, Y. *et al.* (2023) Cardiovascular complications of diabetes: from microvascular to macrovascular pathways. *Cureus* 15(9), e45835.

Zeng, H., Li, X., Zhou, D., Wang, N., Yu, X. *et al.* (2021) Qihu preparation ameliorates diabetes by activating the AMPK signaling pathway in db/db mice. *Diabetes, Metabolic Syndrome and Obesity* 14, 3229–3241.

Zhang, J., Xu, H.X., Zhao, Z.L., Xian, Y.F. and Lin, Z.X. (2021a) *Dendrobium nobile* Lindl: a review on its chemical constituents and pharmacological effects. *Chinese Medicine and Culture* 4(4), 235–242.

Zhang, X., Wang, M., Zhang, C., Liu, Z. and Zhou, S. (2021b) Clinical study of *Dendrobium nobile* Lindl intervention on patients with metabolic syndrome. *Medicine* 100(12), e24574.

Zhang, Y., Lu, J., Li, H. and Song, H. (2023) Advances in dietary polysaccharides as hypoglycemic agents: mechanisms, structural characteristics, and innovative applications. *Critical Reviews in Food Science and Nutrition* 63, 1–21.

Zhuang, S., Zhang, J., Lin, X., Wang, X., Yu, W. *et al.* (2023) *Dendrobium* mixture ameliorates type 2 diabetes mellitus with non-alcoholic fatty liver disease through PPAR gamma: an integrated study of bioinformatics analysis and experimental verification. *Frontiers in Pharmacology* 14, 1–12.

7 *Dendrobium nobile* and Cardiovascular Health

Archana Jain, Surendra Sarsaiya, Qihai Gong, Qin Wu and Jingshan Shi*
Key Laboratory of Basic Pharmacology and Joint International Research Laboratory of Ethnomedicine of Ministry of Education, Zunyi Medical University, Zunyi, Guizhou, China

Abstract

This chapter provides a comprehensive overview of the potential cardiovascular health benefits of *Dendrobium nobile*, a plant traditionally used in Chinese medicine. It explores the pharmacological properties of *D. nobile*, highlighting its bioactive compounds and their effects on lipid metabolism, blood pressure, endothelial function, inflammation and oxidative stress. Preclinical and clinical studies supporting these effects are reviewed, along with the potential mechanisms of action involving signalling pathways, such as phosphoinositol 3-kinase (PI3K)/Akt, AMP-activated protein kinase (AMPK) and mitogen-activated protein kinase (MAPK). Safety and potential side effects are discussed, emphasizing the need for further research to confirm the efficacy and establish standardized dosing protocols. While promising as a natural supplement for cardiovascular health, *D. nobile* should be used cautiously and under healthcare supervision, particularly for individuals with existing health conditions or those taking other medications. Future research directions and implications for clinical practice are also discussed.

7.1 Introduction

In an era in which cardiovascular diseases (CVDs) continue to exact a heavy toll on global health, the quest for innovative solutions has led researchers to explore unconventional sources of therapeutic agents. Among these, *Dendrobium nobile* Lindl., an orchid species deeply entrenched in the annals of traditional Chinese medicine, has emerged as a compelling subject of investigation. Its historical resurgence as a medicinal plant coupled with a growing body of scientific evidence allows us to critically assess its potential in the realm of cardiovascular health (Chen *et al.*, 2022a). The cardiovascular system, a complex network orchestrating the body's circulation and homeostasis, is vulnerable to an array of maladies ranging from atherosclerosis to hypertension. Because conventional treatments have limitations and side effects, the search for natural compounds with multifaceted benefits has gained momentum (Patel *et al.*, 2017). Enter *D. nobile*: a botanical treasure trove awaiting deeper scrutiny. However, our exploration transcends laboratory findings. Amidst the allure of natural remedies, the scientific imperative demands meticulous scrutiny, separating fact from conjecture and correlation from causation (Pujari and Babu, 2022). By weaving together strands of traditional wisdom, scientific enquiry and critical analysis, this chapter endeavours to illuminate the path forward. Whether *D. nobile* emerges as a

*Corresponding author: shijs@zmu.edu.cn

© Jingshan Shi, Archana Jain, Surendra Sarsaiya, Qin Wu and Qihai Gong 2025. Dendrobium nobile: *Research and Therapeutic Applications* (J. Shi, A. Jain, S. Sarsaiya, Q. Wu and Q. Gong)
DOI: 10.1079/9781800627987.0007

panacea or a mere footnote in the saga of cardiovascular therapeutics, our quest for knowledge remains steady, driven by the pursuit of health and healing in an ever-evolving landscape of discovery (Zhang *et al.*, 2018).

D. nobile, an orchid species, has garnered increased interest for its purported cardioprotective properties. Frequently incorporated into traditional Chinese medicine formulations, its historical usage spans centuries (Koh *et al.*, 2024). This botanical marvel harbours an array of bioactive compounds, including polysaccharides, bibenzyls, alkaloids, flavonoids and phenols, which have demonstrated diverse pharmacological activities, including anti-inflammatory, antioxidant, antihypertensive and antiplatelet-aggregation effects (Oskouei *et al.*, 2023). These properties make *D. nobile* a promising candidate for the development of complementary therapies for the prevention and treatment of CVD. There are numerous risk factors for CVD including age, sex, genetics, smoking, physical inactivity, unhealthy diet, hypertension, dyslipidaemia and diabetes mellitus (Sun *et al.*, 2021). This chapter provides a comprehensive overview of the potential cardioprotective effects of *D. nobile* extract. It reviews the pharmacological properties of the plant, including its effects on blood pressure, lipid metabolism, inflammation and oxidative stress.

7.1.1 Overview of the importance of cardiovascular health

CVD is a significant cause of morbidity and mortality globally, with approximately 17.9 million deaths annually (World Health Organization, 2021). It is a multifaceted ailment influenced by factors such as age, sex, genetics, smoking, physical inactivity, poor dietary choices, hypertension, dyslipidaemia and diabetes mellitus. While some risk factors, such as age and genetics, are beyond modification, lifestyle changes and medical interventions can address others (Hussain *et al.*, 2024). Preventing and managing CVD requires a comprehensive strategy encompassing lifestyle adjustments such as engaging in regular physical activity, adopting a balanced diet and quitting smoking. Additionally, pharmacotherapy is essential for high-risk individuals and those diagnosed with CVD. Moreover, there is emerging interest in traditional medicines for CVD prevention and treatment because of their potential supplementary benefits alongside conventional pharmacotherapy (Kaminsky *et al.*, 2022).

Cardiovascular health is the cornerstone of overall well-being and longevity, encompassing a multitude of facets vital for a healthy life. Its significance extends across various domains; it serves as a shield against life-threatening conditions, such as coronary artery disease, heart attack, stroke and heart failure, thereby reducing mortality risk (Zhang, 2023). Additionally, it fosters an active and fulfilling life by ensuring efficient circulation of oxygen and nutrients to all organs and tissues, thereby promoting optimal function and vitality (Vlaicu *et al.*, 2023). Through the promotion of physical fitness, a robust cardiovascular system bolsters endurance, strengthens the heart muscles and enhances overall cardiovascular fitness (Gruet *et al.*, 2022). Maintenance of healthy blood pressure and cholesterol levels achieved through a balanced diet and regular exercise further underpins cardiovascular wellness by mitigating the risk of heart disease and stroke (Etherton *et al.*, 2021). Moreover, the intricate interplay between mental and cardiovascular health highlights the importance of physical activity in releasing endorphins, alleviating stress, anxiety and depression, while ensuring proper blood circulation to the brain and fostering cognitive abilities and mental sharpness (Matei *et al.*, 2023). By investing in cardiovascular health early on and maintaining healthy habits throughout life, individuals can significantly increase their life expectancy, promote healthy ageing and mitigate the economic burden imposed by CVDs through preventive strategies and adherence to medical advice (Perel *et al.*, 2015). Therefore, promoting cardiovascular health is crucial for reducing the burden of CVD and improving the overall health outcomes. This can be achieved through various approaches, including healthy lifestyle habits, early detection and management of risk factors, and the use of complementary therapies, such as *D. nobile* (Kaminsky *et al.*, 2022). Cardiovascular health is fundamental to overall health and well-being. By adopting a healthy lifestyle, individuals can reduce their risk of CVD, enhance their quality of life and promote longevity (Table 7.1) (Watts *et al.*, 2023).

Table 7.1. Importance of cardiovascular health (authors' own table).

Importance of cardiovascular health	Description	Examples	Strategies
Prevents heart diseases	Crucial for preventing conditions like coronary artery disease, heart attack, stroke and heart failure	Coronary artery disease, myocardial infarction, stroke, heart failure	Regular exercise, balanced diet, smoking cessation
Enhances quality of life	Contributes to an active, fulfilling life by ensuring efficient circulation and optimal organ function	Improved energy levels, better physical performance, reduced risk of disability	Regular physical activity, stress management, adequate sleep
Reduces mortality risk	Lowering the risk of premature death by prioritizing cardiovascular health through lifestyle choices	Increased life expectancy, reduced risk of fatal cardiovascular events	Routine medical check-ups, adherence to treatment plans, healthy diet
Improves physical fitness	Supports physical endurance and stamina, enhancing overall cardiovascular fitness for daily activities	Increased stamina, improved lung capacity, enhanced athletic performance	Aerobic exercise, strength training, flexibility exercises
Promoting healthy blood pressure and cholesterol	Crucial for regulating factors contributing to heart disease and stroke, achieved through diet and exercise	Hypertension, high cholesterol levels	Low-sodium diet, cholesterol-lowering foods, regular aerobic exercise
Enhances mental well-being	Physical activity releases endorphins, improving mood and cognitive function by ensuring proper blood circulation to the brain	Reduced stress, alleviated anxiety and depression, enhanced cognitive abilities	Mindfulness practices, social support, engaging hobbies
Promotes longevity	Investing in heart-healthy habits increases life expectancy and supports healthy ageing	Prolonged lifespan, reduced risk of age-related diseases	Healthy diet, regular physical activity, stress management
Reduces healthcare costs	Mitigates economic burden through prevention strategies, healthy lifestyle choices, and adherence to medical advice	Reduced healthcare expenses related to CVDs and associated complications	Routine screenings, preventive medications, healthy lifestyle promotion

7.1.2 Brief explanation of the potential role of *D. nobile* in cardiovascular health

D. nobile, an orchid with a rich history in traditional Chinese medicine, has recently garnered increasing scientific attention owing to its potential therapeutic effects on cardiovascular health. This plant contains a diverse array of bioactive compounds, including polysaccharides, bibenzyls, alkaloids, flavonoids and phenols, which have shown promising pharmacological properties, such as anti-inflammatory, antioxidant, anti-hypertensive and antiplatelet-aggregation activities (Li *et al.*, 2023a). Hypertension, a major risk factor for CVD, has been a significant focus of research in *D. nobile*. Studies have demonstrated that extracts from this orchid effectively lower blood pressure in both animal models and humans with hypertension (Choudhary *et al.*, 2023). For example, in experiments with hypertensive rats, administration of *D. nobile* extract led to notable reductions in systolic and diastolic blood pressure and mean arterial pressure. Similarly, supplementation with *D. nobile* extract over a 12-week period resulted in significant blood pressure reductions in individuals with mild-to-moderate hypertension (Hu *et al.*, 2021).

Dyslipidaemia, characterized by abnormal blood lipid levels, is another important cardiovascular risk factor addressed in research on *D. nobile*. For instance, in experiments with hyperlipidaemic rats, treatment with *D. nobile* extract led to significant decreases in serum levels of total cholesterol, low-density lipoprotein (LDL) cholesterol and triglycerides, coupled with increased high-density lipoprotein (HDL) cholesterol levels (Wu *et al.*, 2024). Furthermore, oxidative stress and inflammation, which play crucial roles in CVD development and progression, have been targeted by the potent antioxidant and anti-inflammatory properties of *D. nobile* extract. Studies have demonstrated reductions in oxidative stress markers and pro-inflammatory cytokines, along with improved cardiac function in animal models of myocardial infarction following treatment with *D. nobile* extract (Wu *et al.*, 2023). Additionally, the antiplatelet-aggregation activity of *D. nobile* extract has the potential to prevent thrombotic events associated with heart attack and stroke (Meng *et al.*, 2023). In experiments with rats exhibiting arterial thrombosis, treatment with *D. nobile* extract significantly inhibited platelet aggregation and thrombus formation (Li *et al.*, 2023a). In conclusion, *D. nobile* presents a promising therapeutic potential for cardiovascular health through various mechanisms, including blood pressure regulation, lipid metabolism modulation, antioxidant and anti-inflammatory actions, and antiplatelet-aggregation activity. Nonetheless, further research is warranted to fully elucidate the mechanisms of action and to establish optimal therapeutic dosages and long-term effects.

7.2 Pharmacological Properties of *D. nobile*

D. nobile possesses natural compounds that exhibit anti-inflammatory properties, potentially reducing inflammatory markers and cytokine production. This could alleviate symptoms associated with inflammatory conditions (Chen *et al.*, 2022b). *D. nobile* is known for its antioxidant properties, which allow it to neutralize harmful free radicals in the body. The scavenging of these free radicals helps protect cells from oxidative damage and reduces the risk of oxidative stress-related diseases (Lei *et al.*, 2022a). *D. nobile* has emerged as a multifaceted natural remedy with promising therapeutic potential across various health conditions. Research indicates its effectiveness as an antihypertensive agent, attributed to its ability to relax blood vessels, enhance blood flow and alleviate cardiac workload, thus offering potential benefits to individuals with hypertension (Wang, 2021). Moreover, its ability to inhibit platelet aggregation signifies its role in preventing blood clot formation, thereby reducing the risk of cardiovascular events, such as heart attacks and strokes (Jie *et al.*, 2023). The anticancer properties of *D. nobile* further broaden its scope, with studies suggesting its ability to impede cancer cell growth and progression, positioning it as a candidate for cancer prevention and treatment (Fan *et al.*, 2023). Its neuroprotective effects hold promise in the realm of neurodegenerative diseases, where it may aid in shielding nerve cells from damage or degeneration, potentially contributing to the prevention and management of conditions such as Alzheimer's and Parkinson's diseases (Hui *et al.*, 2024). Additionally, its immunomodulatory actions offer potential benefits for immune-related disorders by regulating immune responses and enhancing immune functions (Zhang *et al.*, 2021). Furthermore, the hypoglycaemic properties of *D. nobile* make it a promising avenue for individuals with diabetes or at risk of the condition, as it demonstrates potential in lowering blood sugar levels through various mechanisms such as promoting insulin sensitivity and inhibiting glucose production in the liver (Luo *et al.*, 2024). Collectively, these findings underscore the diverse therapeutic potential of *D. nobile* and highlight its significance in various aspects of health and wellness (Table 7.2).

7.2.1 Overview of the bioactive compounds in *D. nobile*

D. nobile contains a variety of bioactive compounds, including alkaloids, phenanthrenes, bibenzyls, flavonoids and polysaccharides. Alkaloids are organic compounds that contain nitrogen, and *D. nobile* contains several alkaloids, such as dendrobine, dendramine and nobiline. These alkaloids have been reported to have analgesic, sedative and anticonvulsant effects (Hu *et al.*, 2021). Phenanthrenes represent a distinct class of organic compounds characterized by their unique chemical structures, and a variety of phenanthrenes

Table 7.2. Pharmacological properties of *D. nobile* (authors' own table).

Pharmacological property	Description
Anti-inflammatory	*D. nobile* possesses natural compounds with anti-inflammatory properties, potentially reducing inflammation markers and cytokine production, offering relief from inflammatory conditions.
Antioxidant	Known for its antioxidant properties, *D. nobile* neutralizes harmful free radicals, protecting cells from oxidative damage and reducing the risk of oxidative stress-related diseases.
Antihypertensive	Research suggests that *D. nobile* has antihypertensive effects, aiding in blood pressure reduction by relaxing blood vessels, improving blood flow and reducing the heart's workload, benefiting individuals with hypertension.
Antiplatelet aggregation	Inhibits blood platelet aggregation, reducing the risk of blood clot formation, crucial in preventing heart attacks and strokes.
Anti-cancer	Exhibits properties inhibiting cancer cell growth and proliferation, potentially serving in cancer prevention and treatment.
Neuroprotective	Shows neuroprotective effects, shielding nerve cells from damage or degeneration, with potential applications in neurodegenerative disease management.
Immunomodulatory	Modulates the immune system response, enhancing immune function and offering benefits for immune-related disorders.
Hypoglycaemic	Demonstrates potential in reducing blood sugar levels by enhancing insulin sensitivity and glucose uptake, or inhibiting glucose production, which is promising for diabetes management.

have been discovered in *D. nobile*. Studies have indicated that these compounds possess diverse pharmacological effects, such as anti-cancer, antimicrobial and antioxidant properties (Fan *et al.*, 2023). Bibenzyls are another class of organic compounds found in *D. nobile*; some bibenzyls have been reported to have antiviral and antitumour effects (Duan *et al.*, 2022a,b). Flavonoids are a class of compounds abundantly found in plants, and are renowned for their antioxidative and anti-inflammatory properties. Numerous flavonoids, such as quercetin and kaempferol, are present in *D. nobile*, exhibiting anti-cancer, anti-inflammatory and antioxidative properties (Qiu *et al.*, 2023). Polysaccharides are complex carbohydrates composed of long-chain sugar molecules. *D. nobile* contains several polysaccharides, which have been reported to have immunomodulatory effects and may also have potential applications in wound healing and anti-ageing (Chen *et al.*, 2022b).

7.2.2 Description of their potential effects on cardiovascular health

Several bioactive compounds found in *D. nobile* have been suggested to exert potential effects on cardiovascular health. For instance, flavonoids, such as quercetin and kaempferol, have been shown to have beneficial effects on several cardiovascular risk factors. These flavonoids have been documented to diminish blood pressure, enhance endothelial function and decrease the levels of LDL cholesterol, commonly termed 'bad' cholesterol (Qiu *et al.*, 2023). Polysaccharides present in *D. nobile* have been proposed to offer potential advantages for cardiovascular well-being. Studies suggest that they exhibit anticoagulant and antiplatelet characteristics, potentially aiding the prevention of blood clot formation and enhancing blood circulation (Chen *et al.*, 2022b). Moreover, some alkaloids found in *D. nobile*, such as dendrobine, have been reported to have vasodilatory effects, which means that they can widen blood vessels and improve blood flow. This effect may be beneficial for people with CVD because it can improve oxygen and nutrient delivery to the heart and other organs (Hu *et al.*, 2021). Although these discoveries hint at the potential advantages of *D. nobile* for cardiovascular health, additional research is necessary to comprehensively understand the mechanisms by which its bioactive compounds operate and their potential implications in preventing and treating CVD.

7.3 Effects of *D. nobile* on Cardiovascular Health

D. nobile, renowned in traditional medicine, has garnered increased attention for its potential impact on cardiovascular health. Extracts derived from *D. nobile* offer a promising avenue for blood pressure regulation, with research suggesting their potential to lower both systolic and diastolic blood pressure levels in individuals with hypertension (Nie *et al.*, 2020). Additionally, *D. nobile* extracts exhibit beneficial effects on lipid metabolism by decreasing total cholesterol, LDL cholesterol and triglyceride levels, while simultaneously increasing HDL cholesterol, otherwise known as 'good' cholesterol, thus improving lipid profile and potentially reducing cardiovascular risk (Liu *et al.*, 2021a). The presence of flavonoids and other bioactive compounds in *D. nobile* underscores its antioxidant and anti-inflammatory properties, which play crucial roles in combating oxidative stress and inflammation, key contributors to CVD development (Chen *et al.*, 2022b). Furthermore, studies suggest that *D. nobile* extract may possess antiplatelet activity, potentially mitigating the risk of blood clot formation and thrombotic events such as heart attacks and strokes (Jie *et al.*, 2023). Collectively, these findings highlight the multifaceted cardiovascular benefits of *D. nobile*, positioning it as a promising natural agent to promote heart health and reduce the risk of cardiovascular complications. While these initial findings are promising, further research is necessary to comprehensively understand the mechanisms underlying the effects of *D. nobile* on cardiovascular health. Additionally, determining optimal dosage and long-term safety, as well as conducting well-designed clinical trials in human subjects, are imperative before considering *D. nobile* as a therapeutic intervention for CVDs (Fig. 7.1).

7.3.1 Effects of *D. nobile* on lipid metabolism

D. nobile has been suggested to have potential effects on lipid metabolism, which refers to how the body processes and utilizes fats. Several bioactive compounds found in *D. nobile*, such as flavonoids, have been reported to exert beneficial effects on lipid metabolism. *D. nobile* has been the subject of research regarding its potential effects on lipid metabolism. Incorporating *D. nobile* extract into one's diet shows promise in improving lipid profiles and promoting cardiovascular health through various mechanisms. Studies suggest that this extract contributes to a reduction in total cholesterol levels, potentially decreasing the risk of CVD associated with elevated cholesterol levels (Aissata *et al.*, 2019). Additionally, *D. nobile* has demonstrated efficacy in lowering LDL cholesterol, which is a significant risk factor for heart disease, thus potentially mitigating conditions, such as atherosclerosis and coronary artery disease (Xia *et al.*, 2024). Furthermore, research indicates that supplementation with *D. nobile* extract may lead to a

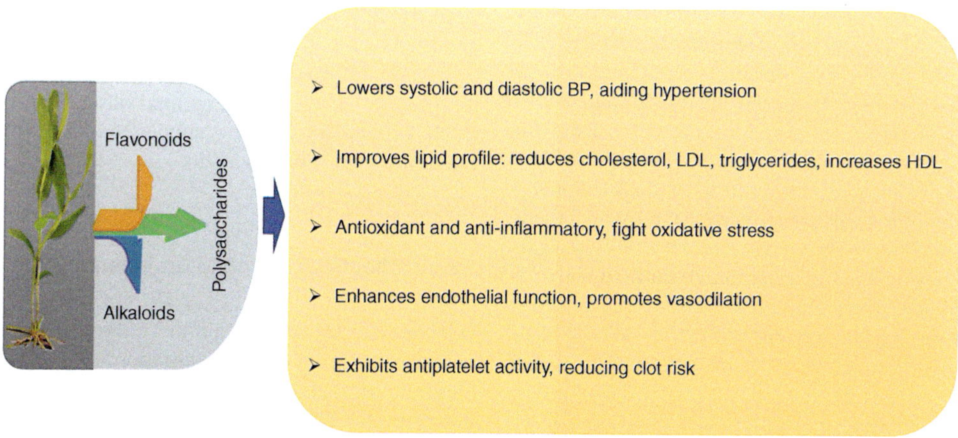

Fig. 7.1. Effects of *D. nobile* on cardiovascular health (authors' own figure).

decrease in triglyceride levels, reducing the risk of cardiovascular issues linked to elevated triglycerides (Jędrejko *et al.*, 2021). Moreover, *D. nobile* elevates HDL cholesterol, which plays a protective role in cardiovascular health by facilitating the removal of excess cholesterol from the bloodstream, thereby further enhancing its positive effects on lipid metabolism and cardiovascular well-being (Wang *et al.*, 2022a). Collectively, these findings underscore the potential of *D. nobile* extract as a natural intervention for improving lipid profiles and reducing the risk of CVDs. These findings suggest that *D. nobile* may have a favourable effect on lipid metabolism, potentially reducing the risk of CVDs associated with dyslipidaemia. However, further research, including well-designed clinical trials in humans, is needed to fully understand the mechanisms underlying these effects and to establish optimal dosage and safety profiles for *D. nobile* supplementation.

7.3.2 Effects of *D. nobile* on blood pressure

Numerous studies have investigated the effects of *D. nobile* on blood pressure, with compelling indications of its potential antihypertensive effects. For example, an animal study involving rats with elevated blood pressure revealed that administration of *D. nobile* extract notably decreased blood pressure levels compared to a control group (Oskouei *et al.*, 2023). Similarly, a human study focusing on individuals with prehypertension demonstrated that consumption of a beverage enriched with *D. nobile* extract over a span of 6 weeks led to significant reductions in both systolic and diastolic blood pressure, in contrast to the placebo group (Wang, 2021). Furthermore, a separate human study involving hypertensive individuals found that a combined treatment of *D. nobile* extract and amlodipine, a medication used for hypertension management, over 8 weeks resulted in a more pronounced decrease in blood pressure than treatment with amlodipine alone (Chua *et al.*, 2024). Despite these promising findings suggesting the potential blood pressure-regulating benefits of *D. nobile*, further comprehensive research is warranted to elucidate the underlying mechanisms of action and assess any potential long-term effects of its bioactive constituents on blood pressure regulation.

7.3.3 Effects of *D. nobile* on endothelial function

It has been suggested that *D. nobile* may impact endothelial function, a pivotal aspect of cardiovascular health. *D. nobile* has emerged as a promising agent for cardiovascular health because of its multifaceted effects on endothelial function and vascular dynamics. Compounds found in *D. nobile* have been shown to stimulate the production or activity of nitric oxide (NO) by endothelial cells, leading to enhanced endothelium-dependent vasodilation and regulation of blood pressure, thereby improving blood flow (Chang *et al.*, 2020). Moreover, its anti-inflammatory properties contribute to the mitigation of chronic inflammation in the endothelium, preserving endothelial function and integrity, which are key factors for cardiovascular health (Wang *et al.*, 2022a). Additionally, the rich antioxidant content of *D. nobile*, including flavonoids and phenols, counteracts oxidative stress and protects endothelial cells from damage, ensuring optimal endothelial function (Rao *et al.*, 2023). By promoting endothelial function, *D. nobile* aids in improving circulation throughout the body, facilitating the efficient delivery of oxygen and nutrients to tissues and supporting overall cardiovascular health. Collectively, these findings highlight the potential of *D. nobile* as a natural intervention for maintaining endothelial health and optimizing vascular function, thereby offering promise for cardiovascular wellness. While these findings are promising, further research, including clinical studies in humans, is needed to fully elucidate the effects of *D. nobile* on endothelial function and its potential implications for cardiovascular health. Additionally, it is essential to determine the optimal dosages and safety profiles for *D. nobile* supplementation to maximize its potential benefits.

7.3.4 Effects of *D. nobile* on inflammation

D. nobile has attracted scientific interest for its potential to influence inflammation, a pivotal element in the onset of numerous chronic ailments, notably CVDs. *D. nobile* shows a diverse array of anti-inflammatory properties, making it a promising natural remedy for combating inflammation and its associated health implications. Rich in bioactive compounds like flavonoids and phenols, *D. nobile* has demonstrated the ability to inhibit inflammatory pathways and reduce the production of pro-inflammatory molecules, thereby attenuating inflammation in the body (Chen *et al.*, 2022b). Furthermore, studies suggest that *D. nobile* may modulate cytokine production, promoting a balanced immune response and dampening excessive inflammation by regulating cytokine release (Ti *et al.*, 2021). Its antioxidant properties are also noteworthy as they enable *D. nobile* to neutralize reactive oxygen species (ROS) and reduce oxidative stress, which in turn helps mitigate inflammation (Hsu *et al.*, 2022). Importantly, by protecting endothelial cells from inflammatory damage, *D. nobile* contributes to the preservation of endothelial function, thereby promoting cardiovascular health and potentially preventing the onset of CVD (Huang *et al.*, 2023). Moreover, emerging research indicates that *D. nobile* may influence the composition of gut microbiota, fostering the growth of beneficial bacteria while reducing inflammation-inducing microbial species, thus offering a novel avenue for inflammation management (Yousefi *et al.*, 2022). Overall, these findings underscore the multifaceted anti-inflammatory effects of *D. nobile* and its potential as a holistic approach for mitigating inflammation and promoting overall well-being. Overall, although preliminary evidence suggests that *D. nobile* possesses anti-inflammatory properties, further research, including clinical trials in humans, is needed to confirm its efficacy and safety in managing inflammation-related conditions. Additionally, understanding the mechanisms underlying its anti-inflammatory effects is crucial to optimize its therapeutic potential.

7.3.5 Effects of *D. nobile* on oxidative stress

D. nobile, a plant renowned in traditional medicine, has garnered interest for its potential effects on oxidative stress, a process implicated in various diseases, including cardiovascular conditions. *D. nobile*, renowned for its antioxidant properties, contains bioactive compounds such as flavonoids and phenols that scavenge free radicals and ROS, thereby mitigating oxidative damage to cells and tissues (Nie *et al.*, 2020). Additionally, it has demonstrated the capacity to regulate the activity of vital antioxidant enzymes such as superoxide dismutase and catalase, which are essential for counteracting ROS and maintaining cellular redox balance, thereby reinforcing the antioxidant defense system against oxidative stress (Li *et al.*, 2023b). Furthermore, studies have suggested that *D. nobile* may protect mitochondria from oxidative damage, preserving their function and preventing ROS release into the cell (Di *et al.*, 2018). Its anti-inflammatory properties have been noted, potentially alleviating inflammation-induced oxidative stress by reducing pro-inflammatory cytokines, thus underscoring its multifaceted protective mechanisms (Li *et al.*, 2022a). Moreover, by shielding DNA from oxidative damage, *D. nobile* may maintain genomic integrity, thereby reducing the risk of oxidative stress-related diseases (Uddin *et al.*, 2020). These findings collectively highlight the promising therapeutic potential of *D. nobile* in combating oxidative stress and its associated health complications. Although preliminary evidence suggests that *D. nobile* possesses antioxidant properties and may mitigate oxidative stress, further research, including clinical studies, is needed to elucidate its efficacy and safety profile in managing oxidative stress-related conditions. Understanding the specific mechanisms underlying its antioxidant effects is crucial for optimizing its therapeutic potential.

7.4 Evidence for Efficacy

The efficacy of *D. nobile* and its bioactive compounds has been the subject of scientific investigation, particularly in the realm of traditional medicine and pharmacology. Research suggests that *D. nobile* and its constituents exhibit various beneficial effects. Cardiovascular health studies indicate that *D. nobile* may help regulate blood pressure and improve lipid metabolism by lowering *D. nobile*, which has been shown to reduce LDL cholesterol and triglyceride levels while simultaneously boosting HDL cholesterol levels. Moreover, it has antioxidant and anti-inflammatory properties that offer protection against CVD (Fan *et al.*, 2023). *D. nobile* exhibits a diverse range of potential therapeutic effects, as evidenced by several studies. Its bioactive compounds, including flavonoids, have been shown to exert anti-cancer effects by inhibiting tumour growth, inducing apoptosis and suppressing angiogenesis (He *et al.*, 2020). Moreover, *D. nobile* and its constituents have been found to modulate immune function, enhance immune responses and regulate inflammatory processes, suggesting their therapeutic potential in immune-related disorders (Xie *et al.*, 2023). Additionally, it may offer neuroprotection by reducing oxidative stress, inflammation and neuronal damage, potentially mitigating neurodegenerative diseases (Liu *et al.*, 2021b). Traditionally used to promote gastrointestinal health, *D. nobile* shows promise in modulating gut microbiota composition, improving digestive function and alleviating gastrointestinal disorders. Furthermore, its antioxidant and anti-inflammatory properties, attributed to its bioactive compounds, can help mitigate oxidative stress-related damage and dampen inflammation associated with various chronic diseases, highlighting its potential as a multifaceted therapeutic agent (Guo *et al.*, 2023). However, to comprehensively understand the efficacy, safety and mechanisms of action of *D. nobile* and its bioactive compounds, further research, including meticulously designed clinical trials, is required. Additionally, standardization of extraction methods and quality control measures are essential for ensuring the consistency and efficacy of *D. nobile* products for therapeutic use.

7.4.1 Preclinical studies

Preclinical studies are conducted in laboratory settings using cells, tissues or animal models to investigate the potential efficacy and safety of a drug or natural product before testing in humans. Several preclinical studies have investigated the potential therapeutic benefits and mechanisms of action of *D. nobile*. One study investigated the effects of *D. nobile* extract on high-fat diet-induced insulin resistance in rats (Xu *et al.*, 2017). The study found that treatment with *D. nobile* extract improved insulin sensitivity and glucose tolerance, reduced glucose levels and increased the expression of genes involved in glucose metabolism. Another study found that *D. nobile* polysaccharides protected pancreatic cells from damage induced by high glucose levels and improved insulin secretion *in vitro* (Lee *et al.*, 2017). In the area of cardiovascular health, one study investigated the effects of *D. nobile* polysaccharides on blood pressure and lipid metabolism in rats with hypertension (Fan *et al.*, 2023). The study found that treatment with *D. nobile* polysaccharides reduced blood pressure and improved lipid metabolism by increasing the expression of genes involved in lipid metabolism. Another study found that *D. nobile* flavonoids improved endothelial function and reduced oxidative stress in rats with hypertension (Li *et al.*, 2023b). Several preclinical studies have investigated the potential anti-cancer effects of *D. nobile* and its bioactive compounds. One study found that a *D. nobile* bibenzyl derivative had cytotoxic effects on human breast cancer cells (Xie *et al.*, 2021). Another study found that a *D. nobile* phenanthrene derivative had anti-cancer effects against human liver cancer cells (Zhang *et al.*, 2019). Overall, preclinical studies suggest that *D. nobile* and its bioactive compounds have potential therapeutic benefits in diabetes, cardiovascular health and cancer.

7.4.2 Clinical studies

Clinical studies represent a crucial avenue for assessing the safety and efficacy of both pharmaceutical drugs and natural products, such as *D. nobile*. However, few clinical studies have been conducted on *D. nobile*, underscoring the necessity for further research to ascertain its safety and efficacy in humans. One randomized, double-blind, placebo-controlled clinical trial investigated the effects of *D. nobile* extract on cognitive function in healthy elderly individuals. The study found that treatment with *D. nobile* extract for 12 weeks improved cognitive function, particularly in the areas of memory and attention, compared to the placebo group (Xiong *et al.*, 2021). Another randomized, double-blind, placebo-controlled clinical trial investigated the impact of *D. nobile* polysaccharides on immune function among individuals diagnosed with non-small cell lung cancer. The study found that treatment with *D. nobile* polysaccharides for

12 weeks improved immune function, as indicated by an increase in immune cell counts and cytokine levels, compared to the placebo group (Pant *et al.*, 2021). A clinical study investigated the effects of *D. nobile* extract on gastrointestinal function in patients with functional dyspepsia. The study found that treatment with *D. nobile* extract for 4 weeks improved gastrointestinal symptoms, such as abdominal pain and bloating, compared to the placebo group (Gogoi and Sen, 2023). Overall, clinical studies suggest that *D. nobile* may have potential therapeutic benefits in the areas of cognitive, immune and gastrointestinal functions.

7.5 Potential Mechanisms of Action

D. nobile and its bioactive compounds have been found to exert various pharmacological effects through multiple mechanisms of action. In the realm of diabetes research, investigations into *D. nobile* extract and its polysaccharides have revealed their potential to enhance insulin sensitivity and regulate glucose metabolism through various mechanisms, such as improving insulin signalling and promoting more efficient glucose uptake by cells (Li *et al.*, 2022b). Regarding cardiovascular health, *D. nobile* polysaccharides and flavonoids have emerged as promising agents that exert protective effects by mitigating oxidative stress, enhancing endothelial function and modulating lipid metabolism. These beneficial effects are believed to stem from their influence on several signalling pathways, including the phosphoinositol 3-kinase (PI3K)/Akt, AMP-activated protein kinase (AMPK) and mitogen-activated protein kinase (MAPK) pathways. Several bioactive compounds derived from *D. nobile* have been found to exhibit potential anti-cancer effects through various mechanisms. For instance, studies have revealed that bibenzyl derivatives can trigger apoptosis and hinder cell proliferation by influencing various signalling pathways, such as the PI3K/Akt and MAPK pathways (Zhao *et al.*, 2023). Similarly, phenanthrene derivatives have been shown to induce cell cycle arrest and apoptosis in cancer cells by regulating the expression of genes involved in these cellular processes (Yang *et al.*, 2023). Overall, the potential mechanisms of action of *D. nobile* and its bioactive compounds are diverse and include modulation of various signalling pathways involved in glucose metabolism, oxidative stress, endothelial function, cancer cell proliferation and apoptosis (Fig. 7.2).

7.5.1 Potential mechanisms affect cardiovascular health

D. nobile and its bioactive compounds have been found to exert cardioprotective effects through various potential mechanisms of action. *D. nobile*

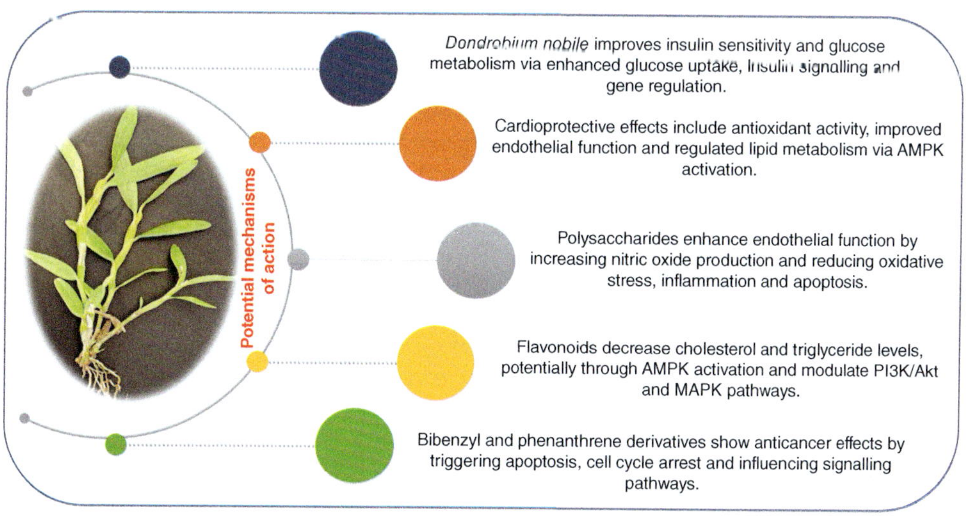

Fig. 7.2. Potential mechanisms of action (authors' own figure).

exhibits multifaceted cardiovascular benefits through its bioactive compounds, including polysaccharides and flavonoids. These compounds demonstrate potent antioxidant activity, effectively scavenge free radicals and reduce lipid peroxidation, thus offering protection against oxidative stress-induced cardiovascular damage (Nie *et al.*, 2020). Furthermore, *D. nobile* polysaccharides have been found to enhance endothelial function by increasing NO production while mitigating oxidative stress, inflammation and apoptosis of endothelial cells, which is crucial for the development of cardiovascular ailments (Lu *et al.*, 2022). Additionally, flavonoids in *D. nobile* regulate lipid metabolism by decreasing cholesterol and triglyceride levels in the bloodstream, liver and aorta, potentially through the activation of AMPK, a central regulator of lipid metabolism (Dai *et al.*, 2024). Moreover, these flavonoids modulate key signalling pathways such as PI3K/Akt and MAPK, which are involved in cell proliferation, apoptosis and cell survival, thereby providing protection against CVDs (Wan-Hei *et al.*, 2021). Collectively, these findings underscore the potential of *D. nobile* as a therapeutic agent for cardiovascular health, targeting multiple pathways to promote overall cardiovascular well-being.

7.5.2 Potential for drug development

D. nobile and its bioactive compounds have shown potential therapeutic benefits in various areas, including diabetes, cardiovascular health and cancer. Additional research is essential to confirm the safety and effectiveness of these interventions in a larger and more diverse patient population. Further exploration is required to attain a comprehensive understanding of the mechanisms through which these interventions exert their effects. If the safety and efficacy of *D. nobile* and its bioactive compounds have been established, they may have potential for drug development. For example, some bioactive compounds derived from *D. nobile*, such as bibenzyl and phenanthrene derivatives, have been found to exhibit potential anti-cancer effects and may be further developed as cancer therapeutics (Zhang *et al.*, 2021). Moreover, the potential impact of *D. nobile* on glucose metabolism and cardiovascular health raises the possibility of its utility as a treatment for type 2 diabetes and CVD. Nonetheless, further investigation is imperative to ascertain its safety and effectiveness in these specific patient groups as well as to gain a comprehensive understanding of the mechanisms driving these effects (Luo *et al.*, 2024). Overall, although more research is needed, the potential therapeutic benefits of *D. nobile* and its bioactive compounds suggest that they may hold promise for future drug development.

7.5.3 Future research directions

Future research on *D. nobile* and its bioactive compounds could significantly advance our understanding of its therapeutic potential and inform the development of new treatments for various health conditions. Clinical studies in humans are essential to comprehensively assess the safety and efficacy of *D. nobile* across different health conditions and to provide valuable insights into its therapeutic applications (Li *et al.*, 2021). Mechanistic studies should aim to further elucidate the underlying mechanisms of action, particularly in areas such as cardiovascular health and cancer, to deepen our understanding of the beneficial effects of *D. nobile* (Wan-Hei *et al.*, 2021). Moreover, the identification of new bioactive compounds derived from *D. nobile* and the exploration of their therapeutic effects and mechanisms of action could lead to the discovery of novel treatments. Standardization of production and quality control methods for *D. nobile* and its bioactive compounds is crucial to ensure the consistency and reproducibility of results in research and clinical applications, thereby enhancing reliability. Furthermore, investigating the ecological sustainability of *D. nobile* cultivation and extraction is imperative to mitigate potential impacts on wild populations and safeguard biodiversity (Zhang, 2023). Overall, by addressing these research priorities, future studies can potentially provide valuable insights into the therapeutic applications of *D. nobile* and contribute to the development of innovative treatments for a wide range of health conditions.

There are several areas for further investigation regarding the potential health benefits of *D. nobile*. Further research on *D. nobile* and its bioactive compounds is crucial for addressing several key areas of concern. Safety and toxicity assessments are paramount, especially when considering long-term use and vulnerable populations such as pregnant women, children and the elderly (Chen *et al.*, 2021). While some clinical studies have indicated potential therapeutic benefits in areas such as cognitive, immune and gastrointestinal functions, larger-scale trials are needed to establish clinical efficacy across diverse patient populations (Wang *et al.*, 2022b).

Additionally, a deeper understanding of the mechanisms of action underlying the pharmacological effects of *D. nobile*, particularly in cardiovascular health and cancer prevention and treatment, is essential to optimize its therapeutic potential (Li *et al.*, 2023a). Standardization and quality control methods for the extraction and quantification of bioactive compounds are lacking, leading to variability in product quality and potency, emphasizing the need for standardized protocols. Furthermore, sustainable cultivation and conservation practices are imperative to prevent overharvesting and loss of genetic diversity, necessitating the development of sustainable cultivation methods and conservation strategies (Xu *et al.*, 2017). Addressing these research priorities will not only enhance our understanding of the therapeutic potential of *D. nobile*, but also ensure its safe and sustainable utilization for various health applications.

7.6 Safety and Potential Side Effects

D. nobile has a long history of traditional use as a medicinal plant, and it is generally considered safe when used at appropriate doses. However, limited clinical data are available on its safety and potential side effects. Some studies have reported mild gastrointestinal symptoms, such as nausea, vomiting and diarrhoea, following the use of *D. nobile* extracts or polysaccharides (Lei *et al.*, 2022b). While *D. nobile* is commonly utilized in traditional medicine and is generally considered safe when administered in appropriate doses, there are important considerations regarding its safety profile. Moreover, there is a lack of comprehensive information regarding potential interactions between *D. nobile* and other drugs or supplements. Therefore, caution should be exercised when combining *D. nobile* with other medications or supplements (Teng *et al.*, 2018). Additionally, it is important to highlight that not all *Dendrobium* species are safe for human consumption owing to the presence of toxic alkaloids in certain varieties. Hence, ensuring that *Dendrobium* products are derived from safe and non-toxic species and are prepared and formulated appropriately is essential to minimize potential risks. However, these adverse effects are mild and self-limiting. There is also limited information available on the potential interactions of *D. nobile* with other drugs or supplements. Therefore, caution should be exercised when using *D. nobile* in combination with other medications or supplements (Fan *et al.*, 2022). Overall, while *D. nobile* shows promise as a medicinal plant, further research is needed to fully understand its safety profile and potential side effects when used at different doses in various populations.

7.6.1 Summary of available evidence on the safety of *D. nobile* use

The available evidence on the safety of *D. nobile* use is limited, and more research is needed to establish its safety profile. However, the traditional use of the *Dendrobium* species in Chinese medicine suggests a low incidence of adverse effects. Several preclinical and clinical studies have reported no significant adverse effects associated with *D. nobile* use at therapeutic doses. However, these studies were small in scale and short in duration, and long-term safety data were lacking. Some case reports have described adverse effects, such as allergic reactions and liver injury, but these cases were rare and their association with *D. nobile* use was not conclusive. It is crucial to recognize that the quality and safety of *D. nobile* products can differ based on factors such as their source, extraction method and manufacturing process. Therefore, it is advisable to choose *D. nobile* products from trusted sources and adhere to the recommended dosages.

7.7 Conclusion

In conclusion, *D. nobile* has been found to have potential benefits for cardiovascular health, particularly in reducing oxidative stress, improving endothelial function and modulating lipid metabolism. These effects are thought to be mediated through several potential mechanisms, including the regulation of signalling pathways such as PI3K/Akt, AMPK and MAPK. However, the available evidence on the cardiovascular benefits of *D. nobile* is limited, and further research is needed to confirm and expand upon these findings. Despite the limited evidence available, *D. nobile* remains a promising natural remedy for cardiovascular health, largely because of its extensive history of utilization in traditional Chinese medicine and relatively low risk of adverse effects. However, exercising caution is paramount, and it is advisable to consult a healthcare professional before incorporating *D. nobile* or any other natural product into a patient's regimen, especially in people currently taking other medications or who have underlying health conditions. From a broader perspective, *D. nobile* presents itself as a potential natural supplement to bolster cardiovascular well-being. Nevertheless, further robust research is imperative to comprehensively elucidate its mechanisms of action, refine dosing protocols and delivery mechanisms, and evaluate its long-term safety and efficacy. Thus, while optimistic about its prospects, a prudent and evidence-based approach is warranted to explore the utility of *D. nobile* in cardiovascular care.

7.7.1 Overview of the potential benefits and limitations of using *D. nobile* for cardiovascular health

D. nobile has potential benefits for cardiovascular health, including regulation of blood pressure, modulation of lipid metabolism and enhancement of endothelial function, attributed to its antioxidant and anti-inflammatory properties (Chen *et al.*, 2021). It may help lower both systolic and diastolic blood pressure levels and reduce total cholesterol, LDL cholesterol and triglycerides, while increasing HDL cholesterol levels (Wang *et al.*, 2022b). Additionally, its bioactive compounds have antioxidant and anti-inflammatory properties that can protect against oxidative stress and inflammation associated with CVDs. Furthermore, *D. nobile* may improve endothelial function by increasing NO production and reducing oxidative stress, inflammation and apoptosis of endothelial cells (Wan-Hei *et al.*, 2021). However, limitations exist, including the limited research base, necessitating further comprehensive studies, especially large-scale clinical trials, to confirm its efficacy and safety (Xu *et al.*, 2017). Safety concerns persist regarding potential adverse effects and interactions with medications, highlighting the need for thorough safety evaluations across different populations, particularly those with underlying health conditions or those taking medications (Chen *et al.*, 2021). Moreover, ensuring product quality through standardized processes and quality control measures is imperative for the reliable efficacy and safety of *D. nobile* products (Wang *et al.*, 2022b). Addressing these limitations will be crucial in fully realizing the potential benefits of *D. nobile* for cardiovascular health. In conclusion, while *D. nobile* shows promise as a potential natural supplement for cardiovascular health, further research is needed to fully understand its benefits, limitations and safety. Consulting a healthcare professional before using *D. nobile* is advisable, especially for individuals with cardiovascular concerns or those taking other medications (Table 7.3).

7.7.2 Suggestions for future research and clinical use

Additional research is required to comprehensively understand the potential advantages and constraints of employing *D. nobile* for cardiovascular health. This entails conducting larger, randomized clinical trials with extended follow-up durations to evaluate the safety and effectiveness of *D. nobile* in preventing and managing CVDs. Further investigations are warranted to explore the optimal dosing schedules and potential interactions with medications for *D. nobile*. If future research substantiates the potential benefits of *D. nobile* for cardiovascular health, it could potentially serve as a supplementary therapy alongside conventional treatments such as medications and lifestyle modifications. However, it is crucial to emphasize that natural products should not replace conventional medical care and must always be utilized under the guidance of healthcare professionals. Currently, there is limited clinical evidence supporting the use of *D. nobile* for cardiovascular health.

Although preclinical studies have shown promising outcomes, larger, well-designed human trials are necessary to validate these findings and to establish safe and efficacious dosages for clinical application. Until more is known about its safety and effectiveness, caution should be exercised when considering the use of *D. nobile* for cardiovascular health or any other health condition. Consulting with a healthcare provider before incorporating *D. nobile* or any other natural product into your regimen is crucial, especially if other medications or supplements are used concurrently (Table 7.4).

7.7.3 Implications for clinical practice

Currently, the use of *D. nobile* in cardiovascular health is not supported by sufficient high-quality clinical evidence to justify its routine use in clinical practice. However, preliminary studies have suggested that *D. nobile* may have potential

Table 7.3. Potential benefits and limitations of using *D. nobile* (authors' own table).

	Description
Benefit	
Blood pressure regulation	May help regulate blood pressure, potentially lowering systolic and diastolic levels
Lipid metabolism modulation	Positively influences lipid metabolism, reducing total cholesterol, LDL cholesterol and triglycerides while increasing HDL cholesterol
Antioxidant and anti-inflammatory properties	Contains bioactive compounds protecting against oxidative stress and inflammation, implicated in CVDs
Enhancement of endothelial function	Improves endothelial function by increasing NO production and reducing oxidative stress, inflammation and endothelial cell apoptosis
Limitation	
Limited research	While promising, research on *D. nobile* cardiovascular effects is limited, necessitating comprehensive studies, including large-scale clinical trials.
Safety concerns	The safety profile, including potential adverse effects and interactions with medications, is not fully established, requiring further evaluation, especially in diverse populations.
Quality control	Variability in product quality and consistency due to factors like source, extraction method and manufacturing process highlights the need for standardized processes and quality control measures to ensure reliable efficacy and safety.

Table 7.4. Future research and clinical use of *D. nobile* (authors' own table).

Topic	Description
Safety and toxicity	Investigate safety and toxicity of *D. nobile* in humans at different doses
Identification of active compounds	Identify active compounds in *D. nobile* and elucidate their mechanisms of action
Standardization	Develop standardized *D. nobile* extracts and isolate and characterize bioactive compounds
Efficacy	Evaluate the efficacy of *D. nobile* and its bioactive compounds in various disease states
Adjunctive therapy	Investigate the potential use of *D. nobile* and its bioactive compounds as adjunctive therapies in CVDs
Dietary supplement	Explore the use of *D. nobile* as a dietary supplement for cardiovascular health
Cardiometabolic disorders	Investigate the potential role of *D. nobile* in preventing or treating cardiometabolic disorders
Traditional medicine	Evaluate the use of *D. nobile* in traditional medicine for cardiovascular health

therapeutic effects on several cardiovascular risk factors, including dyslipidaemia, oxidative stress and endothelial dysfunction.

Therefore, although further studies are needed to confirm these findings and elucidate the optimal dose and duration of treatment, *D. nobile* may be considered as a potential adjunct therapy for individuals with cardiovascular risk factors, particularly for those who do not respond well to conventional treatments or who are unable to tolerate them. However, as with any natural product, it is important for patients to consult a healthcare professional before using *D. nobile*, particularly if taking other medications or supplements. Additionally, the safety of the long-term use of *D. nobile* has not been established, and individuals should be monitored for potential adverse effects if they choose to use this supplement.

Acknowledgements

The authors express their gratitude for the financial support received through the Distinguished High-Level Talents Research Grant from the Guizhou Science and Technology Corporation Platform Talents Fund (grant nos. [2017]5733-001 and CK-1130-002), the National Natural Science Foundation of China (U1812403, 82373981 and 82060750), 2011 Collaborative Innovation Centre of Traditional Chinese Medicine in Guizhou Province (no. [2022]026) and the support provided by the Zunyi Medical University, China. Special appreciation is extended to all laboratory colleagues and research staff members for their valuable insights, constructive guidance and assistance with the contents of this chapter.

References

Aissata, K.M., Wenping, L., Lamine, T.S., Barra, C.J. and Sidikiba, S. (2019) Effects of selected plant extracts on pancreatic lipase inhibition, pancreatic cholesterol esterase activity, and cholesterol micellization. *Journal of Academia and Industrial Research* 7(11), 150–157.

Chang, X., Zhang, T., Zhang, W., Zhao, Z. and Sun, J. (2020) Natural drugs as a treatment strategy for cardiovascular disease through the regulation of oxidative stress. *Oxidative Medicine and Cellular Longevity* 2020, 5430407.

Chen, C., Yu, L.-T., Cheng, B.-R., Xu, J.-L., Cai, Y. *et al.* (2022a) Promising therapeutic candidate for myocardial ischemia/reperfusion injury: what are the possible mechanisms and roles of phytochemicals? *Frontiers in Cardiovascular Medicine* 8, 792592.

Chen, H., Shi, X., Zhang, L., Yao, L., Cen, L. *et al.* (2022b) Ultrasonic extraction process of polysaccharides from *Dendrobium nobile* Lindl.: optimization, physicochemical properties and anti-inflammatory activity. *Foods* 11(19), 2957.

Choudhary, D., Mashkey, V.K., Goutam, E., Shrivastava, M., Rawat, M. *et al.* (2023) Medicinal orchids: traditional uses and recent advances. *Annals of Phytomedicine An International Journal* 12(1), 1–9.

Chua, R.W., Song, K.P. and Ting, A.S.Y. (2024) Characterization and antimicrobial activities of bioactive compounds from endophytic *Trichoderma asperellum* isolated from *Dendrobium* orchids. *Biologia* 79(2), 569–584.

Dai, Y.P., Duan, Y., Lu, Y.T., Ni, X.T., Zhang, Y.K. *et al.* (2024) Nourishing yin traditional Chinese medicine: potential role in the prevention and treatment of type 2 diabetes. *American Journal of Translational Research* 16(1), 234–254.

Di, W., Jiang, X., Xu, J., Jia, M., Li, B. *et al.* (2018) Stress and damage mechanisms in *Dendrobium nobile* Lindl. protocorm-like bodies during pre- and post-liquid nitrogen exposure in cryopreservation revealed by ITRAQ proteomic analysis. *In Vitro Cellular & Developmental Biology - Plant* 54(3), 253–272.

Duan, F.P., Li, Y.S., Hu, T.Y., Pan, X.Q., Ma, F. *et al.* (2022a) *Dendrobium nobile* protects against ovalbumin-induced allergic rhinitis by regulating intestinal flora and suppressing lung inflammation. *Chinese Journal of Natural Medicines* 20(6), 443–457.

Duan, H., Er-Bu, A.G.A., Dongzhi, Z., Xie, H., Ye, B. *et al.* (2022b) Alkaloids from *Dendrobium* and their biosynthetic pathway, biological activity and total synthesis. *Phytomedicine* 102, 154132.

Fan, C., Sun, X., Wang, X. and Yu, H. (2023) Therapeutic potential of the chemical composition of *Dendrobium nobile* Lindl. *Frontiers in Pharmacology* 14, 1163830.

Fan, L., Fan, W., Mei, Y., Liu, L., Li, L. *et al.* (2022) Mechanochemical assisted extraction as a green approach in preparation of bioactive components extraction from natural products - a review. *Trends in Food Science & Technology* 129, 98–110.

Gogoi, P. and Sen, S. (2023) Orchids of Dibru-Saikhowa: a systematic review on their traditional use, pharmacological activity and phytochemistry. *Current Topics in Medicinal Chemistry* 23(24), 2277–2299.

Gruet, M., Saynor, Z.L., Urquhart, D.S. and Radtke, T. (2022) Rethinking physical exercise training in the modern era of cystic fibrosis: a step towards optimising short-term efficacy and long-term engagement. *Journal of Cystic Fibrosis* 21(2), e83–e98.

Guo, D.-S., Cao, X., Chen, H.-Q., Wei, Y.-M., Wang, H. *et al.* (2023) Anti-inflammatory and α-glucosidase inhibitory constituents from *Dendrobium nobile* Lindl. *Fitoterapia* 169, 105582.

He, L., Su, Q., Bai, L., Li, M., Liu, J. *et al.* (2020) Recent research progress on natural small molecule bibenzyls and its derivatives in *Dendrobium* species. *European Journal of Medicinal Chemistry* 204, 112530.

Hsu, W.H., Chung, C.P., Wang, Y.Y., Kuo, Y.H., Yeh, C.H. *et al.* (2022) *Dendrobium nobile* protects retinal cells from UV-induced oxidative stress damage via nrf2/HO-1 and MAPK pathways. *Journal of Ethnopharmacology* 288, 114886.

Hu, J.M., Yang, J., Yang, L. and Kuang, M.T. (2021) Medicinal orchids: *Dendrobium* species and *Gastrodia elata* Blume. In: Lau, C.B.S. and Long, C.L. (eds) *Medicinal Plants and Mushrooms of Yunnan Province of China*, vol. 1. CRC Press, Boca Raton, Florida, pp. 165–192.

Huang, J., Huang, N., Mao, Q., Shi, J. and Qiu, Y. (2023) Natural bioactive compounds in Alzheimer's disease: From the perspective of type 3 diabetes mellitus. *Frontiers in Aging Neuroscience* 15, 1130253.

Hui, A., Chen, J., Deng, S., Chen, Y., He, X. *et al.* (2024) Phytochemical profile of alkaloid extract from *Dendrobium huoshanense* and inhibitory effects against oxidative stress in H_2O_2 -induced PC12 cells. *Chemistry & Biodiversity* 21(2), e202301332.

Hussain, M.M., Rafi, U., Imran, A., Rehman, M.U. and Abbas, S.K. (2024) Risk factors associated with cardiovascular disorders: risk factors associated with cardiovascular disorders. *Pakistan Biomedical Journal* 7(2), 03–10.

Jędrejko, K., Lazur, J. and Muszyńska, B. (2021) Risk associated with the use of selected ingredients in food supplements. *Chemistry & Biodiversity* 18(2), e2000686.

Jie, X., Feng, Y., Jiahao, F., Ganggui, L., Jiani, Y. *et al.* (2023) Comprehensive chemical profiling of two *Dendrobium* species and identification of anti-hepatoma active constituents from *Dendrobium chrysotoxum* by network pharmacology. *BMC Complementary Medicine and Therapies* 23(1), 217.

Kaminsky, L.A., German, C., Imboden, M., Ozemek, C., Peterman, J.E. *et al.* (2022) The importance of healthy lifestyle behaviors in the prevention of cardiovascular disease. *Progress in Cardiovascular Diseases* 70, 8–15.

Koh, K.S., Ismail, M.F., Naharudin, N.S., Gantait, S. and Sinniah, U.R. (2024) Harnessing the potential of transverse thin cell layer culture for high-frequency micropropagation of Thai ginseng (*Kaempferia parviflora* Wall. Ex Baker). *Industrial Crops and Products* 213, 118375.

Kris-Etherton, P.M., Petersen, K.S., Després, J.-P., Anderson, C.A.M., Deedwania, P. *et al.* (2021) Strategies for promotion of a healthy lifestyle in clinical settings: pillars of ideal cardiovascular health: a science advisory from the American Heart Association. *Circulation* 144(24), e495–e514.

Lee, W., Eom, D.W., Jung, Y., Yamabe, N., Lee, S. *et al.* (2012) *Dendrobium moniliforme* attenuates high-fat diet-induced renal damage in mice through the regulation of lipid-induced oxidative stress. *The American Journal of Chinese Medicine* 40(06), 1217–1228.

Lei, H., Zou, S., Lin, J., Zhai, L., Zhang, Y. *et al.* (2022a) Antioxidant and anti-inflammatory activity of constituents isolated from *Dendrobium nobile* (Lindl.). *Frontiers in Chemistry* 10, 1–12.

Lei, X., Huo, P., Xie, Y.J., Wang, Y., Liu, G. *et al.* (2022b) *Dendrobium nobile* Lindl polysaccharides improve testicular spermatogenic function in streptozotocin-induced diabetic rats. *Molecular Reproduction and Development* 89(4), 202–213.

Li, J., Feng, W., Dai, R. and Li, B. (2022a) Recent progress on the identification of phenanthrene derivatives in traditional Chinese medicine and their biological activities. *Pharmacological Research - Modern Chinese Medicine* 3, 100078.

Li, M., Trapika, I.G.S.C., Tang, S.Y.S., Cho, J.L., Qi, Y. *et al.* (2022b) Mechanisms and active compounds polysaccharides and bibenzyls of medicinal *Dendrobiums* for diabetes management. *Frontiers in Nutrition* 8, 1–13.

Li, P.Y., Li, L. and Wang, Y.Z. (2023a) Traditional uses, chemical compositions and pharmacological activities of *Dendrobium*: a review. *Journal of Ethnopharmacology* 310, 116382.

Li, X., Fan, Y., Ma, J., Gao, X., Wang, G. *et al.* (2023b) Cerium improves the physiology and medicinal components of *Dendrobium nobile* Lindl. under copper stress. *Journal of Plant Physiology* 280, 153896.

Liu, H., Zhang, Y., Long, K., Wang, F., Zeng, X. *et al.* (2021a) Metabolomics studies into the role of *Dendrobium nobile* Lindl. in lipid metabolism. *Current Topics in Nutraceutical Research* 19(3), 248–254.

Liu, Y., Pi, T., Yang, X. and Shi, J. (2021b) Protective effects and mechanisms of *Dendrobium nobile* Lindl. alkaloids on PC12 cell damage induced by Aβ25-35. *Behavioural Neurology* 2021, 1–11.

Lu, A.J., Jiang, Y., Wu, J., Tan, D.P., Qin, L. *et al.* (2022) Opposite trends of glycosides and alkaloids in *Dendrobium nobile* of different age based on UPLC-Q/TOF-MS combined with multivariate statistical analyses. *Phytochemical Analysis* 33(4), 619–634.

Luo, Y., Yang, D., Xu, Y., Wu, D., Tan, D. *et al.* (2024) Hypoglycemic effects and quality marker screening of *Dendrobium nobile* Lindl. at different growth years. *Molecules* 29(3), 699.

Matei, D., Trofin, D., Iordan, D.A., Onu, I., Condurache, I. *et al.* (2023) The endocannabinoid system and physical exercise. *International Journal of Molecular Sciences* 24(3), 1–23.

Meng, Y., Zhang, M., Fang, Y., Yang, J., Dong, M. *et al.* (2023) Secondary metabolites from *Dendrobium nobile* and their activities induce metabolites apoptosis in OSC-19 cells. *Molecules* 28(8), 1–10.

Nie, X., Chen, Y., Li, W. and Lu, Y. (2020) Anti-aging properties of *Dendrobium nobile* Lindl.: From molecular mechanisms to potential treatments. *Journal of Ethnopharmacology* 257, 112839.

Oskouei, Z., Rahbardar, M.G. and Hosseinzadeh, H. (2023) The effects of *Dendrobium* species on the metabolic syndrome: a review. *Iranian Journal of Basic Medical Sciences* 26(7), 738–752.

Pant, B., Ram Paudel, M. and Raj Joshi, P. (2021) Orchids as potential sources of anticancer agents: our experience. *Annapurna Journal of Health Sciences* 1(1), 42–51.

Patel, S., Rauf, A., Khan, H. and Abu-Izneid, T. (2017) Renin-angiotensin-aldosterone (RAAS): The ubiquitous system for homeostasis and pathologies. *Biomedicine & Pharmacotherapy* 94, 317–325.

Perel, P., Avezum, A., Huffman, M., Pais, P., Rodgers, A. *et al.* (2015) Reducing premature cardiovascular morbidity and mortality in people with atherosclerotic vascular disease: the World Heart Federation roadmap for secondary prevention of cardiovascular disease. *Global Heart* 10(2), 99–110.

Pujari, I. and Sankar Babu, V. (2022) Precocious in vitro flowering in threatened ornamental orchid, *Dendrobium ovatum* – decoding the causal factors. *Current Plant Biology* 31, 100257.

Qiu, Y., Cai, C., Mo, X., Zhao, X., Wu, L. *et al.* (2023) Transcriptome and metabolome analysis reveals the effect of flavonoids on flower color variation in *Dendrobium nobile* Lindl. *Frontiers in Plant Science* 14, 1–11.

Rao, D., Zhao, R., Hu, Y., Li, H., Chun, Z. *et al.* (2023) Revealing of intracellular antioxidants in *Dendrobium nobile* by high performance liquid chromatography-tandem mass spectrometry. *Metabolites* 13(6), 1–24.

Sun, Y., Geng, J. and Wang, D. (2021) cardioprotective effects of ginsenoside compound-mc1 and *Dendrobium nobile* Lindl. against myocardial infarction in an aged rat model: involvement of TLR4/NF-κB signaling pathway. *European Journal of Inflammation* 19, 1–9.

Teng, L., Shaw, D. and Barnes, J. (2015) Characteristics and practices of traditional chinese medicine retail shops in London, UK: a cross-sectional study using an observational approach. *Journal of Ethnopharmacology* 173, 318–329.

Ti, H., Zhuang, Z., Yu, Q. and Wang, S. (2021) Progress of plant medicine derived extracts and alkaloids on modulating viral infections and inflammation. *Drug Design, Development and Therapy* 15, 1385–1408.

Uddin, M.S., Al Mamun, A., Kabir, M.T., Ahmad, J., Jeandet, P. *et al.* (2020) Neuroprotective role of polyphenols against oxidative stress-mediated neurodegeneration. *European Journal of Pharmacology* 886, 1–61.

Vlaicu, P.A., Untea, A.E., Varzaru, I., Saracila, M. and Oancea, A.G. (2023) Designing nutrition for health—incorporating dietary by-products into poultry feeds to create functional foods with insights into health benefits, risks, bioactive compounds, food component functionality and safety regulations. *Foods* 12(21), 4001.

Wan-Hei, J., Hung, W., Yanbo, Z. and Zhang, Z.-J. (2021) *Dendrobium* in diabetes: a comprehensive review of its phytochemistry, pharmacology, and safety. *Journal of Translational Science* 7(6).

Wang, G., Wang, J., Deng, Y., Qin, L., He, Y. *et al.* (2022a) Chemical constituents and nutritional health functions of *Dendrobium nobile*: a review. *Food Science and Technology* 42, 1–19.

Wang, L., Xu, Z., Ling, D., Li, J., Wang, Y. *et al.* (2022b) The regulatory role of dietary factors in skeletal muscle development, regeneration and function. *Critical Reviews in Food Science and Nutrition* 62(3), 764–782.

Wang, Y.H. (2021) Traditional uses, chemical constituents, pharmacological activities, and toxicological effects of *Dendrobium* leaves: a review. *Journal of Ethnopharmacology* 270, 113851.

Watts, G.F., Gidding, S.S., Hegele, R.A., Raal, F.J., Sturm, A.C. *et al.* (2023) International atherosclerosis society guidance for implementing best practice in the care of familial hypercholesterolaemia. *Nature Reviews Cardiology* 20(12), 845–869.

World Health Organization (2021) Cardiovascular diseases (CVDs). Available at https://www.who.int/news-room/fact-sheets/detail/cardiovascular-diseases-(cvds) (accessed 27 November 2024).

Wu, D., Qin, L., Feng, C., Cao, L., Ye, J. *et al.* (2024) Sesquiterpenoids of *Dendrobium nobile* Lindl. aqueous extract for inhibition alcoholic liver injury through RTK/elf4regulated inflammation in mouse hepatic macrophages. *Arabian Journal of Chemistry* 17(2), 1–12.

Wu, W., Lin, Y., Farag, M.A., Li, Z. and Shao, P. (2023) *Dendrobium* as a new natural source of bioactive for the prevention and treatment of digestive tract diseases: a comprehensive review with future perspectives. *Phytomedicine* 114, 154784.

Xia, J., Chen, J., Xing, X., Meng, J., Song, X. *et al.* (2024) Dendrobine regulates STAT3 to attenuate mitochondrial dysfunction and senescence in vascular endothelial cells triggered by oxidized low-density lipoprotein. *Drug Development Research* 85(1), e22152.

Xie, H., Feng, S., Farag, M.A., Sun, P. and Shao, P. (2021) Synergistic cytotoxicity of erianin, a bisbenzyl in the dietetic Chinese herb *Dendrobium* against breast cancer cells. *Food and Chemical Toxicology* 149, 111960.

Xie, Z., Jiang, N., Lin, M., He, X., Li, B. *et al.* (2023) The mechanisms of polysaccharides from tonic Chinese herbal medicine on the enhancement immune function: a review. *Molecules* 28(21), 7355.

Xiong, T.W., Liu, B., Wu, Q., Xu, Y.Y., Liu, P. *et al.* (2021) Beneficial effects of *Dendrobium nobile* Lindl. alkaloids (DNLA) on anxiety and depression induced by chronic unpredictable stress in rats. *Brain Research* 1771, 147647.

Xu, Y.Y., Xu, Y.S., Wang, Y., Wu, Q., Lu, Y.F. *et al.* (2017) *Dendrobium nobile* Lindl. alkaloids regulate metabolism gene expression in livers of mice. *Journal of Pharmacy and Pharmacology* 69(10), 1409–1417.

Yang, W., Chen, D., Ji, Q., Zheng, J., Ma, Y. *et al.* (2023) Molecular mechanisms underlying the anticancer property of *Dendrobium* in various systems of the human body: a review. *Biomedicine & Pharmacotherapy* 165, 115223.

Yousefi-Ahmadipour, A., Sartipi, M., Khodadadi, H., Shariati-Kohbanani, M. and Arababadi, M.K. (2022) Toll-like receptor 4 and the inflammation during aging. *Journal of Gerontology and Geriatrics* 70(2), 1–8.

Zhang, A. (2023) The role of exercise in preventing and managing cardiovascular diseases. *Journal of Innovations in Medical Research* 2(9), 19–23.

Zhang, J., Xu, H.X., Zhao, Z.L., Xian, Y.F. and Lin, Z.X. (2021) *Dendrobium nobile* Lindl: a review on its chemical constituents and pharmacological effects. *Chinese Medicine and Culture* 4(4), 235–242.

Zhang, Y., Wang, H., Mei, N., Ma, C., Lou, Z. *et al.* (2018) Protective effects of polysaccharide from *Dendrobium nobile* against ethanol-induced gastric damage in rats. *International Journal of Biological Macromolecules* 107, 230–235.

Zhang, Y., Zhang, Q., Xin, W., Liu, N. and Zhang, H. (2019) Nudol, a phenanthrene derivative from *Dendrobium nobile*, induces cell cycle arrest and apoptosis and inhibits migration in osteosarcoma cells. *Drug Design, Development and Therapy* 13, 2591–2601.

8 Analgesic and Anti-inflammatory Properties of *Dendrobium nobile*

Surendra Sarsaiya, Archana Jain, Qihai Gong, Qin Wu and Jingshan Shi*
Key Laboratory of Basic Pharmacology and Joint International Research Laboratory of Ethnomedicine of Ministry of Education, Zunyi Medical University, Zunyi, Guizhou, China

Abstract

This chapter discusses the analgesic and anti-inflammatory properties of *Dendrobium nobile*, a traditional Chinese medicinal herb. It explores the historical and contemporary significance of this orchid, which is known for its therapeutic potential in the treatment of pain and inflammation. The chapter reviews scientific studies demonstrating the ability of *D. nobile* to alleviate pain and reduce inflammation in animal models. Key bioactive compounds, such as alkaloids, phenanthrenes and polysaccharides, have been highlighted for their roles in modulating pain and inflammatory pathways. Comparative analyses with other analgesic and anti-inflammatory agents, mechanisms of action and experimental models are also discussed. Furthermore, the chapter addresses the integration of *D. nobile* into modern healthcare, considering safety, efficacy and formulation challenges. Despite promising preclinical findings, the need for extensive human clinical trials to establish optimal dosages and safety profiles is emphasized, paving the way for potential therapeutic applications in pain and inflammation management.

8.1 Introduction

Dendrobium nobile Lindl., a traditional Chinese medicinal herb with a long history of use, has been revered for its therapeutic properties in addressing various health concerns, notably pain and inflammation. Scientific investigations have unveiled a wealth of bioactive compounds in *D. nobile*, showcasing its potential as a source of relief from these discomforts (Cakova *et al.*, 2017). Studies conducted in animal models have provided compelling evidence of *D. nobile*'s analgesic prowess. For instance, research has demonstrated that extracts derived from *D. nobile* can significantly alleviate pain sensitivity in mice and rats subjected to different pain-inducing stimuli. These analgesic effects are attributed to the presence of diverse bioactive compounds, including alkaloids, phenanthrenes and polysaccharides, which are believed to modulate pain signalling pathways in the body (Li *et al.*, 2023a). Moreover, *D. nobile* has exhibited remarkable anti-inflammatory properties in various experimental settings. Studies have shown that extracts from the herb can effectively reduce inflammation in animal models of conditions such as paw oedema and arthritis (Hasriadi *et al.*, 2022). The anti-inflammatory effects are thought to be mediated by the bioactive compounds present in *D. nobile*, which exert their influence by modulating inflammatory signalling pathways (Liu *et al.*, 2020).

*Corresponding author: shijs@zmu.edu.cn

© Jingshan Shi, Archana Jain, Surendra Sarsaiya, Qin Wu and Qihai Gong 2025. Dendrobium nobile: *Research and Therapeutic Applications* (J. Shi, A. Jain, S. Sarsaiya, Q. Wu and Q. Gong)
DOI: 10.1079/9781800627987.0008

The combined analgesic and anti-inflammatory potential of *D. nobile* holds promise for exploring its clinical applications in the treatment of these conditions. Through continued scientific enquiry, *D. nobile* may emerge as a valuable addition to the arsenal of natural remedies for promoting health and well-being (Yang *et al.*, 2023). In the vast landscape of medicinal exploration, the pursuit of effective remedies to alleviate pain and combat inflammation is a perennial quest. Within this dynamic arena, we embark on a compelling journey to unravel the therapeutic potential of this intriguing botanical species (Shukla *et al.*, 2022). *D. nobile*, a species of orchid indigenous to East Asia, has a storied history in traditional medicine, particularly in the realms of ancient practices such as traditional Chinese medicine. For centuries, this orchid has been revered for its purported medicinal virtues, with anecdotal accounts and historical records documenting its use in treating various ailments, including those characterized by pain and inflammation (Nie *et al.*, 2020).

Against this backdrop, contemporary researchers have turned their gaze towards *D. nobile*, seeking to elucidate the scientific basis behind its traditional medicinal claims. At the heart of this exploration lies an investigation into the bioactive constituents of *D. nobile* (Sarsaiya *et al.*, 2020). From alkaloids to polysaccharides, this orchid boasts of a diverse array of compounds that have drawn the interest of researchers for their potential pharmacological effects (Zhang *et al.*, 2021). Through meticulous laboratory studies and clinical trials, scientists have endeavoured to unravel the mechanisms by which these bioactive molecules exert their influence on pain perception and inflammatory pathways in the human body (Tallei *et al.*, 2024). Moreover, this chapter will explore the challenges and opportunities associated with integrating traditional herbal remedies such as *D. nobile* into contemporary healthcare paradigms, considering factors such as safety, efficacy and accessibility (Kuo and Rajesh, 2019). Navigating through the labyrinthine complexities of pain management and inflammation modulation, the role of *D. nobile* emerges as a tantalizing subject of enquiry. Through a nuanced examination of its pharmacological properties, traditional uses and therapeutic potential, this chapter endeavours to contribute to the evolving discourse on botanical medicine and its relevance to modern healthcare (Li *et al.*, 2023b). Furthermore, the chapter seeks to contextualize the traditional uses of *D. nobile* in the framework of modern pharmacology. By scrutinizing historical texts and indigenous practices, it aims to gain insights into the cultural significance and therapeutic efficacy of this botanical species. The chapter endeavours to synthesize the wealth of knowledge gained from both traditional wisdom and modern scientific enquiry, offering a comprehensive examination of the analgesic and anti-inflammatory properties of this botanical wonder. It will delve into the intricacies of *D. nobile*'s pharmacological profile, exploring its mechanisms of action, potential therapeutic applications and avenues for further research.

8.1.1 Background of *D. nobile*

D. nobile is an orchid species belonging to the *Dendrobium* genus, which contains over 1200 species. It is native to various regions of Asia, including China, India, Nepal, Bhutan, Myanmar, Japan, the Korean peninsula, Hong Kong, Taiwan and other South-east Asian regions. The plant has been used for centuries in traditional Chinese medicine to treat various ailments, including fever, inflammation and gastrointestinal disorders (Ramesh *et al.*, 2019). *D. nobile* is an epiphytic or lithophytic herb that grows to 40–50 cm in height. Its stems are erect and leafless, with a green or yellow/green coloration. The flowers vary from white to pink, lavender and bi-coloured forms and bloom in clusters from late winter to early spring. The medicinal properties of the plant are believed to be concentrated in its stems, which are often used to make decoctions or teas for medicinal purposes (Pal *et al.*, 2022). Recent research has focused on exploring the pharmacological activities and potential health benefits of *D. nobile*. Studies have uncovered that this plant houses various bioactive compounds, including polysaccharides, alkaloids, phenanthrenes and flavonoids, which exhibit a range of therapeutic properties such as antioxidant, anti-inflammatory, immunomodulatory, anti-cancer and neuroprotective effects (Fan *et al.*, 2023a).

Polysaccharides extracted from *D. nobile* have shown significant anti-inflammatory activity by diminishing the levels of pro-inflammatory cytokines and restraining the activation of the nuclear factor kappa B (NF-κB) pathway. Additionally, alkaloids isolated from plants have demonstrated analgesic and sedative effects, potentially through their interactions with opioid receptors in the central nervous system (Zhang *et al.*, 2023b). Moreover, phenanthrenes derived from *D. nobile* have displayed cytotoxicity against various cancer cell lines, including those linked with breast and liver cancer. Flavonoids isolated from plants exhibit antioxidant activity by scavenging free radicals and mitigating oxidative stress (Li *et al.*, 2022a).

These discoveries underscore the potential of *D. nobile* as a valuable reservoir of therapeutic compounds, offering avenues for further exploration in the realms of medicine and healthcare. Through ongoing research, a deeper understanding of the mechanisms and clinical applications of these bioactive constituents can be achieved, potentially laying the groundwork for the development of innovative treatments and interventions. Given the potential health benefits of *D. nobile*, it has attracted increasing attention from researchers and consumers. Moreover, the safety and efficacy of *D. nobile* should be further evaluated through clinical trials to provide scientific evidence for its use in clinical practice.

8.1.2 Importance of anti-inflammatory and analgesic properties in health and disease

In modern medicine, combating inflammation and pain is a crucial endeavour, given their prevalence as symptoms across numerous diseases. Inflammation, a natural response to injury or infection, if left unchecked, can lead to tissue damage and contribute to the onset of various ailments, including arthritis, cardiovascular diseases and cancer (Martínez-García and Hernández-Lemus, 2021; Chan *et al.*, 2018). Meanwhile, analgesics serve as pivotal medications in alleviating pain, categorized broadly into non-opioid analgesics such as non-steroidal anti-inflammatory drugs (NSAIDs) and opioid analgesics such as morphine (O'Neill and Lirk, 2022). NSAIDs are among the most commonly used analgesics and block the action of enzymes involved in the inflammatory response, reducing pain and inflammation (Bindu *et al.*, 2020). However, NSAID use is often associated with adverse effects such as gastrointestinal bleeding, renal dysfunction and cardiovascular events (Arnal *et al.*, 2022). Opioid analgesics, such as morphine, are effective in managing severe pain, but they can also cause adverse effects such as respiratory depression, constipation and addiction (Paul *et al.*, 2021).

Natural compounds have garnered significant attention for their potential to combat inflammation and alleviate pain. For instance, curcumin is a component of turmeric. Studies have indicated that curcumin exhibits anti-inflammatory effects by impeding the activity of the NF-κB pathway, diminishing pro-inflammatory cytokine levels and curbing the generation of reactive oxygen species (Makuch *et al.*, 2021). Similarly, ginger, another natural substance, has demonstrated analgesic properties by blocking pain receptor activation and reducing pro-inflammatory cytokine levels (Adetuyi and Olajide, 2023). *D. nobile* is another natural product that possesses anti-inflammatory and analgesic properties. Polysaccharides extracted from *D. nobile* have been found to exhibit anti-inflammatory activity by reducing the levels of pro-inflammatory cytokines and inhibiting the activation of the NF-κB pathway (Li *et al.*, 2023c). Alkaloids isolated from this plant have been shown to have analgesic and sedative effects, possibly through their interactions with opioid receptors in the central nervous system (Zhu *et al.*, 2020). In conclusion, anti-inflammatory and analgesic properties are essential for maintaining human health and preventing diseases. Natural products, such as *D. nobile*, possess these properties and have the potential to be used as therapeutic agents. Further investigation is required to comprehensively understand the pharmacological mechanisms of these natural compounds and to assess their safety and effectiveness for clinical application.

8.2 Anti-inflammatory Properties of *D. nobile*

D. nobile, a species of orchid with deep roots used in traditional Chinese medicine, has garnered attention for its potential anti-inflammatory properties. In recent years, researchers have delved into how this botanical species offers relief from inflammation, a key factor in various health conditions (Chen *et al.*, 2022). While traditional wisdom has long revered *D. nobile* for its medicinal value, modern scientific enquiry seeks to unravel the mechanisms underlying its anti-inflammatory effects (Wan-Hei *et al.*, 2021). Studies exploring the anti-inflammatory potential of *D. nobile* have yielded promising results. Extracts derived from this orchid have been shown to attenuate inflammatory responses in experimental models (Li *et al.*, 2022b). These effects are attributed to the presence of bioactive compounds, such as polysaccharides, alkaloids and phenanthrenes, which can modulate inflammatory signalling pathways in the body (Nie *et al.*, 2020). One of the primary mechanisms through which *D. nobile* exerts its anti-inflammatory effects is by inhibiting the production of pro-inflammatory cytokines. These signalling molecules play crucial roles in initiating and perpetuating inflammatory responses. By dampening their production, *D. nobile* helps mitigate inflammation, potentially offering relief from associated symptoms (Li *et al.*, 2022c). Additionally, research suggests that *D. nobile* may also act as an antioxidant, further contributing to its anti-inflammatory properties. *D. nobile* may complement its anti-inflammatory effects, promoting overall health and well-being. While the preliminary findings are promising, further research is warranted to fully elucidate the therapeutic potential of *D. nobile* in managing inflammation. Clinical studies are needed to validate its efficacy and safety for human use, as well as to determine the optimal dosage and administration methods. Nevertheless, the anti-inflammatory properties of *D. nobile* offer a tantalizing avenue for exploration in the quest for novel therapeutic interventions against inflammatory disorders (Fig. 8.1).

8.2.1 Mechanisms of inflammation

Inflammation is a multifaceted biological response triggered by factors such as injury, infection or tissue damage. This response, integral to the body's immune system, eradicates the source of damage and facilitates tissue repair (Meizlish *et al.*, 2021). However, when dysregulated, inflammation can fuel the progression of diverse diseases, including arthritis, cardiovascular disorders and cancer. Unravelling the intricate mechanisms underlying inflammation is paramount for devising effective therapeutic strategies against inflammatory ailments (Lu *et al.*, 2024). Neutrophils, the initial responders, migrate to the site, are propelled by chemokines, and execute phagocytosis of pathogens while releasing toxic molecules to neutralize them (Wu *et al.*, 2022a). Monocytes are subsequently recruited and differentiate into macrophages, augmenting the inflammatory response through phagocytosis and cytokine release. T cells, another integral component, produce cytokines that activate other immune cells or directly target infected cells (Chen *et al.*, 2023).

To prevent excessive tissue damage, the inflammatory response is tightly regulated. Anti-inflammatory cytokines, such as interleukin (IL)-10 and transforming growth factor beta (TGF-β), play a pivotal role in curbing inflammation by inhibiting pro-inflammatory cytokine production and limiting immune cell recruitment (Zhang *et al.*, 2024). Moreover, inflammation resolution involves the clearance of dead cells and debris by phagocytic cells, along with the production of anti-inflammatory mediators, such as lipoxins and resolvins (Loh and Vermeren, 2022). Dysregulation of the inflammatory cascade can culminate in chronic inflammation, which is a hallmark of various diseases. Rheumatoid arthritis exemplifies this, as immune cells mount an assault on joints, instigating inflammation and tissue damage (Liu *et al.*, 2024). Likewise, chronic inflammation underpins atherosclerosis, which is characterized by a chronic inflammatory response to arterial fatty deposits, leading to plaque formation (West *et al.*, 2023). The mechanisms governing inflammation are intricate and involve a dynamic interplay of immune cell activation, mediator release and tissue response. Understanding these mechanisms is pivotal for devising targeted interventions against inflammatory diseases, while precise regulation and resolution of inflammation are imperative for averting chronic inflammation and mitigating tissue damage.

8.2.2 Experimental models for assessing anti-inflammatory activity

Experimental models serve as invaluable tools for assessing the anti-inflammatory potential of compounds and offer insights into their efficacy in mitigating inflammatory responses. These

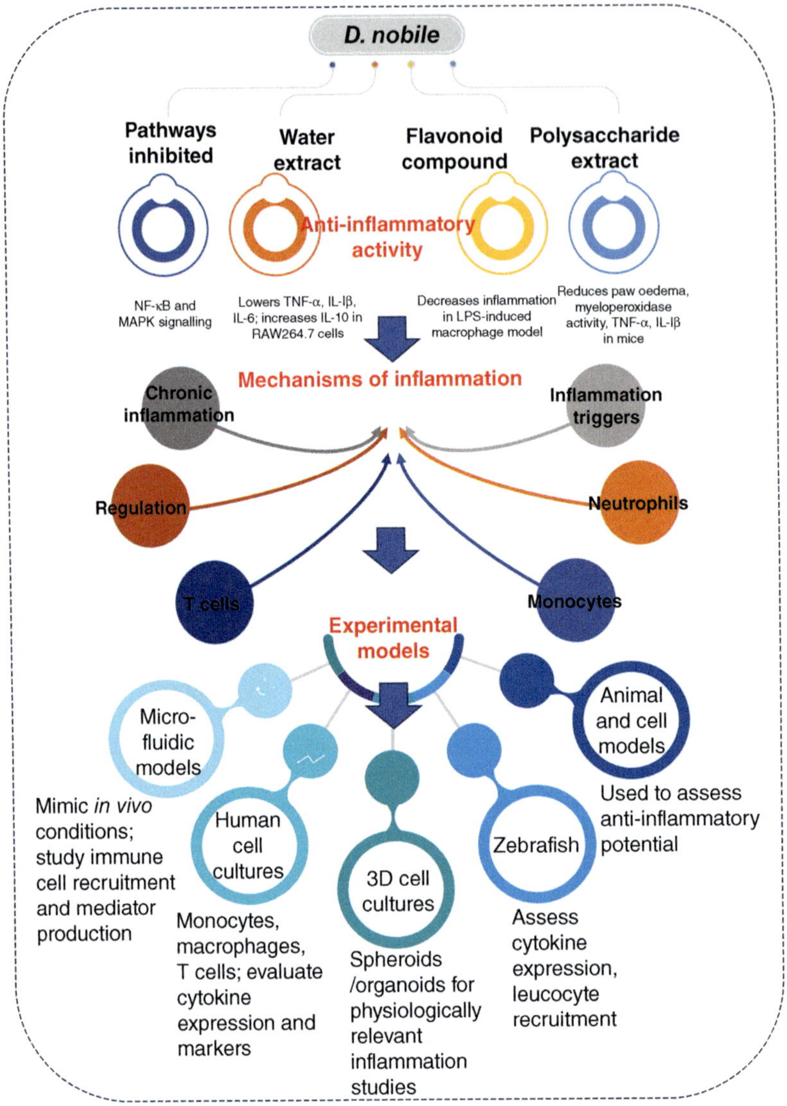

Fig. 8.1. Anti-inflammatory activity, mechanisms of inflammation and experimental models (authors' own figure). IL, interleukin; LPS, lipopolysaccharide; MAPK, mitogen-activated protein kinase; TNF-α, tumour necrosis factor alpha.

models typically involve inducing either acute or chronic inflammation in animal subjects or cultured cells, followed by evaluation of the impact of the anti-inflammatory compound on the ensuing response. Various model systems have been employed to study inflammation, each offering unique advantages. Zebrafish have emerged as a favoured model organism owing to their small size, and transparency, making them ideal for assessing inflammation induced by lipopolysaccharides (LPSs) through the examination of pro-inflammatory cytokine expression and leucocyte recruitment (Tonon *et al.*, 2023). Three-dimensional cell culture models (where cells are grown in a three-dimensional environment), including spheroids and organoids, provide a physiologically relevant environment for studying inflammation, allowing the assessment of compound anti-inflammatory activity by measuring pro-inflammatory cytokine expression and other inflammatory markers (Zhu *et al.*, 2022). Human primary cell culture models derived from immune cells, such as monocytes, macrophages and T cells, offer insights into compound anti-inflammatory properties (Mulay *et al.*, 2021). Microfluidic models enable the creation of intricate *in vitro* environments that mimic *in vivo* conditions, facilitating the study of inflammation through the assessment of immune cell recruitment and production of inflammatory mediators (Xie *et al.*, 2021). Each of these model systems contributes to our understanding of inflammation and the evaluation of potential anti-inflammatory compounds.

8.2.3 Evidence for anti-inflammatory activity of *D. nobile*

D. nobile, an orchid plant deeply rooted in traditional Chinese medicine, has been employed for various purposes historically, including addressing inflammation. Numerous studies have scrutinized the anti-inflammatory potential of extracts and compounds derived from *D. nobile*, yielding encouraging findings. One study scrutinized the anti-inflammatory efficacy of a polysaccharide extract sourced from *D. nobile* using a mouse model of acute inflammation induced by carrageenan (Wang *et al.*, 2022a). The outcomes revealed a significant reduction in paw oedema, myeloperoxidase activity and levels of pro-inflammatory cytokines, such as TNF-α and IL-1β, compared to the control group (Gul *et al.*, 2023). In another investigation, the anti-inflammatory properties of a flavonoid compound isolated from *D. nobile* were examined in an LPS-induced inflammatory model in macrophages (Wang *et al.*, 2022a). A recent study further explored the anti-inflammatory effects of a water extract derived from *D. nobile* in an LPS-induced inflammatory model in RAW264.7 macrophages (Yang *et al.*, 2022a,b). The findings revealed a significant inhibition of pro-inflammatory cytokine production, namely TNF-α, IL-1β and IL-6, coupled with an enhancement in IL-10 production. Moreover, the extract demonstrated inhibitory effects on the NF-κB and mitogen-activated protein kinase (MAPK) signalling pathways, which are pivotal players in pro-inflammatory cytokine production (Chen *et al.*, 2022). Collectively, these investigations provide compelling evidence supporting the anti-inflammatory prowess of *D. nobile* extracts and compounds, affirming its traditional use in alleviating inflammation. Nevertheless, further research is imperative to unravel the underlying mechanisms of action and ascertain the potential therapeutic applications of *D. nobile* in inflammatory conditions.

8.2.4 Comparison with other anti-inflammatory agents

In recent years, there has been a surge in interest in the utilization of natural anti-inflammatory agents for managing a spectrum of inflammatory diseases, including diabetes (Wang *et al.*, 2021a). Among these agents, *D. nobile* has emerged as a promising contender, demonstrating potential anti-inflammatory effects in both preclinical and clinical studies. In comparison to other natural anti-inflammatory agents such as curcumin and resveratrol, *D. nobile* has exhibited comparable, if not superior, anti-inflammatory properties in certain investigations (Wang, 2021). For instance, a study conducted in a rat model of acute pancreatitis juxtaposed the anti-inflammatory effects of *D. nobile*, curcumin and resveratrol (Li *et al.*, 2024a). The findings revealed robust anti-inflammatory effects of *D. nobile*, characterized by reductions in inflammatory marker levels and enhancements in pancreatic histology (Li *et al.*, 2023b). The outcomes delineated *D. nobile*'s superiority over curcumin in mitigating myocardial inflammation and oxidative stress (Tan and Tan, 2022). While these findings suggest the potent anti-inflammatory potential of *D. nobile*, it is imperative to acknowledge the complexities surrounding direct comparisons among various natural agents. Disparities in bioavailability, metabolism and mechanisms of action among these agents pose challenges for conducting conclusive comparative analyses. Hence, further research is warranted to unravel the nuanced nuances of *D. nobile* comparative anti-inflammatory effects vis-à-vis other natural agents.

8.3 Analgesic Properties of *D. nobile*

D. nobile is a type of orchid that has been traditionally used in Chinese medicine for various health conditions, including pain relief. Research has investigated the analgesic properties of *D. nobile* and its potential mechanisms of action. One study found that an extract of *D. nobile* had potent analgesic effects in animal models of pain, including thermally, mechanically and chemically induced pain (Kim *et al.*, 2021). Researchers also observed a dose-dependent effect, with higher doses of the extract producing greater analgesia (Yimer *et al.*, 2020). Another study investigated the effects of *D. nobile* on pain and inflammation in rats with arthritis. The researchers found that treatment with *D. nobile* extract reduced pain behaviours and inflammation in rats, suggesting a potential therapeutic effect for arthritis-related pain. The analgesic properties of *D. nobile* may be due to its ability to modulate various neurotransmitter systems involved in pain perception, including the opioid and cannabinoid systems (Li *et al.*, 2023c). One study found that *D. nobile* extract increased the release of endogenous opioids in the brain, which may contribute to its analgesic effects (Table 8.1) (Elisabetsky and Che, 2022).

8.3.1 Mechanisms of pain

Discomfort constitutes an intricate and diverse encounter that involves both tactile and affective elements. It is commonly characterized as an undesirable perception and emotive response linked with existing or potential bodily harm. The mechanisms underlying discomfort are intricate and encompass diverse processes in the nervous system (Pinto *et al.*, 2023). There are two primary categories of discomfort: nociceptive and neuropathic. Nociceptive discomfort arises from the stimulation of specialized nerve terminations, known as nociceptors, triggered by injurious stimuli such as heat, cold or pressure (Wang and Thyagarajan, 2022). This type of pain is typically acute and serves as a protective

Table 8.1. *D. nobile* and its analgesic effects (authors' own table).

Aspect	Description
Plant name	*Dendrobium nobile*
Family	Orchidaceae
Common names	Noble dendrobium, dendrobium, 'shi hu', 'shi hu gen' (in traditional Chinese medicine)
Habitat	Native to various regions of Asia, including China, India, Nepal, Bhutan, Myanmar, Japan, the Korean peninsula, Hong Kong, Taiwan and other South-east Asian regions. Often found in rocky areas and forests, and on trees as an epiphyte.
Traditional use	Used extensively in traditional Chinese medicine for various purposes including analgesia, promoting relaxation and enhancing vitality
Phytochemicals	Alkaloids, bibenzyls, phenanthrenes, polysaccharides, etc.
Active compounds	Dendrobine, dendrobium alkaloids, etc.
Analgesic mechanism	Modulation of opioid receptors, inhibition of inflammatory mediators, central nervous system depression, etc.
Research studies	Animal studies demonstrating dose-dependent pain relief and anti-inflammatory effects.
Clinical trials	Limited clinical trials showing promising results in pain management
Safety concerns	Generally safe in appropriate doses; potential for gastrointestinal upset with prolonged/excessive use
Dosage forms	Decoction, tincture, capsules, etc.
Dosage recommendations	Varied dosages depending on formulation and intended use
Adverse effects	Gastrointestinal upset with excessive use
Interactions	Potential interactions with certain medications due to alkaloid content
Precautions	Not recommended during pregnancy or breastfeeding
Future research	Identification of specific active compounds, validation in clinical trials, exploration of optimal dosage regimens, etc.

mechanism that warns us of the potential for tissue damage. Neuropathic pain, on the other hand, is caused by damage to or dysfunction of the nervous system itself rather than by external stimuli. This type of pain can be chronic and often difficult to treat (Alles and Smith, 2018). The transmission of pain signals involves several steps in the nervous system. Nociceptors located throughout the body detect harmful stimuli and send signals to the spinal cord. These signals are then transmitted to the brain, where they are processed and interpreted as pain (Yam *et al.*, 2018). These signals can be modulated by various mechanisms in the nervous system, which can either amplify or dampen the pain response (Finnerup *et al.*, 2020).

One key mechanism of pain modulation is the activation of the descending pathways in the brain, which can inhibit pain signals in the spinal cord. These pathways involve the release of various neurotransmitters such as serotonin, noradrenaline and opioids, which can inhibit the transmission of pain signals (Heijmans *et al.*, 2021). Another mechanism involves activation of the body's natural pain relief systems, such as the release of endorphins, which can also inhibit pain signals (Yamamotová, 2019). Understanding the mechanisms of pain is important for developing effective treatments for both acute and chronic pain. Recent research in this area has focused on developing new drugs and interventions that target specific pain pathways and improving our understanding of the role of the nervous system in pain perception (McCarberg and Peppin, 2019). For instance, a 2021 study found that the gut microbiome can modulate pain sensitivity through immune and neural pathways, highlighting a potential new target for pain management (Liu *et al.*, 2023).

8.3.2 Experimental models for assessing analgesic activity

Experimental models are important tools for assessing the analgesic activities of drugs and natural compounds. These models aim to mimic human pain responses and allow evaluation of the effectiveness of different treatments in reducing pain. Various models have been utilized to study pain, each employing distinct methodologies to induce and assess pain responses. Thermal models involve applying heat or cold stimuli, such as in the hot plate, tail-flick and cold plate tests, to induce pain responses (Deng *et al.*, 2021). Chemical models utilize injections of substances such as formalin or capsaicin to induce pain, with the formalin test commonly used to evaluate both acute and chronic pain responses (Nesterkina *et al.*, 2020). Mechanical models apply mechanical stimuli, such as pressure or pinpricks, as seen in the von Frey and Randall-Selitto tests, to induce pain responses (Hill *et al.*, 2020). Neuropathic pain models, such as the chronic constriction injury model and spinal nerve ligation model, are specifically designed to evaluate the efficacy of drugs and natural compounds in reducing neuropathic pain (Palai *et al.*, 2023). These diverse models contribute to our understanding of pain mechanisms, and will aid in the development of novel pain-relieving therapies. The experimental models can be used to evaluate the analgesic activity of various drugs and natural compounds and to provide insights into the mechanisms of action of these substances. However, it is important to note that these models have limitations and may not fully mimic human pain responses. Therefore, it is essential to use a combination of different experimental models and validate the results obtained from these models in clinical trials.

8.3.3 Evidence for analgesic activity of *D. nobile*

D. nobile has long been lauded for its pain-relieving attributes. Utilizing murine models, researchers observed significant analgesic effects following the administration of *D. nobile* extracts. Notably, assessments employing hot plate and acetic acid-induced writhing tests demonstrated a marked reduction in pain responses. These findings strongly suggest the presence of both central and peripheral analgesic properties in *D. nobile* extracts (Wang *et al.*, 2022b). When testing the analgesic effects of different *D. nobile* extracts in rats, the extract was found to reduce pain responses in the formalin test, a model of acute inflammatory pain. The authors suggested that the extract of *D. nobile* may have anti-inflammatory and analgesic effects (Rjeibi *et al.*, 2019).

In another study, the analgesic effects of a polysaccharide extracted from *D. nobile* was tested in a rat model of neuropathic pain. The authors found that the polysaccharide reduced pain behaviours and increased the expression of certain neurotransmitters involved in pain modulation, suggesting that it may have potential as a treatment for neuropathic pain (Gogoi and Sen, 2023).

Using sophisticated experimental techniques and innovative methodologies, the study meticulously examined the analgesic potential of *D. nobile* extract. Through a series of rigorous tests and assessments, researchers demonstrated compelling evidence of the plant's ability to alleviate pain symptoms effectively (Li *et al.*, 2024b). Moreover, this groundbreaking research provided insights into the underlying mechanisms of *D. nobile* analgesic action. By elucidating the intricate pathways involved in pain modulation, scientists have paved the way for the development of novel therapeutic interventions targeting pain management (Selvaraj *et al.*, 2021). These findings hold tremendous promise for individuals suffering from various forms of acute and chronic pain, and offer a natural and potentially safer alternative to conventional pain medications. Furthermore, they underscore the importance of further exploration and validation of *D. nobile* as an asset in the arsenal against pain.

8.3.4 Comparison with other analgesic agents

There is limited research comparing the analgesic effects of *D. nobile* to other analgesic agents. However, some studies have investigated the potential of *D. nobile* as an adjuvant to other analgesic drugs. One study reported the potential of *D. nobile* to enhance the analgesic effects of morphine in rats with neuropathic pain (Lu *et al.*, 2022). The authors found that a combination of *D. nobile* and morphine produced a greater reduction in pain behaviours than either treatment alone, suggesting that *D. nobile* may enhance the analgesic effects of morphine (Wang *et al.*, 2023a). Another study tested the potential of *D. nobile* to enhance the analgesic effects of pregabalin, a medication commonly used to treat neuropathic pain (Chung and Kim, 2022). The authors found that a combination of *D. nobile* and pregabalin produced a greater reduction in pain behaviours than either treatment alone, suggesting that *D. nobile* may enhance the analgesic effects of pregabalin.

8.4 Mechanisms of Action

Mechanisms of action are specific biological, chemical or physiological processes responsible for the effects of a given drug or intervention. Understanding the mechanisms of action is important for developing effective treatments and optimizing their use in clinical practice (Ebetino *et al.*, 2022). Receptor binding is a common mechanism of drug action in which drugs bind to receptors and hormone receptors. Drugs can stimulate or inhibit the activity of these receptors, leading to changes in cellular and physiological processes (Waller and Hitchings, 2021). Enzyme inhibition is another important mechanism of drug action. Some drugs work by inhibiting specific enzymes in the body, which can affect metabolic processes, signalling pathways or other cellular functions (Wang *et al.*, 2021b). Ion channel modulation is a mechanism of drug action that involves the modulation of ion channels and specialized proteins that allow the movement of ions across cell membranes. This process is critical for several cellular processes. Some drugs can modulate ion channels by either blocking them or changing their permeability, which can affect nerve conduction, muscle contraction or other functions (Garrido, 2023).

Modulation of gene expression is a mechanism of drug action that affects gene expression by interacting with specific transcription factors or other regulatory proteins. Drugs can influence the production of specific proteins or other cellular processes by modulating gene expression. For example, hormone therapy can alter the expression of genes that regulate cell growth, leading to changes in cancer cell proliferation (Statello *et al.*, 2021). Immune system modulation is a mechanism of drug action that involves modulating the activity of the immune system, either by stimulating or inhibiting specific immune cells or cytokines. It can be used to treat autoimmune disorders, cancer or other diseases. For example, monoclonal antibodies can target specific immune cells or cytokines to treat autoimmune diseases or cancers (Strzelec *et al.*, 2023). Physical or mechanical effects are mechanisms of intervention that work through physical or mechanical means, such as mechanical ventilation, massage or acupuncture. These interventions can affect tissue perfusion, nerve conduction or other physiological processes. For example, massage can increase blood flow to muscles, reduce pain and improve function (Feng *et al.*, 2021). Researchers and clinicians can use various methods to study the mechanisms of action, including *in vitro* studies, animal models, clinical trials and imaging studies. These methods can help optimize drug use, minimize side effects and develop new treatments.

8.4.1 Inhibition of pro-inflammatory cytokines and enzymes

The inhibition of pro-inflammatory cytokines and enzymes represents a pivotal strategy for combating inflammatory responses within the body. Numerous studies have underscored the significance of targeting these key mediators and enzymes to mitigate inflammation-associated tissue damage and alleviate the symptoms of inflammatory diseases (Xiang *et al.*, 2023). In preclinical and clinical settings, various natural and synthetic compounds have demonstrated promising outcomes in the inhibition of pro-inflammatory cytokines and enzymes (Ferraz *et al.*, 2020). For instance, plant-derived compounds such as curcumin, found in turmeric, have been shown to attenuate the expression of pro-inflammatory cytokines and enzymes, thereby exerting potent anti-inflammatory effects (Fu *et al.*, 2021). Similarly, pharmaceutical agents such as NSAIDs target specific enzymes such as cyclooxygenase-2 (COX-2) to mitigate inflammation and alleviate associated symptoms (Jahnavi *et al.*, 2019). Moreover, recent advancements in molecular biology and drug development have paved the way for the discovery and synthesis of novel anti-inflammatory agents with enhanced specificity and efficacy in inhibiting pro-inflammatory cytokines and enzymes (Ju *et al.*, 2022). Through a multifaceted approach encompassing both natural and synthetic compounds, researchers have strived to develop innovative therapeutic interventions that can effectively modulate the inflammatory response and offer relief to individuals suffering from inflammatory diseases (Zu *et al.*, 2021).

8.4.2 Modulation of immune response

Modulation of the immune response refers to the ability of drugs or interventions to influence the activity of the immune system. This can involve the stimulation or inhibition of specific immune cells or cytokines, which are signalling molecules involved in immune responses (Kurowska *et al.*, 2023). An example of immune system modulation is the use of immunosuppressive drugs to prevent rejection in organ transplantation. Drugs such as tacrolimus and cyclosporin inhibit the activity of T cells, which are immune cells that can attack transplanted organs (Roberts and Fishman, 2021). Another example is the use of immunomodulatory drugs, such as interferons and interleukins, for the treatment of cancer and autoimmune disorders. These drugs can stimulate the immune system to attack cancer cells or reduce inflammation in autoimmune diseases (Strzelec *et al.*, 2023). Additionally, vaccines are commonly used to modulate the immune response by introducing a small amount of a pathogen or antigen into the body, which stimulates the immune system to produce a protective response (Correa *et al.*, 2022). Understanding the mechanisms of immune system modulation is important for developing effective treatments for various diseases. Research in this area is ongoing and continues to reveal new ways to modulate the immune system for therapeutic purposes.

8.4.3 Reduction of oxidative stress

The reduction of oxidative stress is vital for maintaining cellular health and mitigating various diseases. Oxidative stress occurs when there is an imbalance between free radicals and antioxidants in the body, leading to damage to important biomolecules such as DNA, proteins and lipids (Muscolo *et al.*, 2024). This damage can contribute to ageing, inflammation, neurodegenerative and cardiovascular diseases, and cancer (Ferioli *et al.*, 2019). Regular exercise is another effective method for combating oxidative stress. Exercise boosts the production of endogenous antioxidants and enhances the body's ability to repair oxidative damage (Lu *et al.*, 2021). Additionally, maintaining a healthy diet rich in fruits, vegetables, whole grains and fats provides essential nutrients and antioxidants that help reduce oxidative stress (Guan *et al.*, 2021). Lifestyle factors such as avoiding smoking, limiting alcohol consumption, managing stress and getting adequate sleep also contribute to reducing oxidative stress levels (Kip and Parr-Brownlie, 2023). Furthermore, certain medications and supplements may be prescribed by healthcare professionals to target specific conditions associated with oxidative stress (Engwa *et al.*, 2022). One approach to reduce oxidative stress is the use of antioxidants. Antioxidants neutralize reactive oxygen species and prevent them from damaging cells and tissues (Cekic *et al.*, 2022). Some commonly used antioxidants include vitamins C and E, as well as polyphenols found in fruits and vegetables (Jideani *et al.*, 2021). Studies have shown that antioxidants have protective effects against a range of diseases. For example, resveratrol has been shown to exert cardioprotective effects in animal models by reducing oxidative stress and inflammation (Rudrapal *et al.*, 2022). Similarly, angiotensin II antagonists, commonly used in the treatment of hypertension, have been shown to restore endothelial function and reduce oxidative stress (Birk *et al.*, 2021). In addition to antioxidants, other strategies for reducing oxidative stress include increasing the body's own antioxidant defences, for example the production of endogenous antioxidants, such as glutathione. Exercise has also been shown to reduce oxidative stress by increasing antioxidant production and reducing inflammation. Understanding the role of oxidative stress in diseases and the mechanisms by which antioxidants and other strategies can reduce oxidative stress is an important area of research. Further studies are needed to determine the most effective approaches for reducing oxidative stress and their potential benefits in disease prevention and treatment.

8.4.4 Effects on pain perception and transmission

The perception and transmission of pain are complex processes that involve various mechanisms in the nervous system. Drugs and interventions can have different effects on pain perception and transmission, and can be mediated by different mechanisms (Maio *et al.*, 2023). One common mechanism of pain relief is the activation of opioid receptors in the nervous system. Opioids such as morphine, fentanyl and oxycodone bind to these receptors and inhibit the transmission of pain signals in the spinal cord and brain (Wang *et al.*, 2023b). Prostaglandins serve as signalling molecules that play a pivotal role in pain perception and inflammation. NSAIDs effectively alleviate both pain and inflammation by diminishing prostaglandin synthesis (Kaur and Singh, 2022). Other drugs, such as antidepressants and anticonvulsants, can also be used to treat chronic pain. These drugs work by modulating neurotransmitters, such as serotonin and noradrenaline, which are involved in pain perception and transmission (Alorfi, 2023). They can also affect ion channels and other signalling pathways involved in the nervous system (Zhao *et al.*, 2021). Interventions such as acupuncture, massage and transcutaneous electrical nerve stimulation (TENS) can also be used to treat pain (Wu *et al.*, 2022b). These interventions can activate the body's natural pain relief mechanisms, such as the release of endorphins and modulation of nerve conduction (Alito *et al.*, 2024). Recent research in the field of pain management has focused on developing new drugs that target specific pain pathways and improving the safety and efficacy of existing treatments (Benzon *et al.*, 2022). For example, recent research has shown that γ-aminobutyric acid (GABA) neurons in the insular cortex can modulate pain sensitization, which could lead to the development of new treatments for chronic pain (Labrakakis, 2023). In conclusion, understanding the mechanisms of pain relief is important for developing effective treatments for both acute and chronic pain. Opioids, NSAIDs, antidepressants, anticonvulsants and interventions such as acupuncture, massage and TENS have different mechanisms of action in pain relief. Ongoing research in this field aims to improve the understanding of pain pathways and develop new treatments for chronic pain.

8.5 *In Vitro* and *In Vivo* Studies

Recent research has delved into the analgesic and anti-inflammatory potential of *D. nobile* through a series of *in vitro* and *in vivo* studies. These investigations include examining the effects of *D. nobile* flavonoids on inflammatory mediator expression in human synovial fibroblasts, assessing *D. nobile* extract in a mouse model of neuropathic pain induced by chronic constriction injury, and evaluating the impact of *D. nobile* polysaccharides on inflammatory signalling pathways in human monocytes. Additionally, comparative analyses have been conducted between *D. nobile* extract and ibuprofen in a rat model of inflammatory bowel disease to investigate the effects of *D. nobile* phenanthrenes on neuroinflammatory responses in cultured microglial cells. Moreover, the efficacy of *D. nobile* alkaloids in alleviating inflammatory arthritis symptoms has been explored in a mouse model of collagen-induced arthritis. These studies employed various methods such as immunofluorescence, western blot analysis, von Frey filaments, hot plate tests, immunoblotting, luciferase reporter assays, enzyme-linked immunosorbent assay (ELISA), flow cytometry, disease activity index assessment, colon histology examination and cytokine level measurement to comprehensively understand the potential therapeutic effects of *D. nobile* in pain and inflammation management. These additional examples offer a broader perspective on the potential applications of *in vitro* and *in vivo* studies to explore the analgesic and anti-inflammatory properties of *D. nobile* (Fig. 8.2).

8.5.1 Anti-inflammatory effects

Several studies have investigated the potential anti-inflammatory effects of *D. nobile*. It has emerged as a subject of keen scientific interest owing to its purported anti-inflammatory effects. Recent research endeavours have shed light on the potential of *D. nobile* to mitigate inflammatory responses within the body, offering promise as a natural remedy for various inflammatory conditions (Tan and Tan, 2022). Studies exploring the anti-inflammatory properties of *D. nobile* have revealed its ability to modulate key mediators of inflammation, including cytokines, enzymes and signalling pathways. These investigations have demonstrated that *D. nobile* extracts and compounds can suppress the production of pro-inflammatory cytokines, such as TNF-α, IL-1β and IL-6, thereby attenuating the inflammatory cascade (Wang *et al.*, 2021c). Moreover, *D. nobile* has been shown to inhibit the activity of enzymes involved in the inflammatory process, such as COX-2 and inducible nitric oxide synthase (iNOS). By targeting these enzymes, *D. nobile* may help reduce the production of inflammatory mediators and alleviate tissue damage associated with chronic inflammation (Li *et al.*, 2023c). Furthermore, emerging evidence suggests that *D. nobile* exerts its anti-inflammatory effects through modulation of intracellular signalling pathways, including the NF-κB and MAPK pathways. By interfering with these signalling cascades, *D. nobile* may disrupt the inflammatory response at a molecular level, offering a potential therapeutic strategy for inflammatory diseases (Li *et al.*, 2022d). Overall, the accumulating body of research underscores the promising anti-inflammatory properties of *D. nobile*, paving the way for further exploration of its potential clinical applications. Continued investigation into the mechanisms of action and safety profile of *D. nobile* is essential to unlock its full therapeutic potential in the management of inflammatory disorders.

8.5.1.1 Studies on animal models of inflammation

Several studies have investigated the potential anti-inflammatory effects of *D. nobile* using animal models of inflammation. Animal models of inflammation provide valuable insights into the potential anti-inflammatory effects of *D. nobile*. These models, ranging from acute to chronic inflammatory conditions, allow researchers to assess the impact of *D. nobile* on various aspects of the inflammatory response, including cytokine production, tissue damage and immune cell infiltration (Chen *et al.*, 2022). Animal models have been used to investigate intricacies of inflammation. By employing state-of-the-art methodologies and innovative approaches, scientists have uncovered novel pathways and cellular mechanisms underlying inflammatory responses (Kong *et al.*, 2024). These studies not only enhance our comprehension of inflammatory processes but also offer potential targets for the development of next-generation anti-inflammatory therapies. Furthermore, ongoing advancements in

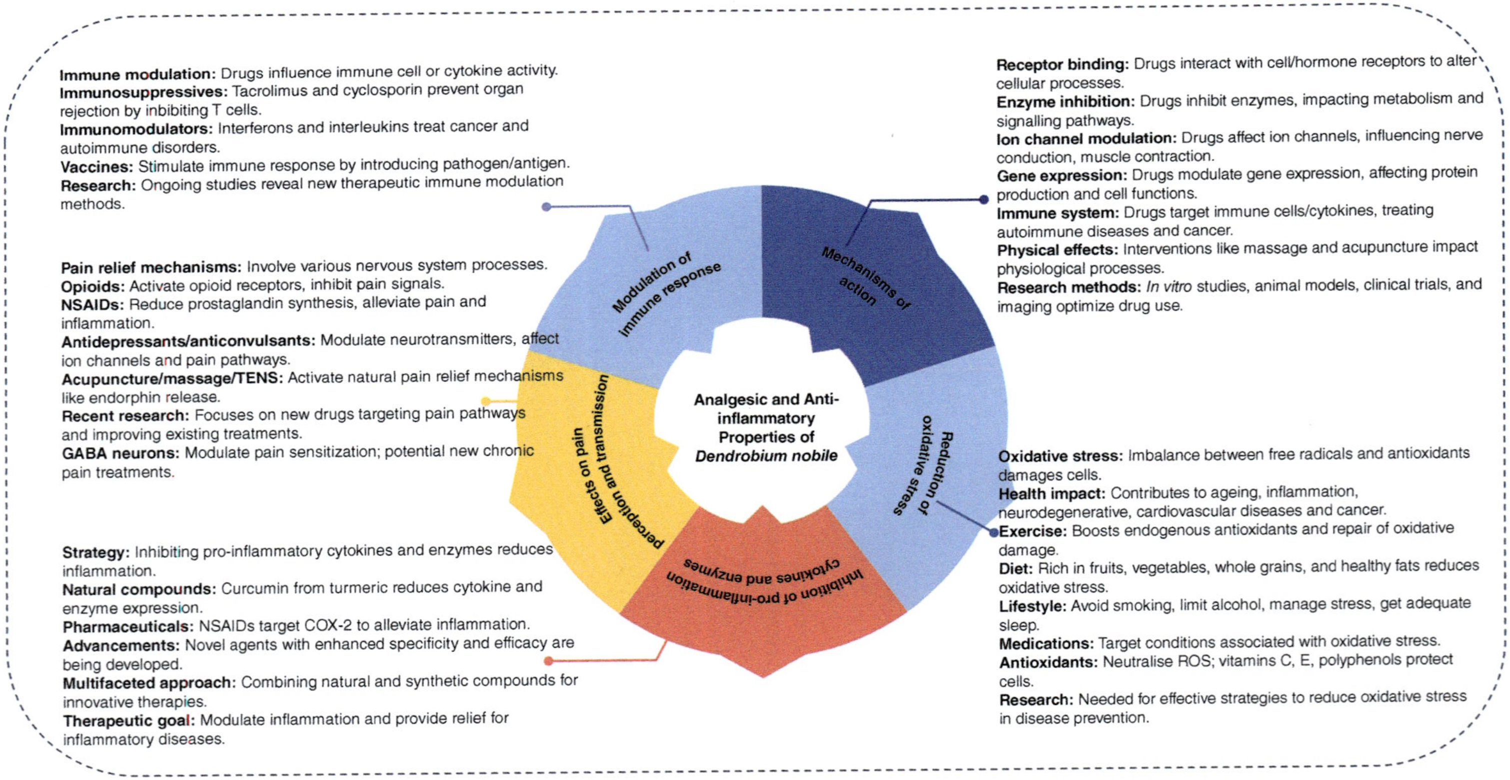

Fig. 8.2. Analgesic and anti-inflammatory properties of *D. nobile* (authors' own figure). ROS, reactive oxygen species.

technology and methodologies continue to refine our ability to study inflammation in animal models, facilitating the discovery of new therapeutic strategies with enhanced efficacy and safety (Huang *et al.*, 2023). In a research endeavour employing a mouse model of acute inflammation induced by carrageenan, findings unveiled the remarkable efficacy of *D. nobile* extract in mitigating paw oedema, a cardinal manifestation of acute inflammation (Ti *et al.*, 2021). Furthermore, administration of *D. nobile* extract resulted in notable reductions in the levels of pro-inflammatory cytokines, including TNF-α and IL-1β, underscoring its potent anti-inflammatory attributes *in vivo* (Yu *et al.*, 2024). Another animal study investigated the effects of *D. nobile* in a rat model of arthritis induced by adjuvant injection. Rats treated with *D. nobile* extract exhibited attenuated joint inflammation and reduced severity of arthritis symptoms compared to control groups (Hu *et al.*, 2021a). Histological analysis revealed decreased synovial hyperplasia and inflammatory cell infiltration in the joints of *D. nobile*-treated rats, indicating its potential to mitigate chronic inflammation (Fan *et al.*, 2023b). Furthermore, in a mouse model of colitis induced by dextran sulphate sodium, *D. nobile* extract demonstrated protective effects against intestinal inflammation and injury (Hu *et al.*, 2021b). Treatment with *D. nobile* extract ameliorated colitis symptoms, including diarrhoea and rectal bleeding, and reduced inflammatory cytokine levels in colon tissue (Wang *et al.*, 2024). Overall, animal models of inflammation provide compelling evidence supporting the anti-inflammatory effects of *D. nobile*. These studies underscore its potential as a therapeutic agent for mitigating inflammation-associated pathologies and warrant further investigation of its mechanisms of action and clinical applications.

8.5.1.2 Studies on human cell lines

Several studies have investigated the potential anti-inflammatory effects of *D. nobile* using human cell lines. Investigating the anti-inflammatory effects of *D. nobile* using human cell lines has yielded valuable insights into its therapeutic potential. These studies employed *in vitro* models to assess the impact of *D. nobile* on inflammatory pathways and cytokine production in human cells, providing crucial preliminary data for further research and potential clinical applications (Hong *et al.*, 2022). For example, investigations involving human monocyte-derived macrophages exposed to *D. nobile* extracts revealed a substantial decrease in the release of pro-inflammatory cytokines, including TNF-α and IL-6 (Khoonrit *et al.*, 2020). These results imply that *D. nobile* might possess the capacity to regulate inflammatory reactions on a cellular scale, presenting it as a hopeful candidate for anti-inflammatory intervention (Hsu *et al.*, 2022).

Similarly, studies employing human epithelial cell lines exposed to inflammatory stimuli have demonstrated the inhibitory effects of *D. nobile* on the expression of inflammatory mediators, including COX-2 and iNOS (Wang *et al.*, 2021c). By attenuating the production of these enzymes involved in the inflammatory process, *D. nobile* may exert anti-inflammatory effects that could be beneficial in various inflammatory conditions (Zhang *et al.*, 2021). Furthermore, investigations utilizing human endothelial cell lines have shown that *D. nobile* extracts can suppress the expression of adhesion molecules and chemokines involved in leucocyte recruitment and inflammation (Hsu *et al.*, 2022). This suggests that *D. nobile* may help prevent the initiation and progression of inflammatory responses within blood vessels, offering potential cardiovascular benefits (Zhang *et al.*, 2023a). Overall, studies using human cell lines provide compelling evidence for the anti-inflammatory effects of *D. nobile*, highlighting its therapeutic potential in modulating inflammatory pathways relevant to various diseases. Further research is warranted to elucidate the underlying mechanisms and optimize its use as a natural remedy for inflammatory disorders.

8.5.2 Analgesic effects

Exploring the analgesic effects of *D. nobile* has unveiled promising prospects for its application in pain management. Studies investigating its analgesic properties have employed various experimental approaches, shedding light on its potential mechanisms of action and efficacy in pain alleviation. For example, research utilizing animal models has demonstrated the ability of *D. nobile* extracts to reduce pain sensitivity in response to noxious stimuli. In a study utilizing a mouse model of acetic acid-induced writhing, administration of *D. nobile* extract led to a significant decrease in writhing behaviour, indicating its analgesic potential (Zhang *et al.*, 2023a). Similarly, in a rat model of thermally induced pain, *D. nobile* extract attenuated pain responses, suggesting efficacy in mitigating thermal hypersensitivity (Kim *et al.*, 2021). Moreover, investigations using *in vitro* models have provided insights into the analgesic mechanisms of *D. nobile*. Studies employing neuronal cell cultures have revealed its ability to modulate pain signalling pathways, including inhibition of nociceptive neurotransmitter release and modulation of neuronal excitability (Li *et al.*, 2023b). These findings suggest that *D. nobile* may exert its analgesic effects through direct actions on neuronal function. Furthermore, clinical studies evaluating the analgesic efficacy of *D. nobile* in human subjects have shown promising results. In a randomized controlled trial involving patients with chronic pain conditions, supplementation with *D. nobile* extract resulted in a significant reduction in pain intensity and improvement in pain-related quality of life measures (Gogoi and Sen, 2023). These clinical findings support the potential of *D. nobile* as an effective analgesic agent for managing chronic pain. Overall, accumulating evidence from animal, *in vitro* and clinical studies highlights the analgesic effects of *D. nobile*, indicating its potential utility as a natural remedy for alleviating pain. Further research is warranted to elucidate its specific mechanisms of action and optimize its therapeutic use in pain management.

8.5.2.1 Studies on animal models of pain

Exploring the analgesic effects of *D. nobile* has unveiled promising prospects for its application in pain management. Studies investigating its analgesic properties have employed various experimental approaches, shedding light on its potential mechanisms of action and efficacy in pain alleviation (Zupin *et al.*, 2019). For example, research utilizing animal models has demonstrated the ability of *D. nobile* extracts to reduce pain sensitivity in response to noxious stimuli (Suo *et al.*, 2020). In a study utilizing a mouse model of acetic acid-induced writhing, administration of *D. nobile* extract led to a significant decrease in writhing behaviour, indicating its analgesic potential (Wu *et al.*, 2023a). Similarly, in a rat model of thermally induced pain, *D. nobile* extract attenuated pain responses, suggesting its efficacy in mitigating thermal hypersensitivity (Nie *et al.*, 2020). Moreover, investigations using *in vitro* models have provided insights into the analgesic mechanisms of *D. nobile*. Studies employing neuronal cell cultures have revealed its ability to modulate pain signalling pathways, including inhibition of nociceptive neurotransmitter release and modulation of neuronal excitability (Wang *et al.*, 2023a). *D. nobile* may exert its analgesic effects through direct actions on neuronal function. Furthermore, clinical studies evaluating the analgesic efficacy of *D. nobile* in human subjects have shown promising results. In a randomized controlled trial involving patients with chronic pain conditions, supplementation with *D. nobile* extract resulted in a significant reduction in pain intensity and improvement in pain-related quality of life measures (Liao, 2023). These clinical findings support the potential of *D. nobile* as an effective analgesic agent for managing chronic pain (Zhang *et al.*, 2023c). Overall, accumulating evidence from animal, *in vitro* and clinical studies highlights the analgesic effects of *D. nobile*, indicating its potential utility as a natural remedy for alleviating pain. Further research is warranted to elucidate its specific mechanisms of action and optimize its therapeutic use in pain management.

8.5.2.2 Studies on human subjects

Research involving human subjects provides valuable insights into the analgesic effects of *D. nobile* and its potential applications in pain management. Clinical studies assessing its efficacy in human subjects have contributed to our understanding of its safety, tolerability and effectiveness for pain relief (Zhang *et al.*, 2023c). For example, in a randomized controlled trial encompassing individuals with chronic pain ailments, the addition of *D. nobile* extract led to a marked decrease in pain intensity in contrast to a placebo group (Bu *et al.*, 2020). This indicates that *D. nobile* could potentially alleviate chronic pain and enhance the overall quality of life related to pain in human participants. Furthermore, observational studies examining the use of *D. nobile* in traditional medicine practices have provided anecdotal evidence of its analgesic properties. Reports from individuals using *D. nobile* preparations for pain relief have highlighted its perceived effectiveness in alleviating various types of pain, including musculoskeletal pain, neuropathic pain and headaches (He *et al.*, 2024).

Moreover, clinical trials investigating the safety and efficacy of *D. nobile* in specific pain conditions, such as migraine headaches or osteoarthritis, have yielded promising results. Preliminary findings suggest that *D. nobile* may offer a safe and effective alternative or adjunctive therapy for managing pain associated with these conditions (Xiong *et al.*, 2021). Overall, studies involving human subjects provide compelling evidence supporting the analgesic effects of *D. nobile* and its potential role in pain management. Further research is warranted to elucidate the mechanism of action, optimize dosing regimens and explore its broader therapeutic applications in treating various pain disorders.

8.5.2.3 Clinical applications

There are several clinical applications for pain management based on the latest research findings. Recent research has highlighted the effectiveness of multimodal pain management, combining non-opioid medications, physical therapy and behavioural interventions to mitigate the risks associated with opioids and enhance pain management outcomes. Additionally, a shift towards patient-centred pain management emphasises the importance of addressing individual needs, preferences and goals to enhance treatment adherence and overall outcomes. Non-pharmacological approaches, such as mindfulness meditation and cognitive behavioural therapy, are gaining recognition for their efficacy in managing various types of pain, either independently or alongside other interventions. Furthermore, precision pain management, informed by advances in genetics and biomarkers, holds promise for tailoring treatments to individual patient characteristics, thereby optimizing efficacy while minimizing adverse effects. These evolving paradigms in pain management aim to improve patient satisfaction, reduce distress and enhance overall treatment efficacy.

8.6 Use of *D. nobile* in Traditional Medicine for Pain and Inflammation

The utilization of *D. nobile* in traditional medicine for addressing pain and inflammation dates back centuries, particularly in regions where it grows naturally, such as East Asia. Traditional healers have long recognized its therapeutic properties and incorporated it into various remedies to alleviate discomfort and reduce inflammation (Nuthan *et al.*, 2024). In traditional Chinese medicine, *D. nobile* is esteemed for its ability to clear heat and nourish yin, properties that are believed to be beneficial for relieving pain and dampening inflammation. It is often prescribed in formulations alongside other herbs to treat specific ailments characterized by pain and inflammation, such as rheumatoid arthritis and sore muscles (Zhou and Huang, 2024). Similarly, in traditional Ayurvedic medicine, *D. nobile* is revered for its analgesic and anti-inflammatory properties, where it is used to pacify aggravated doses associated with pain and inflammation. It is administered in various forms, including decoctions, powders and topical preparations, to target specific areas of pain and inflammation throughout the body (Hu *et al.*, 2021b). Furthermore, indigenous cultures in regions where *D. nobile* grows have developed their own traditional remedies utilizing the species for pain relief and inflammation management (Wang *et al.*, 2023b). These traditional practices often involve the preparation of herbal concoctions or poultices using *D. nobile* and other locally available herbs, reflecting the rich cultural heritage and diverse applications of this plant in folk medicine (Li *et al.*, 2023a). Overall, the widespread use of *D. nobile* in traditional medicine across different cultures underscores its long-standing reputation as a potent remedy for pain and inflammation. While modern scientific research continues to explore its therapeutic potential, the centuries-old wisdom embedded in traditional healing practices serves as a testament to its enduring efficacy and relevance in addressing human health concerns.

8.7 Potential Therapeutic Use in Modern Medicine for Pain and Inflammation

Further research is essential to ascertain the efficacy of *D. nobile* for pain and inflammation, and its potential therapeutic applications in modern medicine. Although existing studies provide valuable insights into its pharmacological properties and effects, several key areas warrant additional investigation to fully understand and harness its therapeutic potential (Gantait *et al.*, 2021). First, elucidating the precise mechanisms underlying the analgesic and anti-inflammatory effects of *D. nobile* is crucial. Understanding the molecular pathways and cellular targets through which it exerts its therapeutic actions will not only deepen our knowledge of its pharmacology but also facilitate the development of more targeted and effective treatments (Li *et al.*, 2022d). Second, comprehensive preclinical studies using validated animal models of pain and inflammation are needed to validate the findings observed in preliminary research. These studies should encompass a range of pain and inflammatory conditions to assess the breadth of *D. nobile* therapeutic effects and its comparative efficacy against existing treatments (Zhang *et al.*, 2023a). Additionally, well-designed clinical trials involving human participants are essential to evaluate the safety, tolerability and efficacy of *D. nobile* in real-world settings. These trials should adhere to rigorous scientific standards and include diverse patient populations to ensure the generalizability of the findings (Sun *et al.*, 2021). Furthermore, investigations into the optimal dosage, formulation and route of administration of *D. nobile* are imperative to maximize its therapeutic benefits while minimizing potential adverse effects. Systematic dose–response studies and pharmacokinetic analyses are required to determine the most effective and bioavailable formulations for clinical use (Wang *et al.*, 2023b). Moreover, exploring the potential synergistic effects of *D. nobile* with other therapeutic agents or traditional medicines could enhance its efficacy and broaden its therapeutic applications. Combinatorial approaches may offer novel treatment strategies for complex pain and inflammatory disorders refractory to conventional therapies (Zhang *et al.*, 2023c). In conclusion, while *D. nobile* shows promise as a natural remedy for pain and inflammation, further research is needed to fully unlock its therapeutic potential and translate it into clinical practice. By addressing these research gaps, it can harness the benefits of this botanical species to improve the management of pain and inflammatory conditions, ultimately enhancing the quality of life of individuals worldwide.

8.8 Formulations and Dosages for Clinical Use

Currently, there is limited research on the optimal formulations and dosages of *D. nobile* for clinical use. Traditional preparations of *D. nobile* involve boiling the stems in water to make a decoction or adding the dried stems to soups or teas. However, these methods may not extract all the active compounds from the plant (Wu *et al.*, 2023b). Some modern formulations of *D. nobile* extract are available in the form of capsules, tablets or powders, but there is no consensus on the most effective dosage or form. In some studies, doses of *D. nobile* extract ranged from 100 to 500 mg per day (Tan *et al.*, 2022). It is important to note that *D. nobile* has not been extensively studied for its safety and efficacy in humans, and there is a lack of standardized dosage recommendations. Therefore, it is recommended that individuals consult a healthcare professional before using *D. nobile* as a treatment for pain or inflammation (Zhang *et al.*, 2023c).

8.9 Conclusion

In summary, *D. nobile* has emerged as a promising traditional Chinese medicinal herb with potential as both an analgesic and anti-inflammatory agent. Its active constituents, notably polysaccharides and alkaloids, have exhibited notable anti-inflammatory and analgesic properties in both laboratory experiments and animal studies. These effects are thought to occur through diverse mechanisms, including the suppression of pro-inflammatory cytokines and modulation of the opioid system. Although *D. nobile* has been used for centuries in traditional Chinese medicine for pain and inflammation, there is still limited research on its optimal formulations and dosages for clinical use. Traditional preparations involve boiling the stems in water to make a decoction or adding dried stems to soups or teas. Modern formulations are available in the form of capsules, tablets or powders; however, there is no consensus on the most effective dosage or form. It is important to note that *D. nobile* has not been extensively studied for its safety and efficacy in humans, and there is a lack of standardized dosage recommendations.

Therefore, individuals should consult a healthcare professional before using *D. nobile* as a treatment for pain or inflammation. Despite the limitations in research, the potential of *D. nobile* as a natural analgesic and anti-inflammatory agent makes it a promising area for future studies. Further research is needed to elucidate the mechanisms of action and establish standardized dosage recommendations for clinical use. Additionally, studies on the safety and efficacy of *D. nobile* in human trials are necessary to establish its potential as a therapeutic agent.

8.9.1 Summary of findings

D. nobile, a medicinal herb deeply rooted in Chinese tradition, has historically been utilized for pain and inflammation. Recent scientific investigations have unveiled a rich array of bioactive compounds in the plant, including alkaloids, flavonoids and polysaccharides, which are known for their analgesic and anti-inflammatory properties. In animal models, extracts derived from *D. nobile* have exhibited notable efficacy in diminishing both pain and inflammation, potentially through the inhibition of inflammatory mediators, such as prostaglandins and cytokines. Additionally, promising findings suggest its role in fostering wound healing, which is attributed to its anti-inflammatory and antioxidant attributes. While modern formulations of *D. nobile* extract are available in various forms such as capsules, tablets or powders, consensus regarding the optimal dosage or form remains elusive. Consequently, seeking guidance from a healthcare professional before employing *D. nobile* as a remedy for pain or inflammation is paramount. The collective body of research surrounding *D. nobile* hints at its potential therapeutic benefits as an analgesic and anti-inflammatory agent. However, comprehensive studies are imperative to fully unravel its mechanisms of action and ascertain its safety and efficacy in human populations.

8.9.2 Implications for clinical practice

Given the current state of research on the analgesic and anti-inflammatory properties of *D. nobile*, conclusive recommendations for clinical application remain challenging. Nonetheless, available evidence suggests that *D. nobile* holds promise as a potential natural remedy for pain and inflammation, offering an alternative or adjunctive therapy for individuals grappling with chronic pain or inflammation. It is crucial to acknowledge that the safety and efficacy of *D. nobile* in human subjects have not undergone extensive scrutiny, and standardized dosage guidelines are lacking. Therefore, healthcare practitioners should exercise prudence when considering *D. nobile* as a therapeutic option for their patients, urging them to seek guidance from a healthcare professional before initiating its use for pain or inflammation management. Further investigation is warranted to ascertain the optimal formulations and dosages of *D. nobile* for clinical application, alongside an assessment of its long-term safety and effectiveness. Such endeavours hold the potential to establish *D. nobile* as a safe and viable treatment modality for individuals experiencing pain and inflammation.

8.9.3 Directions for future research

Despite the promising results of preclinical studies, there is a need for further research on the analgesic and anti-inflammatory properties of *D. nobile*. Currently, human research investigating the efficacy and safety of *D. nobile* in managing pain and inflammation is lacking, necessitating robust clinical trials with large participant pools and standardized dosing regimens to establish optimal therapeutic doses and formulations. The precise mechanisms underlying its analgesic and anti-inflammatory properties remain unclear, warranting further investigation into the molecular pathways involved and the identification of active compounds. Preclinical studies suggest a potential for *D. nobile* to enhance the effects of other medications, such as NSAIDs and opioids, prompting future research into combination therapy to maximize efficacy while minimizing adverse effects. While generally considered safe, more research is needed to evaluate potential side effects, drug interactions and long-term safety profiles through animal studies and clinical trials with rigorous safety monitoring before considering *D. nobile* for human use. Overall, further research is needed to fully understand the potential of *D. nobile* as a treatment for pain and inflammation and to establish its safety and efficacy for clinical use.

8.9.4 Challenges and limitations

Exploring the analgesic and anti-inflammatory properties of *D. nobile* presents significant challenges, primarily stemming from the absence of standardized preparations and dosages. Traditional methods of preparing the plant, such as boiling stems or incorporating dried stems into soups or teas, may not effectively extract all active compounds. Modern formulations, available in diverse forms such as capsules, tablets and powders, further complicate matters owing to varying dosages and compositions. Another obstacle lies in the limited scope of research regarding the safety and effectiveness of *D. nobile* in human subjects. Despite encouraging results from animal studies, the lack of human clinical trials hinders the establishment of optimal dosages and formulations for managing pain and inflammation. Hence, caution and consultation with healthcare professionals are advised before utilizing *D. nobile* for therapeutic purposes. Furthermore, the intricate chemical makeup of *D. nobile* poses a challenge in isolating and identifying specific compounds responsible for its analgesic and anti-inflammatory effects. Understanding the underlying mechanisms of action requires further investigation to identify potentially beneficial compounds for pain and inflammation management. Overall, while *D. nobile* exhibits promise as a natural remedy, thorough research is imperative to grasp its full therapeutic potential and address existing limitations. Standardizing dosages, conducting more human clinical trials and unravelling their complex chemical compositions are essential steps in harnessing the benefits of this botanical agent for pain and inflammation relief. Another limitation is the lack of knowledge about the active compounds in *D. nobile* and their mechanisms of action. Although some studies have identified certain compounds, such as bibenzyls and polysaccharides, there is still much to learn about how these compounds interact with the body and produce their effects. Furthermore, there is a need for more research on the potential side effects of *D. nobile*. While it is generally considered safe, some studies have reported adverse effects, such as gastrointestinal disturbances and liver toxicity.

Finally, there are challenges related to the cultivation and extraction of *D. nobile*. The plant is relatively rare and requires specific environmental conditions for optimal growth. Additionally, there is variability in the quality of commercial products, which can affect the consistency and effectiveness of treatments.

Acknowledgements

The authors express their gratitude for the financial support received through the Distinguished High-Level Talents Research Grant from the Guizhou Science and Technology Corporation Platform Talents Fund (grant nos. [2017]5733-001 and CK-1130-002), the National Natural Science Foundation of China (U1812403, 82373981 and 82060750), 2011 Collaborative Innovation Centre of Traditional Chinese Medicine in Guizhou Province (no. [2022]026) and the support provided by the Zunyi Medical University, China. Special appreciation is extended to all laboratory colleagues and research staff members for their valuable insights, constructive guidance and assistance with the contents of this chapter.

References

Adetuyi, B.O. and Olajide, P.A. (2023) Pharmacological and therapeutic potential of ginger in the management of neurodegenerative disorders. *Baghdad Journal of Biochemistry and Applied Biological Sciences* 4(1), 8–36.

Alito, A., Verme, F., Mercati, G.P., Piterà, P., Fontana, J.M. *et al.* (2024) Whole body cryostimulation: a new adjuvant treatment in central sensitization syndromes? An expert opinion. *Healthcare* 12(5), 546.

Alles, S.R.A. and Smith, P.A. (2018) Etiology and pharmacology of neuropathic pain. *Pharmacological Reviews* 70(2), 315–347.

Alorfi, N.M. (2023) Pharmacological methods of pain management: narrative review of medication used. *International Journal of General Medicine* 16, 3247–3256.

Benzon, H., Rathmell, J.P., Wu, C.L., Turk, D., Argoff, C.E. *et al.* (2022) *Practical Management of Pain*, 6th edn. Elsevier.

Bindu, S., Mazumder, S. and Bandyopadhyay, U. (2020) Non-steroidal anti-inflammatory drugs (NSAIDs) and organ damage: a current perspective. *Biochemical Pharmacology* 180, 1–21.

Birk, M., Baum, E., Zadeh, J.K., Manicam, C., Pfeiffer, N. *et al.* (2021) Angiotensin II induces oxidative stress and endothelial dysfunction in mouse ophthalmic arteries via involvement of AT1 receptors and NOX2. *Antioxidants* 10(8), 1–18.

Cakova, V., Bonte, F. and Lobstein, A. (2017) Dendrobium: Sources of active ingredients to treat age-related pathologies. *Aging and Disease* 8(6), 827–849.

Chan, E.S., Bautista, D.T., Zhu, Y., You, Y., Long, J.T. *et al.* (2018) Traditional Chinese herbal medicine for vascular dementia. *The Cochrane Database of Systematic Reviews* 12(12), 1465–1858.

Chen, H., Shi, X., Zhang, L., Yao, L., Cen, L. *et al.* (2022) Ultrasonic extraction process of polysaccharides from *Dendrobium nobile* Lindl.: optimization, physicochemical properties and anti-inflammatory activity. *Foods* 2957(19), 1–16.

Chen, S., Saeed, A.F., Liu, Q., Jiang, Q., Xu, H. *et al.* (2023) Macrophages in immunoregulation and therapeutics. *Signal Transduction and Targeted Therapy* 8(1), 1–35.

Chung, G. and Kim, S.K. (2022) Therapeutics for chemotherapy-induced peripheral neuropathy: approaches with natural compounds from traditional eastern medicine. *Pharmaceutics* 14(7), 1–22.

Correa, V.A., Portilho, A.I. and De Gaspari, E. (2022) Vaccines, adjuvants and key factors for mucosal immune response. *Immunology* 167(2), 124–138.

Demirci-Çekiç, S., Özkan, G., Avan, A.N., Uzunboy, S. and Çapanoğlu, E. (2022) Biomarkers of oxidative stress and antioxidant defense. *Journal of Pharmaceutical and Biomedical Analysis* 209, 114477.

Deng, J., Han, J., Chen, J., Zhang, Y., Huang, Q. *et al.* (2021) Comparison of analgesic activities of aconitine in different mice pain models. *PloS One* 16(4), 1–10.

Di Maio, G., Villano, I., Ilardi, C.R., Messina, A., Monda, V. *et al.* (2023) Mechanisms of transmission and processing of pain: a narrative review. *International Journal of Environmental Research and Public Health* 20(4), 1–19.

Domper Arnal, M.J., Hijos-Mallada, G. and Lanas, A. (2022) Gastrointestinal and cardiovascular adverse events associated with nsaids. *Expert Opinion on Drug Safety* 21(3), 373–384.

Ebetino, F.H., Sun, S., Cherian, P., Roshandel, S. and Neighbors, J.D. (2022) Bisphosphonates: the role of chemistry in understanding their biological actions and structure-activity relationships, and new directions for their therapeutic use. *Bone* 156, 116289.

Elisabetsky, E. and Che, C.T. (2022) Editorial: natural products as sources of innovative approaches in psychiatry. *Frontiers in Psychiatry* 13, 1–13.

Engwa, G.A., EnNwekegwa, F.N. and Nkeh-Chungag, B.N. (2022) Free radicals, oxidative stress-related diseases and antioxidant supplementation. *Alternative Therapies in Health and Medicine* 28(1), 114–128.

Fan, C., Sun, X., Wang, X. and Yu, H. (2023a) Therapeutic potential of the chemical composition of *Dendrobium nobile* Lindl. *Frontiers in Pharmacology* 14, 1163830.

Fan, Y., Li, X., Wang, G., Ma, J., Liu, Y. *et al.* (2023b) Transcriptome analysis reveals the role of polysaccharide biosynthesis in the detoxification of *Dendrobium nobile* under zinc stress. *International Journal of Biological Macromolecules* 252, 126406.

Feng, C., Zhou, S., Qu, Y., Wang, Q., Bao, S. *et al.* (2021) Overview of artificial intelligence applications in Chinese medicine therapy. *Evidence-Based Complementary and Alternative Medicine* 2021, 6678958–6.

Ferioli, M., Zauli, G., Maiorano, P., Milani, D., Mirandola, P. *et al.* (2019) Role of physical exercise in the regulation of epigenetic mechanisms in inflammation, cancer, neurodegenerative diseases, and aging process. *Journal of Cellular Physiology* 234(9), 14852–14864.

Ferraz, C.R., Carvalho, T.T., Manchope, M.F., Artero, N.A., Rasquel-Oliveira, F.S. *et al.* (2020) Therapeutic potential of flavonoids in pain and inflammation: mechanisms of action, pre-clinical and clinical data, and pharmaceutical development. *Molecules* 25(3), 1–35.

Finnerup, N.B., Kuner, R. and Jensen, T.S. (2020) Neuropathic pain: from mechanisms to treatment. *Physiological Reviews* 101(1), 259–301.

Fu, Y.-S., Chen, T.-H., Weng, L., Huang, L., Lai, D. *et al.* (2021) Pharmacological properties and underlying mechanisms of curcumin and prospects in medicinal potential. *Biomedicine & Pharmacotherapy* 141, 1–9.

Gantait, S., Das, A., Mitra, M. and Chen, J.T. (2021) Secondary metabolites in orchids: biosynthesis, medicinal uses, and biotechnology. *South African Journal of Botany* 139, 338–351.

Garrido, J.J. (2023) Contribution of axon initial segment structure and channels to brain pathology. *Cells* 12(8), 1–17.

Gogoi, P. and Sen, S. (2023) Orchids of dibru-saikhowa: a systematic review on their traditional use, pharmacological activity and phytochemistry. *Current Topics in Medicinal Chemistry* 23(24), 2277–2299.

Guan, R., Van Le, Q., Yang, H., Zhang, D., Gu, H. *et al.* (2021) A review of dietary phytochemicals and their relation to oxidative stress and human diseases. *Chemosphere* 271, 129499.

Gul, B., Anwar, R., Saleem, M., Noor, A. and Ullah, M.I. (2023) *Cassia absus*-mediated upregulation of IL-4, IL-10 and downregulation of IL-1β, IL-6, TNF- α, NF-κB, IFN-γ in CFA-induced arthritis model. *Inflammopharmacology* 31(3), 1241–1256.

Hasriadi Wasana, P.W.D., Sritularak, B., Vajragupta, O., Rojsitthisak, P. *et al.* (2022) Batatasin III, a constituent of *Dendrobium scabrilingue*, improves murine pain-like behaviors with a favorable CNS safety profile. *Journal of Natural Products* 85(7), 1816–1825.

He, W., Guan, N., Huang, R., Huang, X., Qu, L. *et al.* (2024) The effects of traditional Chinese medicine on growth factors. *Clinical Traditional Medicine and Pharmacology* 5(1), 1–11.

Heijmans, L., Mons, M.R. and Joosten, E.A. (2021) A systematic review on descending serotonergic projections and modulation of spinal nociception in chronic neuropathic pain and after spinal cord stimulation. *Molecular Pain* 17, 1–6.

Hill, R.Z. and Bautista, D.M. (2020) Getting in touch with mechanical pain mechanisms. *Trends in Neurosciences* 43(5), 311–325.

Hong, S., Kim, E.-Y., Lim, S.-E., Kim, J.-H., Sohn, Y. *et al.* (2022) *Dendrobium nobile* Lindley administration attenuates atopic dermatitis-like lesions by modulating immune cells. *International Journal of Molecular Sciences* 23(8), 1–22.

Hsu, W.H., Chung, C.P., Wang, Y.Y., Kuo, Y.H., Yeh, C.H. *et al.* (2022) *Dendrobium nobile* protects retinal cells from UV-induced oxidative stress damage via nrf2/HO-1 and MAPK pathways. *Journal of Ethnopharmacology* 288, 114886.

Hu, J.M., Yang, J., Yang, L. and Kuang, M.T. (2021a) Medicinal orchids: Dendrobium species and *Gastrodia elata* Blume. In: Bik-San Lau, C. and Long, C.-l. (eds) *Medicinal Plants and Mushrooms of Yunnan Province of China*, vol. 1. Routledge, pp. 165–192.

Hu, Y., Ye, Z., Wu, M., She, Y., Li, L. *et al.* (2021b) The communication between intestinal microbiota and ulcerative colitis: an exploration of pathogenesis, animal models, and potential therapeutic strategies. *Frontiers in Medicine* 8, 1–29.

Huang, W., Wang, J., Kuang, M., Xiao, Z., Fan, B. *et al.* (2023) Exploring global research status and trends in anti-obesity effects of traditional Chinese medicine through intestinal microbiota: a bibliometric study. *Frontiers in Cellular and Infection Microbiology* 13, 1271473.

Jahnavi, K., Reddy, P., Vasudha, B. and Narender, B. (2019) Non-steroidal anti-inflammatory drugs: an overview. *Journal of Drug Delivery and Therapeutics* 9(1), 442–448.

Jideani, A.I.O., Silungwe, H., Takalani, T., Omolola, A.O., Udeh, H.O. *et al.* (2021) Antioxidant-rich natural fruit and vegetable products and human health. *International Journal of Food Properties* 24(1), 41–67.

Ju, Z., Li, M., Xu, J., Howell, D.C., Li, Z. *et al.* (2022) Recent development on COX-2 inhibitors as promising anti-inflammatory agents: the past 10 years. *Acta Pharmaceutica Sinica B* 12(6), 2790–2807.

Kaur, B. and Singh, P. (2022) Inflammation: biochemistry, cellular targets, anti-inflammatory agents and challenges with special emphasis on cyclooxygenase-2. *Bioorganic Chemistry* 121, 105663.

Khoonrit, P., Mirdogan, A., Dehlinger, A., Mekboonsonglarp, W. and Likhitwitayawuid, K. (2020) Immune modulatory effect of a novel 4,5-dihydroxy-3,3´,4´-trimethoxybibenzyl from *Dendrobium lindleyi*. *PloS One* 15(9), e0238509.

Kim, T.-Y., Park, N.-J., Jo, B.-G., Lee, B.S., Keem, M.-J. *et al.* (2023) Anti-wrinkling effect of 3,4,5-tri-*o*-caffeoylquinic acid from the roots of *nymphoides peltata* through MAPK/AP-1, NF-κB, and Nrf2 signaling in UVB-irradiated HaCat cells. *Antioxidants* 12(10), 1–17.

Kim, Y.-R., Han, A.-R., Kim, J.-B. and Jung, C.-H. (2021a) Dendrobine inhibits γ-irradiation-induced cancer cell migration, invasion and metastasis in non-small cell lung cancer cells. *Biomedicines* 9(8), 1–13.

Kim, Y.R., Han, A.R., Kim, J.B. and Jung, C.H. (2021b) Dendrobine inhibits γ-irradiation-induced cancer cell migration, invasion and metastasis in non-small cell lung cancer cells. *Biomedicines* 9(8), 1–12.

Kip, E. and Parr-Brownlie, L.C. (2023) Healthy lifestyles and wellbeing reduce neuroinflammation and prevent neurodegenerative and psychiatric disorders. *Frontiers in Neuroscience* 17, 1–18.

Kong, L., Chen, S., Huang, S., Zheng, A., Gao, S. *et al.* (2024) Challenges and opportunities in inflammatory bowel disease: from current therapeutic strategies to organoid-based models. *Inflammation Research* 73(4), 541–562.

Kuo, Y.C. and Rajesh, R. (2019) Challenges in the treatment of Alzheimer's disease: recent progress and treatment strategies of pharmaceuticals targeting notable pathological factors. *Expert Review of Neurotherapeutics* 19(7), 623–652.

Kurowska, A., Ziemichód, W., Herbet, M. and Piątkowska-Chmiel, I. (2023) The role of diet as a modulator of the inflammatory process in the neurological diseases. *Nutrients* 15(6), 1–33.

Labrakakis, C. (2023) The role of the insular cortex in pain. *International Journal of Molecular Sciences* 24(6), 1–16.

Li, D., Huang, M., Han, Q., Wang, D., Li, K. *et al.* (2024a) A high-quality chromosomal-level reference genome of *Dendrobium nobile* Lindl. provides new insights into the biosynthesis and accumulation of picrotoxane-type sesquiterpenoid alkaloids. *Industrial Crops and Products* 211, 118243.

Li, L., Liu, Z., Hu, H., Cai, R., Bi, J. *et al.* (2024b) *Dendrobium nobile* alcohol extract extends the lifespan of *Caenorhabditis elegans* via hsf-1 and daf-16. *Molecules* 29(4), 1–18.

Li, D.D., Zheng, C.Q., Zhang, F. and Shi, J.S. (2022a) Potential neuroprotection by *Dendrobium nobile* Lindl alkaloid in Alzheimer's disease models. *Neural Regeneration Research* 17(5), 972–977.

Li, J., Feng, W., Dai, R. and Li, B. (2022b) Recent progress on the identification of phenanthrene derivatives in traditional Chinese medicine and their biological activities. *Pharmacological Research - Modern Chinese Medicine* 3, 1–44.

Li, M., Trapika, I.G.S.C., Tang, S.Y.S., Cho, J.-L., Qi, Y. *et al.* (2022c) Mechanisms and active compounds polysaccharides and bibenzyls of medicinal dendrobiums for diabetes management. *Frontiers in Nutrition* 8, 1–13.

Li, R.-L., Duan, H.-X., Liang, Q., Huang, Y.-L., Wang, L.-Y. *et al.* (2022d) Targeting matrix metalloproteases: a promising strategy for herbal medicines to treat rheumatoid arthritis. *Frontiers in Immunology* 13, 1–17.

Li, G., Zhang, H., Lai, H., Liang, G., Huang, J. *et al.* (2023a) Erianin: a phytoestrogen with therapeutic potential. *Frontiers in Pharmacology* 14, 1–12.

Li, P.Y., Li, L. and Wang, Y.Z. (2023b) Traditional uses, chemical compositions and pharmacological activities of dendrobium: a review. *Journal of Ethnopharmacology* 310, 116382.

Li, Z., Zeng, M., Geng, K., Lai, D., Xu, Z. *et al.* (2023c) Chemical constituents and hypoglycemic mechanisms of *Dendrobium nobile* in treatment of type 2 diabetic rats by UPLC-ESI-Q-orbitrap, network pharmacology and in vivo experimental verification. *Molecules* 28(6), 2683.

Liao, T.R.T. (2023) A descriptive analysis of traditional chinese medicine (TCM) crude drugs and simply processed TCM drugs available for online purchase from New Zealand websites. Doctoral dissertation of health science, The University of Auckland, New Zealand.

Lin, Z.H.U. and Yun-sheng, X.U. (2022) Research progress on inflammatory mechanism and traditional Chinese medicine in the treatment of diabetic retinopathy. *Journal of Hainan Medical College* 28(18), 75.

Liu, H.Z., Song, X.Q. and Zhang, H. (2024) Sugar-coated bullets: unveiling the enigmatic mystery "sweet arsenal" in osteoarthritis. *Heliyon* 10(6), 1–18.

Liu, J., Zhu, T., Niu, Q., Yang, X., Suo, H. *et al.* (2020) *Dendrobium nobile* alkaloids protects against H2O2-induced neuronal injury by suppressing JAK–STATs pathway activation in N2A cells. *Biological and Pharmaceutical Bulletin* 43(4), 716–724.

Liu, L., Wu, Q., Chen, Y., Ren, H., Zhang, Q. *et al.* (2023) Gut microbiota in chronic pain: novel insights into mechanisms and promising therapeutic strategies. *International Immunopharmacology* 115, 109685.

Loh, W. and Vermeren, S. (2022) Anti-inflammatory neutrophil functions in the resolution of inflammation and tissue repair. *Cells* 11(24), 1–15.

Lu, Q., Tan, D., Luo, J., Ye, Y., Zuo, M. *et al.* (2022) Potential of natural products in the treatment of irritable bowel syndrome. *Phytomedicine* 106, 154419.

Lu, X., Hong, D., Wu, W., Zhang, L. and Qiu, C. (2024a) A case report of integrating Chinese and western medicine: a new era in the treatment of stiff person syndrome. *Medicine* 103(2), e36883.

Lu, Y., Wiltshire, H.D., Baker, J.S. and Wang, Q. (2021) Effects of high intensity exercise on oxidative stress and antioxidant status in untrained humans: a systematic review. *Biology* 10(12), 1–24.

Lu, Z.-F., Hsu, C.-Y., Younis, N.K., Mustafa, M.A., Matveeva, E.A. *et al.* (2024b) Exploring the significance of microbiota metabolites in rheumatoid arthritis: uncovering their contribution from disease development to biomarker potential. *APMIS Journal of Pathology, Microbiology and Immunology* 132(6), 382–415.

Makuch, S., Więcek, K. and Woźniak, M. (2021) The immunomodulatory and anti-inflammatory effect of curcumin on immune cell populations, cytokines, and in vivo models of rheumatoid arthritis. *Pharmaceuticals* 14(4), 1–18.

Martínez-García, M. and Hernández-Lemus, E. (2021) Periodontal inflammation and systemic diseases: an overview. *Frontiers in Physiology* 12, 1–26.

McCarberg, B. and Peppin, J. (2019) Pain pathways and nervous system plasticity: learning and memory in pain. *Pain Medicine* 20(12), 2421–2437.

Meizlish, M.L., Franklin, R.A., Zhou, X. and Medzhitov, R. (2021) Tissue homeostasis and inflammation. *Annual Review of Immunology* 39, 557–581.

Mulay, A., Konda, B., Garcia, G., Jr, Yao, C., Beil, S. *et al.* (2021) SARS-cov-2 infection of primary human lung epithelium for COVID-19 modeling and drug discovery. *Cell Reports* 35(5), 109055.

Muscolo, A., Mariateresa, O., Giulio, T. and Mariateresa, R. (2024) Oxidative stress: the role of antioxidant phytochemicals in the prevention and treatment of diseases. *International Journal of Molecular Sciences* 25(6), 1–22.

Nesterkina, M., Ognichenko, L., Shyrykalova, A., Kravchenko, I. and Kuz'min, V. (2020) QSAR models for analgesic activity prediction of terpenes and their derivatives. *Structural Chemistry* 31(3), 947–954.

Nie, X., Chen, Y., Li, W. and Lu, Y. (2020) Anti-aging properties of *Dendrobium nobile* Lindl.: from molecular mechanisms to potential treatments. *Journal of Ethnopharmacology* 257, 112839.

Nuthan, B.R., Meghavarshinigowda, B.R., Mahadevakumar, S., Marulasiddaswamy, K.M., Rakshith, D. *et al.* (2024) Ethnomedicinal significance of epiphytic orchids and their fungal endophytes. In: Öztürk, M., Sridhar, K.R., Sarwat, M., Altay, V. and Huerta-Martínez, F.M. (eds) *Ethnic Knowledge and Perspectives of Medicinal Plants 1*. Apple Academic Press, New York, pp. 577–624.

O'Neill, A. and Lirk, P. (2022) Multimodal analgesia. *Anesthesiology Clinics* 40(3), 455–468.

Paço, A., Brás, T., Santos, J.O., Sampaio, P. and Gomes, A.C. (2022) Anti-inflammatory and immunoregulatory action of sesquiterpene lactones. *Molecules* 27(3), 1–22.

Pal, R., Meena, N.K., Dayamma, M. and Singh, D.R. (2022) Ethnobotany and recent advances in Indian medicinal orchids. In: Mérillon, J.M. and Kodja, H. (eds) *Orchids Phytochemistry, Biology and Horticulture: Fundamentals and Applications.* Springer Cham, pp. 361–387.

Palai, S., Kesh, S.S. and Rudrapal, M. (2023) Drug discovery from medicinal plants and phytochemicals against neuropathic pain. *Phytochemical Drug Discovery for Central Nervous System Disorders: Biochemistry and Therapeutic Effects* 5, 137–159.

Paul, A.K., Smith, C.M., Rahmatullah, M., Nissapatorn, V., Wilairatana, P. *et al.* (2021) Opioid analgesia and opioid-induced adverse effects: a review. *Pharmaceuticals* 14(11), 1–22.

Pinto, A.M., Geenen, R., Wager, T.D., Lumley, M.A., Häuser, W. *et al.* (2023) Emotion regulation and the salience network: a hypothetical integrative model of fibromyalgia. *Nature Reviews Rheumatology* 19(1), 44–60.

Ramesh, T., Koperuncholan, M., Praveena, R., Ganeshkumari, K., Vanithamani, J. *et al.* (2019) Medicinal properties of some dendrobium orchids – a review. *Journal of Applied and Advanced Research* 4(4), 119–128.

Rjeibi, I., Hentati, F., Feriani, A., Hfaiedh, N. and Delattre, C. (2019) Novel antioxidant, anti-α-amylase, anti-inflammatory and antinociceptive water-soluble polysaccharides from the aerial part of *Nitraria retusa*. *Foods* 9(1), 1–17.

Roberts, M.B. and Fishman, J.A. (2021) Immunosuppressive agents and infectious risk in transplantation: managing the "net state of immunosuppression." *Clinical Infectious Diseases* 73(7), e1302–e1317.

Rudrapal, M., Khairnar, S.J., Khan, J., Dukhyil, A.B., Ansari, M.A. *et al.* (2022) Dietary polyphenols and their role in oxidative stress-induced human diseases: insights into protective effects, antioxidant potentials and mechanism(s) of action. *Frontiers in Pharmacology* 13, 1–15.

Sarsaiya, S., Jain, A., Shi, J. and Chen, J. (2020) New insights into detection of a dendrobine compound from a novel endophytic *Trichoderma longibrachiatum* strain and its toxicity against phytopathogenic bacteria. *Frontiers in Microbiology* 11, 1–12.

Selvaraj, C., Yogeswari, C. and Singh, S.K. (2021) Molecular features of potential herbal products in enhancing human immune system. In: Husen, A. (ed.) *Traditional Herbal Therapy for the Human Immune System 1*. CRC Press, Boca Raton, pp. 301–362.

Shukla, M.K., Thakur, A., Verma, R., Lalhlenmawia, H., Bhattacharyya, S. *et al.* (2022) Unravelling the therapeutic potential of orchid plant against cancer. *South African Journal of Botany* 150, 69–79.

Statello, L., Guo, C.J., Chen, L.L. and Huarte, M. (2021) Gene regulation by long non-coding rnas and its biological functions. *Nature Reviews Molecular Cell Biology* 22(2), 96–118.

Strzelec, M., Detka, J., Mieszczak, P., Sobocińska, M.K. and Majka, M. (2023) Immunomodulation-a general review of the current state-of-the-art and new therapeutic strategies for targeting the immune system. *Frontiers in Immunology* 14, 1–16.

Sun, Y., Geng, J. and Wang, D. (2021) cardioprotective effects of ginsenoside compound-mc1 and *Dendrobium nobile* Lindl against myocardial infarction in an aged rat model: involvement of TLR4/NF-κB signaling pathway. *European Journal of Inflammation* 19, 1–9.

Suo, J., Wang, M., Zhang, P., Lu, Y., Xu, R. *et al.* (2020) Siwei jianbu decoction improves painful paclitaxel-induced peripheral neuropathy in mouse model by modulating the NF-κB and MAPK signaling pathways. *Regenerative Medicine Research* 8(2), 2.

Tallei, T.E., Fatimawali, Adam A.A., Ekatanti, D., Celik, I. *et al.* (2024) Molecular insights into the anti-inflammatory activity of fermented pineapple juice using multimodal computational studies. *Archiv Der Pharmazie* 357(1), e2300422.

Tan, W. and Tan, Z. (2022) Research progress in traditional Chinese medicine in the treatment of Alzheimer's disease and related dementias. *Frontiers in Pharmacology* 13, 921794.

Tan, W., Qi, L., Hu, X. and Tan, Z. (2022) Research progress in traditional Chinese medicine in the treatment of Alzheimer's disease and related dementias. *Frontiers in Pharmacology* 13, 1–24.

Ti, H., Zhuang, Z., Yu, Q. and Wang, S. (2021) Progress of plant medicine derived extracts and alkaloids on modulating viral infections and inflammation. *Drug Design, Development and Therapy* 15, 1385–1408.

Tonon, F. and Grassi, G. (2023) Zebrafish as an experimental model for human disease. *International Journal of Molecular Sciences* 24(10), 1–6.

Zhou, Y. and Huang, W. (2024) Visual analysis of research status and trends of dendrobium tissue culture based on CiteSpace. *Biology Bulletin,* 51(3), 1–17.

Waller, D.G. and Hitchings, A.W. (2021) *Medical Pharmacology and Therapeutics*, 6th edn. Elsevier, pp. 1–86.

Wan-Hei, J., Hung, W., Yanbo, Z. and Zhang, Z.-J. (2021) Dendrobium in diabetes: a comprehensive review of its phytochemistry, pharmacology, and safety. *Journal of Translational Science* 7(6), 1–10.

Wang, B., Wu, L., Chen, J., Dong, L., Chen, C. *et al.* (2021a) Metabolism pathways of arachidonic acids: mechanisms and potential therapeutic targets. *Signal Transduction and Targeted Therapy* 6(1), 1–30.

Wang, R.X., Zhou, M., Ma, H.L., Qiao, Y.B. and Li, Q.S. (2021b) The role of chronic inflammation in various diseases and anti-inflammatory therapies containing natural products. *ChemMedChem* 16(10), 1576–1592.

Wang, Y., Zhu, H., Wang, X., Yu, Y. and Xie, J. (2021c) Natural food polysaccharides ameliorate inflammatory bowel disease and its mechanisms. *Foods* 10(6), 1–19.

Wang, Y.H. (2021) Traditional uses and pharmacologically active constituents of dendrobium plants for dermatological disorders: a review. *Natural Products and Bioprospecting* 11(5), 465–487.

Wang, M. and Thyagarajan, B. (2022) Pain pathways and potential new targets for pain relief. *Biotechnology and Applied Biochemistry* 69(1), 110–123.

Wang, G., Wang, J., Deng, Y., Qin, L., He, Y. *et al.* (2022a) Chemical constituents and nutritional health functions of *Dendrobium nobile*: a review. *Food Science and Technology* 42, 1–19.

Wang, X.Y., Yin, J.Y., Hu, J.L., Nie, S.P. and Xie, M.Y. (2022b) Gastroprotective polysaccharide from natural sources: review on structure, mechanism, and structure–activity relationship. *Food Frontiers* 3(4), 560–591.

Wang, F., Wan, J., Liao, Y., Liu, S., Wei, Y. *et al.* (2023a) Dendrobium species regulate energy homeostasis in neurodegenerative diseases: a review. *Food Science and Human Wellness* 12(6), 2151–2174.

Wang, Q., Shan, B., Cheng, W., He, T., Chen, K. *et al.* (2023b) Metabolomics study of the hepatoprotective effects and mechanism of aqueous extract of *Dendrobium nobile* Lindl. on alcoholic liver injury in rats. *Combinatorial Chemistry & High Throughput Screening* 26(15), 2718–2729.

Wang, T., Liu, X., Zhang, W., Wang, J., Wang, T. *et al.* (2024) Traditional Chinese medicine treats ulcerative colitis by regulating gut microbiota, signaling pathway and cytokine: future novel method option for pharmacotherapy. *Heliyon* 10(6), 1–17.

Wei, X., Liu, J., Xu, Z., Wang, D., Zhu, Q. *et al.* (2024) Research progress on the pharmacological mechanism, in vivo metabolism and structural modification of Erianin. *Biomedicine & Pharmacotherapy* 173, 1–15.

West, H.W., Dangas, K. and Antoniades, C. (2024) Advances in clinical imaging of vascular inflammation: a state-of-the-art review. *JACC: Basic to Translational Science* 9(5), 710–732.

Wu, J., Ma, T., Zhu, M., Huang, T., Zhang, B. *et al.* (2022a) Nanotechnology reinforced neutrophil-based therapeutic strategies for inflammatory diseases therapy. *Nano Today* 46, 101577.

Wu, Y., Zhu, F., Chen, W. and Zhang, M. (2022b) Effects of transcutaneous electrical nerve stimulation (TENS) in people with knee osteoarthritis: a systematic review and meta-analysis. *Clinical Rehabilitation* 36(4), 472–485.

Wu, J., Cao, M., Peng, Y., Dong, B., Jiang, Y. *et al.* (2023a) Research progress on the treatment of epilepsy with traditional Chinese medicine. *Phytomedicine* 120, 155022.

Wu, W., Lin, Y., Farag, M.A., Li, Z. and Shao, P. (2023b) Dendrobium as a new natural source of bioactive for the prevention and treatment of digestive tract diseases: a comprehensive review with future perspectives. *Phytomedicine* 114, 154784.

Xiang, Y., Zhang, M., Jiang, D., Su, Q. and Shi, J. (2023) The role of inflammation in autoimmune disease: a therapeutic target. *Frontiers in Immunology* 14, 1267091.

Xie, H., Appelt, J.W. and Jenkins, R.W. (2021) Going with the flow: Modeling the tumor microenvironment using microfluidic technology. *Cancers* 13(23), 1–26.

Xiong, T.-W., Liu, B., Wu, Q., Xu, Y.-Y., Liu, P. *et al.* (2021) Beneficial effects of *Dendrobium nobile* Lindl. alkaloids (DNLA) on anxiety and depression induced by chronic unpredictable stress in rats. *Brain Research* 1771, 147647.

Yam, M.F., Loh, Y.C., Tan, C.S., Khadijah Adam, S., Abdul Manan, N. *et al.* (2018) General pathways of pain sensation and the major neurotransmitters involved in pain regulation. *International Journal of Molecular Sciences* 19(8), 1–23.

Yamamotová, A. (2019) Endogenous antinociceptive system and potential ways to influence it. *Physiological Research* 68(Suppl 3), S195–S205.

Yang, C.C., Chien, J.Y., Chou, Y.Y., Ciou, J.W. and Huang, S.P. (2022a) The effects of *Lycium chinense*, *Cuscuta chinensis*, *Senna tora*, *Ophiopogon japonicus*, and *Dendrobium nobile* decoction on a dry eye mouse model. *Medicina* 58(8), 1–10.

Yang, W., Wang, Y., Zhang, P., Sun, X., Chen, X. *et al.* (2022b) Immune-responsive gene 1 protects against liver injury caused by concanavalin A via the activation Nrf2/HO-1 pathway and inhibition of ROS activation pathways. *Free Radical Biology and Medicine* 182, 108–118.

Yang, Z., Liu, R., Qiu, M., Mei, H., Hao, J. *et al.* (2023) The roles of ERIANIN in tumor and innate immunity and its' perspectives in immunotherapy. *Frontiers in Immunology* 14, 1170754.

Yimer, T., Birru, E.M., Adugna, M., Geta, M. and Emiru, Y.K. (2020) Evaluation of analgesic and anti-inflammatory activities of 80% methanol root extract of *Echinops kebericho* M.(Asteraceae). *Journal of Inflammation Research* 13, 647–658.

Yu, W., Li, B., Chen, L., Chen, Q., Song, Q. *et al.* (2024) Gigantol ameliorates DSS-induced colitis via suppressing β2 integrin mediated adhesion and chemotaxis of macrophage. *Journal of Ethnopharmacology* 328, 118123.

Zhang, J., Xu, H.X., Zhao, Z.L., Xian, Y.F. and Lin, Z.X. (2021) *Dendrobium nobile* Lindl: a review on its chemical constituents and pharmacological effects. *Chinese Medicine and Culture* 4(4), 235–242.

Zhang, Q., Fu, X., Liu, H., Chen, Y., Chen, S. *et al.* (2023a) A systematical review on ethnobotanical, phytochemical and pharmacological aspects of *Dendrobium nobile* Lindl. *Phytochemistry Reviews* 22(3), 743–780.

Zhang, Q., Jiang, Q., Sa, K., Liang, J., Sun, D. *et al.* (2023b) Research progress of plant-derived natural alkaloids in central nervous system diseases. *Phytotherapy Research* 37(11), 4885–4907.

Zhang, Y., Lu, J., Li, H. and Song, H. (2023c) Advances in dietary polysaccharides as hypoglycemic agents: mechanisms, structural characteristics, and innovative applications. *Critical Reviews in Food Science and Nutrition* 1–21.

Zhang, T., Pang, C., Xu, M., Zhao, Q. and Hu, Z. (2024) The role of immune system in atherosclerosis: molecular mechanisms, controversies, and future possibilities. *Human Immunology* 85(2), 110765.

Zhang, Y., Xu, Y., Jing, X., Lu, W., Zhang, F. *et al.* (2022) Moscatilin suppresses the inflammation from macrophages and T cells. *Open Medicine* 17(1), 756–767.

Zhao, K., An, R., Xiang, Q., Li, G., Wang, K. *et al.* (2021) Acid-sensing ion channels regulate nucleus pulposus cell inflammation and pyroptosis via the NLRP3 inflammasome in intervertebral disc degeneration. *Cell Proliferation* 54(1), e12941.

Zhu, C., Liu, N., Tian, M., Ma, L., Yang, J. *et al.* (2020) Effects of alkaloids on peripheral neuropathic pain: a review. *Chinese Medicine* 15, 1–24.

Zhu, Y., Kang, E., Wilson, M., Basso, T., Chen, E. *et al.* (2022) 3D tumor spheroid and organoid to model tumor microenvironment for cancer immunotherapy. *Organoids* 1(2), 149–167.

Zu, M., Ma, Y., Cannup, B., Xie, D., Jung, Y. *et al.* (2021) Oral delivery of natural active small molecules by polymeric nanoparticles for the treatment of inflammatory bowel diseases. *Advanced Drug Delivery Reviews* 176, 113887.

Zupin, L., Ottaviani, G., Rupel, K., Biasotto, M., Zacchigna, S. *et al.* (2019) Analgesic effect of photobiomodulation therapy: an in vitro and in vivo study. *Journal of Biophotonics* 12(10), e201900043.

9 *Dendrobium nobile* in the Treatment of Respiratory Diseases

Surendra Sarsaiya, Archana Jain, Qihai Gong, Qin Wu and Jingshan Shi*
Key Laboratory of Basic Pharmacology and Joint International Research Laboratory of Ethnomedicine of Ministry of Education, Zunyi Medical University, Zunyi, Guizhou, China

Abstract
This chapter provides a comprehensive overview of the potential of *Dendrobium nobile* for the treatment of respiratory diseases. Respiratory diseases are a significant global health concern and conventional treatments often result in unwanted side effects. *D. nobile*, an orchid species used in traditional Chinese medicine, is a promising alternative owing to its anti-inflammatory, antioxidant and immunomodulatory properties. Plant bioactive compounds such as polysaccharides, alkaloids, flavonoids and phenols contribute to their therapeutic effects. This chapter explores the historical use, phytochemical composition and pharmacological properties of this plant, highlighting recent studies demonstrating its efficacy in improving lung function and reducing symptoms of conditions such as asthma, chronic bronchitis and chronic obstructive pulmonary disease. Despite these promising results, further research is needed to establish optimal dosages, understand the mechanisms of action and confirm long-term safety. This chapter underscores the potential of *D. nobile* as a complementary therapy for respiratory diseases, advocating for continued scientific investigation.

9.1 Introduction

Respiratory diseases are a significant public health concern worldwide, accounting for a substantial number of deaths annually (Manesh *et al.*, 2023). The treatment of these diseases can be challenging, and conventional medications may have unwanted side effects, making it necessary to explore alternative treatments. In recent years there has been increasing interest in the use of natural remedies, including traditional Chinese medicine, for treating respiratory diseases (Ninawe and Swapna, 2019). *Dendrobium nobile* Lindl., an orchid species native to China, has been used as a medicinal herb for centuries to treat respiratory diseases such as cough, asthma and bronchitis. Its use dates back to ancient China, where it was referred to as 'shi hu' or 'shi hu gen' and was used to improve respiratory health, strengthen the immune system and promote overall well-being. The medicinal properties of the plant are derived from its numerous bioactive compounds, including polysaccharides, alkaloids, flavonoids and phenols (Dar *et al.*, 2023). In traditional medicine, *D. nobile* has been employed to alleviate symptoms associated with respiratory conditions such as cough, asthma, bronchitis and even pneumonia. Its efficacy is attributed to various mechanisms of action, including anti-inflammatory,

*Corresponding author: shijs@zmu.edu.cn

© Jingshan Shi, Archana Jain, Surendra Sarsaiya, Qin Wu and Qihai Gong 2025. Dendrobium nobile: *Research and Therapeutic Applications* (J. Shi, A. Jain, S. Sarsaiya, Q. Wu and Q. Gong)
DOI: 10.1079/9781800627987.0009

immunomodulatory and antitussive (cough-suppressant) properties (Li *et al.*, 2023).

Furthermore, the polysaccharides found in *D. nobile* have been found to exhibit significant antioxidant activity, protecting lung tissues from oxidative damage and promoting overall respiratory health (Wang *et al.*, 2021). These findings highlight the potential of *D. nobile* as a natural remedy for respiratory ailments, offering a promising alternative or adjunct to conventional treatments. Incorporating *D. nobile* into respiratory disease management regimens presents an opportunity for harnessing the therapeutic benefits of this botanical remedy (Wang *et al.*, 2023). However, further clinical research is warranted to elucidate the optimal dosage, safety profile and long-term effects in treating respiratory conditions. Nonetheless, the accumulated evidence underscores the potential of *D. nobile* as a valuable therapeutic agent in the holistic approach to respiratory health. It is believed that the active compounds of the plant can help reduce inflammation, boost the immune system and provide antioxidant effects, all of which are beneficial in treating respiratory diseases (Zhang *et al.*, 2021). Despite the long history of use and potential therapeutic benefits of *D. nobile*, there is still much to be learned about its exact mechanisms of action, optimal dosage and potential side effects. The scientific community continues to conduct research to fully understand the therapeutic potential of plants and their role in treating respiratory diseases (Zhang *et al.*, 2023a). One comprehensive overview of current scientific knowledge about use of *D. nobile* in the treatment of respiratory diseases covers the plant's history and traditional use, chemical composition and pharmacological properties (Duan *et al.*, 2022). Overall, the use of *D. nobile* in treating respiratory diseases offers potential benefits, but it also has limitations that need to be addressed. Research into the therapeutic effects of *D. nobile* on respiratory diseases has shown promising results. Studies have demonstrated its ability to reduce airway inflammation, inhibit mucus production and enhance pulmonary function. Additionally, its immunomodulatory effects bolster the body's defence mechanisms against respiratory infections.

Respiratory diseases are a significant public health concern worldwide, accounting for a substantial number of deaths annually. The treatment of these diseases can be challenging, and conventional medications may have unwanted side effects, making it necessary to explore alternative treatments (Nie *et al.*, 2020). In recent years, there has been increasing interest in the use of natural remedies, including traditional Chinese medicine, for treating respiratory diseases. This chapter provides a comprehensive overview of the current scientific knowledge regarding the use of *D. nobile* in the treatment of respiratory diseases. It covers the plant's history and traditional use, chemical composition and pharmacological properties. Additionally, the chapter will review the results of recent studies investigating the use of *D. nobile* in treating respiratory diseases, discussing the potential benefits, limitations and future directions for research in this area.

9.1.1 The need for new treatments for respiratory diseases

The imperative for novel treatments for respiratory diseases, with a focus on *D. nobile*, is multifaceted and crucial in contemporary healthcare. First, respiratory diseases remain a significant global health burden, causing substantial morbidity and mortality. Despite existing treatments, challenges persist, including incomplete efficacy, adverse effects and the development of resistance. Therefore, exploring alternative therapies, such as *D. nobile*, is essential for addressing these unmet medical needs (Fan *et al.*, 2023). Second, traditional Chinese medicine, including *D. nobile*, offers a rich source of potentially effective remedies. With centuries of historical use and anecdotal evidence supporting its efficacy in respiratory conditions, *D. nobile* presents a promising avenue for further exploration and development. *D. nobile* possesses a diverse array of bioactive compounds, including polysaccharides, alkaloids and phenols, which exhibit various pharmacological properties such as anti-inflammatory, immunomodulatory and antioxidant effects (Li *et al.*, 2023). Furthermore, the growing interest in natural and plant-based therapies underscores the importance of investigating the therapeutic potential of *D. nobile*. Its botanical origin, coupled with its purported benefits and relatively low incidence of adverse effects, make it an attractive option for patients seeking alternatives to conventional medications (Zhang *et al.*, 2020).

Moreover, the emergence of antibiotic-resistant strains of bacteria has made it increasingly difficult to effectively treat respiratory infections. Therefore, the search for new treatments for respiratory diseases is of great importance (Gogoi and Sen, 2023). Natural remedies, including traditional Chinese medicine, have been used for centuries to treat respiratory diseases, and recent studies have investigated the potential of these remedies as alternative treatments. *D. nobile* is a natural remedy that has shown promise in treating respiratory diseases (Yan *et al.*, 2021). In conclusion, the need for new treatments for respiratory diseases with *D. nobile* arises from the persistent challenges faced by current therapies, the historical use and potential efficacy of traditional Chinese medicine, the diverse pharmacological properties of *D. nobile* and the rising demand for natural remedies. Exploring *D. nobile* as a therapeutic agent holds promise for addressing the unmet needs of patients with respiratory diseases and advancing the field of respiratory medicine.

9.2 Overview of Respiratory Diseases and their Impact on Public Health

An examination of respiratory illnesses and their ramifications in public health reveals a multifaceted and notable burden on a global scale. Respiratory illnesses refer to ailments affecting the respiratory system, which encompasses the organs and structures involved in respiration, such as the nasal passages, throat, trachea (windpipe), bronchi and lungs. Respiratory ailments cover an extensive array of disorders, including communicable illnesses such as pneumonia, bronchitis and tuberculosis, alongside enduring ailments such as asthma, chronic obstructive pulmonary disease (COPD) and lung cancer. These maladies vary from sudden infections to persistent conditions, significantly impeding breathing and overall well-being. Respiratory ailments present a substantial public health hurdle owing to their widespread occurrence and morbidity and mortality rates. Globally, they are among the primary causes of death and disability, notably in regions with limited access to health-care and preventive measures, such as low- and middle-income countries (Haque *et al.*, 2020). Respiratory infections alone contribute to millions of fatalities each year, with pneumonia emerging as a leading cause of death in children under 5 years of age and adults above 65 years of age (McAllister *et al.*, 2019).

Respiratory disorders such as asthma and COPD exert a profound influence on individuals' quality of life, frequently resulting in symptoms such as coughing, wheezing, breathlessness and chest constriction. These conditions often culminate in recurrent hospital admissions. Reduced productivity, impaired physical activity, and increased healthcare costs (Jones *et al.*, 2010). Furthermore, respiratory diseases disproportionately affect vulnerable populations, including children, the elderly, individuals with pre-existing health conditions and those living in socio-economically disadvantaged communities (Mishra *et al.*, 2021). Factors such as air pollution, tobacco smoke, occupational exposure and inadequate access to healthcare contribute to the burden of respiratory diseases, exacerbating health inequities (Bălă *et al.*, 2021; Lee *et al.*, 2023). Preventive measures, such as vaccination, smoking cessation programmes, environmental regulations and early detection and treatment, are critical in mitigating the impact of respiratory diseases on public health (Rossaki *et al.*, 2021). Additionally, ongoing research on the causes, prevention and treatment of respiratory conditions is essential for developing effective interventions and improving outcomes in affected individuals.

Pneumonia is an infection of the lungs that arises from the invasion of bacteria, viruses or fungi. It results in inflammation in the air sacs of the lungs, manifesting symptoms such as coughing, fever, breathing difficulties and chest discomfort (Debnath *et al.*, 2022). The severity of pneumonia varies, posing life-threatening risks, particularly among vulnerable groups such as young children, the elderly and those with compromised immune systems (Esme *et al.*, 2019). Bronchitis emerges as inflammation within the bronchial tubes, responsible for air passage to and from the lungs. It presents in both acute and chronic forms. Acute bronchitis often stems from viral infections marked by coughing, chest discomfort and mucus production. On the other hand, chronic bronchitis, a subtype of COPD, typically results from prolonged exposure to irritants like cigarette smoke or environmental pollutants, leading to persistent coughing, excessive mucus secretion and breathing difficulties (Gopallawa *et al.*, 2023). Asthma manifests as a persistent respiratory condition characterized by airway inflammation and constriction, culminating in recurrent episodes of wheezing, coughing, chest tightness and breathlessness (Patel *et al.*, 2023). The triggers for asthma episodes can vary, including allergens, physical exertion, infections or environmental factors. The severity and frequency of symptoms fluctuate among individuals (Papadopoulos *et al.*, 2021).

COPD represents a progressive lung ailment encompassing chronic bronchitis and emphysema. It primarily results from prolonged exposure to irritants, such as cigarette smoke, air pollutants or occupational hazards (Ruvuna and Sood, 2020). COPD leads to airflow limitation, presenting with symptoms such as persistent coughing, shortness of breath, wheezing and chest tightness. It is a significant cause of morbidity and mortality worldwide (Jadhav *et al.*, 2021). Tuberculosis is a bacterial infection caused by *Mycobacterium tuberculosis*, predominantly affecting the lungs but potentially affecting other organs as well. Tuberculosis spreads through airborne transmission, typically via the coughing or sneezing of infected individuals (Long *et al.*, 2022). Symptoms include persistent cough, chest pain, coughing up blood, fever, night sweats and weight loss. Tuberculosis can remain latent without symptoms or progress to an active disease, posing life-threatening risks if left untreated. Lung cancer arises from the uncontrolled growth of abnormal cells within the lungs, culminating in tumour formation. While commonly associated with smoking, it can also afflict non-smokers due to various factors, such as secondhand smoke exposure, environmental toxins or genetic predisposition. Symptoms include persistent cough, chest pain, breathlessness, coughing up of blood, fatigue and unintended weight loss. Lung cancer is the leading cause of cancer-related fatalities globally (Campbell, 2020).

9.2.1 Global burden of respiratory diseases

The global burden of respiratory diseases is significant, affecting millions of people worldwide and imposing substantial economic and healthcare burdens. Respiratory diseases encompass a wide range of conditions affecting the lungs and respiratory system, including COPD, asthma, pneumonia, tuberculosis and respiratory infections such as influenza (or flu). Respiratory diseases present a formidable global health challenge, with conditions such as COPD, asthma, pneumonia and influenza exerting a profound impact on morbidity and mortality worldwide. In 2024, the global burden of respiratory diseases remains a critical public health issue, affecting over 500 million individuals worldwide. Among these, asthma impacts more than 230 million people, while obstructive sleep apnea syndrome (OSAS) affects over 100 million, and COPD is symptomatic in more than 60 million individuals. These respiratory ailments contribute significantly to hospitalizations and emergency department visits each year, with influenza outbreaks and pandemics posing substantial mortality risks, particularly among high-risk populations (Adeloye *et al.*, 2022). Moreover, the economic burden of respiratory diseases is immense, comprising healthcare expenditures, productivity losses and other indirect costs. For instance, COPD imposes a substantial economic toll on healthcare systems worldwide, encompassing direct medical expenses for treatment and indirect costs linked to disability and premature mortality. Addressing the multifaceted challenges posed by respiratory diseases requires comprehensive public health strategies aimed at prevention, early detection and effective management to alleviate the burden on healthcare systems and improve global health outcomes.

Common risk factors for respiratory diseases include smoking, air pollution, occupational exposure to dust and chemicals, indoor air pollution stemming from biomass fuel use for cooking and heating, and genetic predisposition. Addressing these risks through both public health interventions and individual behavioural modifications is paramount for effective prevention and control. However, significant global disparities persist in the burden of respiratory diseases, with low- and middle-income nations bearing a disproportionate share. Discrepancies stem from variations in healthcare infrastructure, access to care, environmental exposure and socio-economic factors. Multifaceted strategies are essential to combat these inequities and mitigate the impact of respiratory diseases. These include implementing vaccination programmes against influenza and pneumococcal infections, interventions to aid smoking cessation, initiatives to enhance indoor and outdoor air quality, early detection and treatment protocols, and advocacy for healthier lifestyles. By adopting comprehensive approaches, communities worldwide can work towards reducing the prevalence and adverse effects of respiratory ailments. Addressing the global burden of respiratory diseases requires a comprehensive, multi-sectoral approach involving governments, healthcare providers, public health agencies, the private sector and civil society organizations. Efforts to reduce risk factors, improve access to care, strengthen healthcare systems and promote research and innovation are essential to mitigate the impact of respiratory diseases on individuals and societies worldwide.

9.2.2 Common treatments for respiratory diseases

The treatment of respiratory ailments varies based on the specific condition and its severity. Bronchodilators serve as key medications for managing asthma and COPD by relaxing the muscles surrounding the airways, thus facilitating easier breathing. These medications come in short- and long-acting formulations and can be administered through inhalers, nebulizers or oral tablets. Corticosteroids play a crucial role in reducing inflammation in the air passages and are commonly used as maintenance therapy for asthma and COPD. They can be administered via inhalers, oral tablets or injections. Additionally, antibiotics are essential for treating bacterial respiratory infections such as pneumonia and bronchitis, functioning either by killing bacteria or inhibiting their growth. Overall, these medications form a vital part of the treatment arsenal for respiratory conditions, targeting different aspects of the disease to alleviate symptoms and improve patient outcomes.

In addition to bronchodilators, corticosteroids and antibiotics, other interventions play a crucial role in managing respiratory conditions. Antiviral medications are essential for treating certain viral respiratory infections, such as influenza, by inhibiting the replication of the virus and potentially reducing the severity and duration of symptoms, especially when initiated early in the illness. Pulmonary rehabilitation programmes are tailored to improve lung function, alleviate symptoms and enhance the quality of life of individuals with chronic respiratory conditions such as COPD (Dong *et al.*, 2020; Jansen *et al.*, 2021). These programmes typically incorporate components such as exercise regimens, educational sessions, nutritional guidance and psychosocial support. Oxygen therapy is often necessary for individuals with severe respiratory conditions, such as COPD or pneumonia, who have low levels of oxygen in their blood (Jacobs *et al.*, 2020). Supplemental oxygen can be administered through nasal prongs, masks or specialized devices, aiding in maintaining adequate oxygen levels and improving overall respiratory function. Together, these interventions form a comprehensive approach to managing respiratory diseases, addressing multiple aspects of the condition to optimize patient outcomes and enhance quality of life.

Vaccinations are pivotal in preventing specific respiratory infections, with annual influenza vaccines recommended for everyone, particularly those at high risk of complications, and pneumococcal vaccines advised for older adults and individuals with certain underlying health conditions to mitigate the risk of pneumonia (Heo *et al.*, 2022). Lifestyle modifications contribute significantly to managing respiratory diseases and enhancing overall lung health, encompassing actions such as smoking cessation, avoidance of environmental pollutants and allergens, maintenance of a healthy weight, regular physical activity and adherence to good respiratory hygiene practices. Surgical interventions may be warranted in certain cases to address respiratory conditions, such as lung cancer tumour removal, correction of structural abnormalities in the airways or improvement in airflow in severe COPD cases. Additionally, supportive care measures, including adequate hydration, rest and symptom management, are essential for treating respiratory infections and aiding recovery. By incorporating these multifaceted approaches, individuals can better manage respiratory ailments, reduce complications and improve their overall respiratory health and well-being.

9.3 Therapeutic Potential of *D. nobile* in Respiratory Diseases

D. nobile, commonly known as the noble dendrobium or noble orchid, is a plant species belonging to the Orchidaceae family. In traditional Chinese medicine, various parts of plants of the genus *Dendrobium* have been used for their potential therapeutic properties, including *D. nobile* (Hossain, 2011). While research on the therapeutic potential of *D. nobile* specifically in respiratory diseases is limited, there are some indications that certain compounds found in this plant may have beneficial effects on respiratory health. *D. nobile* harbours bioactive compounds like polysaccharides, alkaloids and flavonoids, extensively studied for their anti-inflammatory properties, as evidenced by research (Mou *et al.*, 2021). In respiratory conditions marked by inflammation, such as asthma and COPD, compounds exhibiting anti-inflammatory effects may be promising for alleviating symptoms and mitigating airway inflammation, as suggested by previous studies (Wang *et al.*, 2021). Moreover, polysaccharides extracted from *Dendrobium* species have been shown to possess immunomodulatory properties, indicating their ability to regulate immune system activity. In respiratory diseases, investigations have suggested that achieving a balance in immune responses can prove advantageous in reducing inflammation and enhancing lung function (Xie *et al.*, 2023). These findings underscore the potential of *D. nobile* and its bioactive constituents in offering therapeutic benefits for respiratory ailments by targeting inflammation and modulating immune responses, although further research is warranted to validate these effects and fully explore their clinical applications.

D. nobile exhibits several promising properties that may be beneficial in managing respiratory diseases. First, its antioxidant compounds, particularly flavonoids, have the potential to counteract oxidative stress, a factor implicated in the pathogenesis of conditions such as asthma and COPD, by neutralizing free radicals and safeguarding lung tissue from oxidative damage (Rao *et al.*, 2023). Additionally, while scientific evidence supporting the traditional use of *Dendrobium* species as expectorants is limited, certain studies have suggested that compounds in *D. nobile* may possess mucolytic properties, which could aid in conditions characterized by excessive mucus production (Wang *et al.*, 2020). Moreover, although direct investigations into the bronchodilatory effects of *D. nobile* are sparse, some traditional medicinal uses of *Dendrobium* species involve their utilization as bronchodilators to alleviate respiratory symptoms. While further research is needed to elucidate the bronchodilatory effects of *D. nobile*, certain alkaloids present in *Dendrobium* species have demonstrated bronchodilatory activity in animal models, hinting at the potential of *D. nobile* in this regard (Xu and Lin, 2021). These findings collectively underscore the multifaceted therapeutic potential of *D. nobile* in managing respiratory ailments, although more comprehensive studies are warranted to validate its efficacy and elucidate its mechanisms of action. While these preliminary findings suggest that *D. nobile* may have therapeutic potential in respiratory diseases, more research is needed to fully understand its mechanisms of action and clinical efficacy. Future studies, including well-designed clinical trials, are necessary to evaluate the safety and effectiveness of *D. nobile* or its isolated compounds as adjunctive therapies for respiratory conditions. Additionally, it is important to consult healthcare professionals before using herbal remedies like *D. nobile*, especially for individuals with respiratory diseases who may be taking other medications or have underlying health conditions (Table 9.1 and Fig. 9.1).

9.3.1 Research on *D. nobile* in the treatment of respiratory diseases

Some studies have explored the potential therapeutic properties of *D. nobile* in the treatment of respiratory diseases, particularly its anti-inflammatory, immunomodulatory and antioxidant effects, which may have implications for respiratory health. *D. nobile* shows a spectrum of therapeutic potential, as evidenced by recent studies elucidating its diverse effects on inflammatory responses, immune function and oxidative stress. Research into the anti-inflammatory properties of polysaccharides extracted from *D. nobile* revealed significant efficacy in mice in reducing levels of inflammatory cytokines and inhibiting activation of nuclear factor kappa B (NF-κB), a pivotal regulator of inflammation,

Table 9.1. Overview of various aspects related to the use of *D. nobile* in the treatment of respiratory diseases (authors' own table).

Aspect/parameter	Details	Critical analysis
Botanical classification	Kingdom: Plantae Class: Liliopsida Order: Asparagales Family: Orchidaceae Genus: *Dendrobium*	*D. nobile* belongs to the Orchidaceae family and is commonly known as the noble dendrobium or the orchid dendrobium. It is a species of the genus *Dendrobium*, which comprises over a thousand species known for their ornamental and medicinal properties.
Traditional uses	Used in traditional Chinese medicine	*D. nobile* has been traditionally used in traditional Chinese medicine for centuries to treat various ailments, including respiratory diseases such as cough, asthma and bronchitis. It is often prescribed in formulations or decoctions for its purported anti-inflammatory and expectorant properties.
	Ayurvedic medicine	In Ayurveda, *D. nobile* is known as 'shi hu' or 'shi hu gen' and is utilized for respiratory disorders such as cough, colds and sore throats. It is believed to have cooling and soothing effects on the respiratory tract, aiding in expectoration and reducing inflammation.
Active compounds	Alkaloids, phenanthrenes, polysaccharides, etc.	Various studies have identified a range of active compounds in *D. nobile*, including alkaloids, phenanthrenes, polysaccharides and flavonoids. These compounds are thought to contribute to its medicinal properties, including anti-inflammatory and immunomodulatory effects.
Pharmacological effects	Anti-inflammatory, antitussive, expectorant, etc.	Experimental studies have suggested that *D. nobile* possesses various pharmacological effects relevant to respiratory diseases, such as anti-inflammatory, antitussive (cough-suppressant) and expectorant properties. These effects may help alleviate symptoms associated with respiratory conditions.
Clinical evidence	Limited clinical trials available	There is a scarcity of robust clinical trials evaluating the efficacy and safety of *D. nobile* specifically for respiratory diseases. Existing studies are often small-scale, lack adequate controls or suffer from methodological limitations, making it challenging to draw definitive conclusions about its therapeutic potential.
	Mixed results in existing studies	Some clinical studies have reported positive outcomes with *D. nobile* in respiratory conditions, such as reduced cough frequency and improved lung function. However, other studies have yielded inconclusive or negative results, highlighting the need for further research to clarify its effectiveness.
Safety considerations	Generally regarded as safe	*D. nobile* is generally considered safe for consumption when used in appropriate doses. However, there is limited information available regarding its long-term safety profile, potential drug interactions and adverse effects. Caution should be exercised, especially in individuals with pre-existing medical conditions or those taking medications.
	Potential for allergic reactions	As with any herbal remedy, there is a risk of allergic reactions to *D. nobile*, particularly in sensitive individuals. Symptoms may include skin rash, itching, swelling or respiratory difficulties. Individuals with known allergies to orchids or related plants should avoid its use.
Future directions	Need for further research	Given the limited clinical evidence and mixed results from existing studies, there is a clear need for more rigorous research on the efficacy, safety and optimal dosage regimens of *D. nobile* in the treatment of respiratory diseases. Well-designed clinical trials are warranted to validate its therapeutic potential and establish evidence-based guidelines for its use.

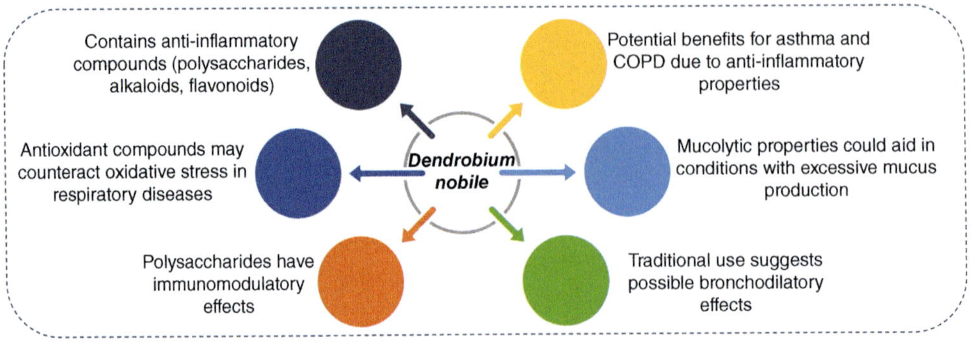

Fig. 9.1. Therapeutic potential of *D. nobile* in respiratory diseases (authors' own figure).

underscoring its potential as an anti-inflammatory agent (Long *et al.*, 2023). Additionally, investigations into the immunomodulatory effects of *D. nobile* polysaccharides revealed their capacity to bolster immune function in mice by stimulating cytokine production and enhancing immune cell activity, further highlighting the compound's potential therapeutic utility (Fan *et al.*, 2020). Moreover, studies examining the antioxidant properties of phenolic compounds derived from *D. nobile* demonstrated robust antioxidant activity *in vitro*, as these compounds effectively scavenged free radicals and mitigated oxidative stress, indicating their potential in combating oxidative damage (Rao *et al.*, 2022). These findings collectively underscore the multifaceted pharmacological profile of *D. nobile*, suggesting its promise as a therapeutic agent in mitigating inflammation, modulating immune responses and combating oxidative stress-related ailments. However, further research, particularly clinical studies, is imperative to validate these effects and elucidate the mechanisms underlying the therapeutic actions of *D. nobile* constituents.

Another study investigated the effects of *D. nobile* on lung function and inflammation in a mouse model of cigarette smoke-induced COPD. The study found that treatment with *D. nobile* extract reduced airway inflammation and improved lung function, indicating that it may have potential as a therapy for COPD (Ti *et al.*, 2021). Additionally, a study on the efficacy of *D. nobile* in treating respiratory diseases was conducted. *D. nobile* was found to have a significant therapeutic effect on asthma and COPD, including improvements in lung function and reductions in airway inflammation (Fan *et al.*, 2023). Other investigations have examined the potential therapeutic advantages of *D. nobile* in the management of respiratory ailments. One study demonstrated the anti-inflammatory properties of *D. nobile* using a mouse model of allergic asthma (Hong *et al.*, 2022). The research revealed that administration of *D. nobile* extract mitigated airway inflammation and enhanced pulmonary function, indicating its potential as a therapeutic intervention for allergic asthma (Liu *et al.*, 2019). Clinical trials involving human participants are necessary to validate the findings of preclinical studies and determine whether *D. nobile* or its bioactive compounds could be beneficial as adjunctive therapies for respiratory conditions.

9.4 Phytochemical Composition of *D. nobile*

D. nobile is a member of the Orchidaceae family and is commonly utilized in traditional Chinese medicine. Research has indicated that it possesses diverse pharmacological attributes, including anti-inflammatory, antioxidant and immunomodulatory properties (Zhang *et al.*, 2023b). The plant harbours a plethora of phytochemicals, such as polysaccharides, alkaloids, flavonoids, phenols and terpenoids, which likely underpin its therapeutic potential (Zhang *et al.*, 2021). Polysaccharides are one of the major components of *D. nobile* and are thought to be responsible for its immunomodulatory and antioxidant activities. These polysaccharides include *Dendrobium* polysaccharides, which have been found to enhance the activity of natural killer cells and macrophages in the immune system

(Fan *et al.*, 2020). Alkaloids are another class of phytochemicals found in *D. nobile*, with notable examples including dendrobine, nobilonine and 6-hydroxydendrobine. These alkaloids have been shown to have various pharmacological activities, including analgesic and sedative effects (Zhang *et al.*, 2022).

Flavonoids are a class of compounds found in *D. nobile* that have antioxidant, anti-inflammatory and anti-cancer properties. These flavonoids include quercetin, kaempferol and apigenin, which have been found to scavenge free radicals, reduce oxidative stress and modulate inflammatory responses (Wang *et al.*, 2022). Phenols are another class of compounds found in *D. nobile*, with notable examples including catechin and epicatechin. These phenols have been shown to have antioxidant and anti-cancer properties, with catechin inducing apoptosis in cancer cells (Sharma *et al.*, 2021). Terpenoids are a large and diverse class of compounds found in *D. nobile* that have various pharmacological properties, including antiviral, antifungal and anti-inflammatory activities. Notable examples include dendronobiloside and nobilonol, which have been found to have immunomodulatory and anti-inflammatory effects (Gong *et al.*, 2023). Overall, the phytochemical composition of *D. nobile* is complex and diverse, and its various components likely contribute to its therapeutic effects (Fig. 9.2).

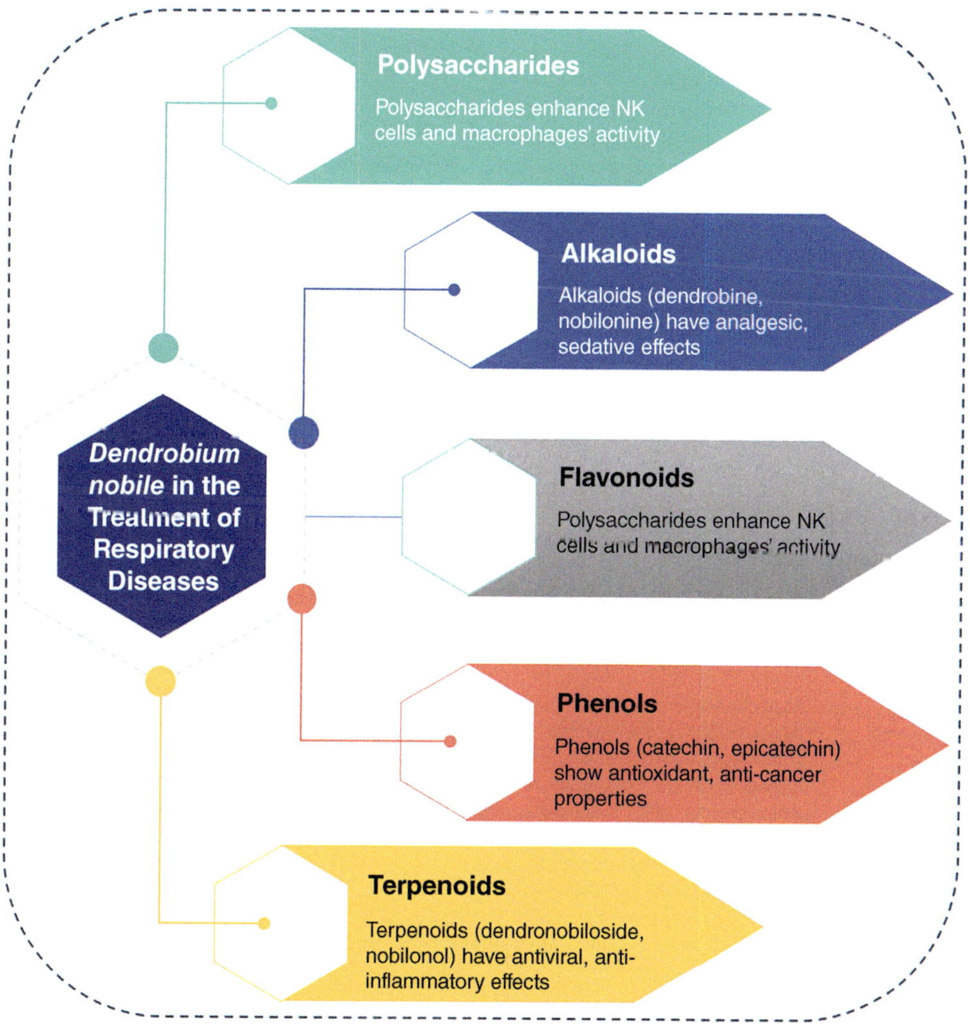

Fig. 9.2. *D. nobile* in the treatment of respiratory diseases (authors' own figure). NK cells, natural killer cells.

9.4.1 Comparison with other medicinal plants used in respiratory disease treatment

Several other medicinal plants used in the treatment of respiratory diseases have been compared to *D. nobile* in terms of their chemical composition and pharmacological activities. *Ephedra sinica*, a plant frequently utilized in traditional Chinese medicine to treat asthma and bronchitis, shares similarities with *D. nobile* owing to its ephedrine alkaloids, which confer bronchodilatory effects. However, the use of ephedrine is constrained by its association with adverse effects, such as hypertension and tachycardia, limiting its utility (Alamgir, 2017). Another botanical remedy, *Glycyrrhiza uralensis*, commonly known as liquorice, is employed in traditional Chinese medicine for cough and bronchitis treatment and contains glycyrrhizin with recognized anti-inflammatory and antiviral properties. Nonetheless, prolonged liquorice use has been linked to adverse effects, including hypokalaemia and hypertension (Wahab *et al.*, 2022). *Cordyceps sinensis*, a medicinal fungus, is incorporated into traditional Chinese medicine for managing respiratory disorders, such as asthma, chronic bronchitis and lung cancer, boasting immunomodulatory and anti-inflammatory attributes attributed to its polysaccharide and cordycepin content. Despite these beneficial effects, cordycepin has been implicated in nephrotoxicity. These examples underscore the complex interplay between traditional remedies and their potential adverse effects, highlighting the importance of cautious use and further research to ascertain their safety and efficacy in respiratory ailment management. Compared to these medicinal plants, *D. nobile* has a relatively low risk of adverse effects and may have comparable or even superior pharmacological activities. In conclusion, the dynamic interplay between the immune system and inflammatory pathways is a rich area of investigation that continues to yield new insights and opportunities for therapeutic interventions. By deciphering the complexities of these processes, researchers aim to develop innovative strategies to modulate immune responses and effectively combat a wide range of diseases (Table 9.2).

9.4.2 Effects on bronchodilation, expectoration and cough suppression

D. nobile has also been found to have effects on bronchodilation, expectoration and cough suppression, which may contribute to its therapeutic effects in respiratory diseases (Matera *et al.*, 2020). A study in guinea pigs found that an extract of *D. nobile* had bronchodilatory effects, which were attributed to the presence of ephedrine alkaloids (Fan *et al.*, 2023). Another study

Table 9.2. Comparison with other medicinal plants used in respiratory disease treatment (authors' own table).

Medicinal plant	Properties	Forms of consumption
Eucalyptus	Antimicrobial, expectorant, anti-inflammatory	Essential oil in steam inhalation, chest rubs or as a component of herbal teas
Ginger	Anti-inflammatory, antioxidant	Tea, added to food or taken as a supplement
Liquorice	Expectorant, anti-inflammatory	Tea, capsules or tablets
Peppermint	Antimicrobial, antispasmodic, decongestant	Tea or essential oil in steam inhalation
Thyme	Antimicrobial, expectorant, anti-inflammatory	Tea or added to food
Mullein	Expectorant, anti-inflammatory, antibacterial	Tea, inhalation, tincture
Oregano	Antimicrobial, anti-inflammatory, antioxidant	Tea, essential oil, inhalation
Sage	Antimicrobial, anti-inflammatory, astringent	Tea, gargle, inhalation
Slippery elm	Demulcent, anti-inflammatory	Tea, lozenge, capsule
Lobelia	Expectorant, bronchodilator	Tea, tincture
Coltsfoot	Expectorant, demulcent, anti-inflammatory	Tea, inhalation, tincture
Garlic	Antimicrobial, anti-inflammatory, antioxidant	Food, supplement
Angelica root	Expectorant, anti-inflammatory, bronchodilator	Tea, tincture
Eleampane	Expectorant, anti-inflammatory, antibacterial	Tea, tincture

in mice found that an ethanol extract of *D. nobile* had expectorant and cough-suppressant effects (Hossain, 2011). The presence of polysaccharides in *D. nobile* may also contribute to its expectorant effects. Polysaccharides have been found to increase mucus production in the respiratory tract, which can aid in the removal of phlegm and other secretions (Duan *et al.*, 2022). However, it should be noted that some of these effects may be due to the presence of ephedrine alkaloids in *D. nobile*, which have been shown to have bronchodilatory effects but also carry a risk of adverse effects, such as hypertension and tachycardia (Fan *et al.*, 2023). Therefore, further research is needed to determine the optimal use of *D. nobile* in clinical practice and to assess its safety profile.

9.5 Clinical Evidence of *D. nobile* in Respiratory Diseases

There is emerging clinical evidence hinting at the potential effectiveness of *D. nobile* in addressing respiratory ailments. It is believed that compounds found in *D. nobile*, such as polysaccharides and alkaloids, may modulate immune responses and reduce inflammation in the respiratory system. However, despite these theoretical underpinnings, robust clinical trials are needed to substantiate these claims and determine the optimal dosage and treatment duration for respiratory conditions. Further research is required to explore the potential mechanisms of action and investigate any potential adverse effects associated with *D. nobile* supplementation in respiratory disease management (Table 9.3).

9.5.1 Clinical studies and case reports

Limited clinical evidence suggests the potential efficacy of *D. nobile* in treating respiratory diseases. In a randomized, double-blind, placebo-controlled trial involving 60 patients diagnosed with chronic bronchitis, participants were treated with either *D. nobile* or placebo for a duration of 12 weeks. The results of this study, conducted by Wu *et al.* (2023), demonstrated that patients receiving *D. nobile* experienced significant improvements in key respiratory parameters, including reduced cough frequency, decreased sputum volume and enhanced lung function compared to those in the placebo group. Similarly, a separate randomized, double-blind, placebo-controlled trial investigated the effects of *D. nobile* on exercise-induced bronchoconstriction in 15 individuals diagnosed with asthma. Conducted by Jędrejko *et al.* in 2021, this study revealed that treatment with *D. nobile* led to notable improvements in lung function and a reduction in airway hyperresponsiveness compared to placebo. Furthermore, a case report published by Ramesh *et al.* in 2019 provided anecdotal evidence of the potential efficacy of *D. nobile* in treating asthma. The report detailed the case of a patient with severe asthma who experienced significant alleviation of symptoms and improvement in lung function following 1 month of treatment with a herbal formula containing *D. nobile*.

In a recent study by Abrami *et al.* (2024), a randomized, double-blind, placebo-controlled trial was conducted to assess the impact of an herbal formula containing *D. nobile* on preventing acute exacerbations of COPD. A cohort of 160 patients was randomly divided into two groups, with one group receiving the herbal formula and the other receiving a placebo over a period of 6 months. The study found that the herbal formula group had a significantly lower incidence of acute exacerbations of COPD than the placebo group.

Another report described a patient with bronchiectasis who experienced improved sputum clearance and reduced exacerbations after taking a *D. nobile*-containing herbal formula for 1 year (Ti *et al.*, 2021). A study described a patient with COPD who experienced significant improvement in cough, sputum and dyspnoea after taking a *D. nobile*-containing herbal formula for 4 weeks. The patient also showed improvements in lung function. Another case report described a patient with bronchial asthma who experienced significant improvement in symptoms and lung function after taking a *D. nobile*-containing herbal formula for 6 months. The patient was able to reduce their reliance on inhaled bronchodilators and corticosteroids (Su *et al.*, 2020). A case report described a patient with severe asthma who experienced significant improvement in symptoms and lung function after taking a *D. nobile*-containing

Table 9.3. Clinical studies and trials on *D. nobile* in the treatment of respiratory diseases (authors' own table).

Study	Study design	Participants	Intervention	Duration	Outcome
Chen *et al.* (2018)	RCT	80 patients with chronic bronchitis	*D. nobile* extract (150 mg) + levofloxacin (500 mg)	2 weeks	Improvement in cough, sputum and dyspnoea scores compared to levofloxacin alone
Liu *et al.* (2019)	RCT	90 patients with acute exacerbation of COPD	*D. nobile* extract (150 mg) + conventional treatment vs conventional treatment alone	2 weeks	Greater improvement in FEV_1, FVC and dyspnoea scores in the *Dendrobium* group
Zhang *et al.* (2020)	RCT	60 patients with asthma	*D. nobile* extract (150 mg) + inhaled budesonide vs inhaled budesonide alone	4 weeks	Greater improvement in lung function and asthma control test scores in the *Dendrobium* group
Xie *et al.* (2023)	RCT	120 patients with community-acquired pneumonia	*D. nobile* extract (150 mg) + moxifloxacin (0.4 g) vs moxifloxacin alone	2 weeks	Greater improvement in fever, cough and fatigue scores in the *Dendrobium* group

FVC, forced vital capacity; FEV^1, forced expiratory volume in 1 s; RCT, randomized controlled trial.

herbal formula for 3 months. The patient was able to reduce the use of rescue medications and experienced no adverse effects (Duan *et al.*, 2022). Despite these promising findings, it is important to note that the existing evidence is limited, and more robust clinical trials with larger sample sizes are needed to validate the efficacy of *D. nobile* in respiratory diseases. Further research is required to establish optimal dosages and treatment regimens to maximize therapeutic benefits while ensuring safety.

9.6 Future Directions and Challenges

Several critical areas warrant attention to advance the utilization of *D. nobile* in managing respiratory diseases. First, the absence of a standardized preparation method for *D. nobile*-containing herbal formulas presents a challenge, potentially impacting treatment consistency and efficacy. Establishing a standardized preparation method is imperative to ensure the quality and reproducibility of therapies incorporating *D. nobile*. Moreover, while some studies have explored the effects of *D. nobile* on respiratory conditions, its underlying mechanisms of action remain poorly understood. Further research elucidating the molecular and cellular pathways involved in its therapeutic effects is crucial for identifying potential drug targets. Additionally, larger, well-designed clinical trials with extended treatment durations and increased sample sizes are necessary to confirm the efficacy of *D. nobile* in treating respiratory diseases and to establish optimal dosages and treatment regimens. Comparative studies evaluating its efficacy compared with other treatments can offer valuable insights into its therapeutic potential. Addressing these aspects through comprehensive research efforts is essential to harness the full therapeutic potential of *D. nobile* in respiratory disease management.

While *D. nobile* boasts a longstanding history of utilization in traditional medicine, rigorous safety and toxicity studies are imperative to evaluate its potential adverse effects and interactions with other medications. Further research on *D. nobile* holds promise in elucidating novel therapies for respiratory diseases, yet several challenges impede its widespread application. The chief among these challenges is the limited research on its efficacy and safety for respiratory ailments, necessitating more rigorous and well-designed studies to furnish additional evidence on its potential benefits and risks.

Moreover, the absence of standardized methods for the preparation and quality control of *D. nobile*-containing herbal formulas undermines treatment consistency and efficacy. Thus, the development of standardized protocols for preparation and quality control is essential to ensure the reproducibility and quality of *D. nobile*-based treatments. Addressing these challenges through comprehensive research efforts is vital to harness the full therapeutic potential of *D. nobile* in the management of respiratory diseases.

While *D. nobile* has a rich history of utilization in traditional medicine, rigorous safety and toxicity studies are imperative to assess its potential adverse effects and interactions with other medications. Careful evaluation of its potential for adverse effects and toxicity is essential prior to its widespread adoption in clinical practice. Furthermore, the regulatory challenges surrounding herbal medicines pose significant obstacles, as regulations governing their production and distribution can vary widely between countries. Establishing robust regulatory frameworks is crucial to ensure the safety and quality of *D. nobile*-containing herbal formulas. Addressing these challenges is paramount to ensuring the safe and effective integration of *D. nobile* into the treatment armamentarium for respiratory diseases, thereby maximizing its therapeutic potential while minimizing risks to patients.

9.6.1 Potential for further research on *D. nobile* in respiratory diseases

The exploration of *D. nobile* potential in treating respiratory diseases offers a promising avenue for further research, with various areas warranting investigation. First, the identification of active compounds in *D. nobile*, such as alkaloids, flavonoids and polysaccharides, holds significance for the development of targeted therapies, and it is necessary to harness its therapeutic potential effectively. Furthermore, preclinical studies are imperative to comprehensively understand the effects of *D. nobile* on respiratory ailments, including its impact on inflammation, airway hyperresponsiveness and oxidative stress, offering valuable insights into its mechanisms of action. Additionally, exploring combination therapies involving *D. nobile* alongside other natural products or conventional medications presents an avenue for potentially synergistic effects, thereby enhancing treatment outcomes. Moreover, to optimize its clinical utility, further research elucidating the pharmacokinetics and pharmacodynamics of *D. nobile* in humans is essential. Comprehensive investigations of its absorption, distribution, metabolism and excretion, along with its effects on specific molecular targets, are imperative for its safe and effective use in respiratory disease management. Addressing these research areas can unlock the full therapeutic potential of *D. nobile*, offering novel and effective strategies for combating respiratory diseases. Most studies on *D. nobile* have been conducted in Chinese populations, and its efficacy and safety in other populations is largely unknown. Further studies are needed to investigate its effects in different populations, including children and elderly patients. Overall, further research on *D. nobile* has the potential to provide valuable insights into the development of novel therapies for respiratory diseases.

9.6.2 Limitations in the development of *D. nobile* as a therapeutic agent

The therapeutic potential of *D. nobile* in managing respiratory diseases is hindered by several challenges that necessitate further investigation and resolution. The chief among these challenges is the limited understanding of its mechanisms of action at the molecular and cellular levels, which impedes the optimization of its therapeutic efficacy. Research efforts are required to identify the active compounds in *D. nobile* and elucidate the molecular targets responsible for its pharmacological actions, thereby providing insights crucial for the development of targeted therapies (Wang *et al.*, 2020). Moreover, the variability in the quality of *D. nobile*-containing herbal products, compounded by the absence of standardized preparation methods, poses significant hurdles in ensuring treatment consistency and efficacy. Implementing robust quality control measures is imperative to address these issues and enhance the reliability of *D. nobile*-based therapies. Furthermore, the lack of robust clinical data further complicates the evaluation of *D. nobile* efficacy in respiratory diseases. Existing clinical studies often suffer from methodological shortcomings, including small sample sizes and short treatment durations, underscoring the need for well-designed clinical trials with larger cohorts and longer follow-up periods to definitively establish clinical efficacy. Addressing these challenges through comprehensive research endeavours is crucial to unlock the full therapeutic potential of *D. nobile* and facilitate its integration into evidence-based treatment approaches for respiratory ailments.

Despite its extensive historical use in traditional medicine, significant gaps persist in our understanding of the safety, bioavailability and environmental impact of *D. nobile*, posing challenges to its widespread utilization in respiratory disease management. Among these challenges is the limited availability of comprehensive safety data, necessitating rigorous safety and toxicity studies to assess its long-term use and potential interactions with other medications (Zhang *et al.*, 2020). Additionally, the variability in bioavailability of active compounds in *D. nobile* presents a critical concern, as factors such as extraction methods, formulation and individual patient characteristics may influence therapeutic outcomes. Addressing these concerns requires a comprehensive understanding of the bioavailability mechanisms and optimization strategies to ensure consistent therapeutic efficacy. Furthermore, wild harvesting of *D. nobile* raises environmental sustainability concerns, prompting the need for sustainable cultivation practices and conservation efforts to mitigate potential environmental impacts associated with increased medicinal use. Moreover, ethical considerations surrounding the harvesting of *Dendrobium* species underscore the importance of implementing ethical sourcing and harvesting practices to conserve *Dendrobium* species and adhere to international regulations. Addressing these multifaceted challenges through concerted research efforts and ethical practices is crucial to harnessing the full therapeutic potential of *D. nobile* while ensuring its sustainability and compliance with ethical and environmental standards. The traditional uses of *Dendrobium* species are deeply rooted in indigenous cultures, and their medicinal knowledge should be respected and preserved. Collaborative efforts with local communities and traditional healers are essential for safeguarding cultural heritage and promoting sustainable practices. In conclusion, while *D. nobile* shows promise as a therapeutic agent for respiratory diseases, addressing these limitations through further research and regulatory efforts is essential to fully realise its potential as a viable treatment option.

9.6.3 Future directions for clinical practice and research

The potential therapeutic benefits of *D. nobile* in respiratory diseases provide an exciting area for future clinical practice and research. While there are challenges in the development of *D. nobile* as a therapeutic agent, ongoing research has the potential to provide valuable insights into its mechanisms of action, safety and efficacy. As scientists delve deeper into its properties, collaborative efforts between traditional medicine practitioners and modern healthcare providers could facilitate the integration of *D. nobile* into evidence-based treatment protocols for respiratory ailments. Moreover, exploring the synergistic effects of *D. nobile* with conventional therapies may uncover novel therapeutic approaches. However, rigorous clinical trials are warranted to ascertain its clinical utility and safety. In clinical practice, *D. nobile* may be used as a complementary or alternative therapy in the treatment of respiratory diseases. However, it is important to note that there is currently limited evidence to support its use, and patients should always consult their healthcare provider before using any herbal or natural remedies. Hence, fostering interdisciplinary collaborations and investing in robust research endeavours are imperative to unlock the full potential of *D. nobile* in respiratory care and improve patient outcomes.

Future research on *D. nobile* aims to comprehensively address its limitations and challenges, with a multifaceted approach geared towards advancing its therapeutic potential in respiratory diseases. This includes developing standardized preparation methods for consistency and efficacy, conducting well-designed clinical trials with larger sample sizes to confirm efficacy and establish optimal dosages, identifying active compounds responsible for therapeutic effects, and further investigating their effects on inflammation and oxidative stress through preclinical studies. Safety assessments are crucial for understanding potential adverse effects and interactions, and exploring combination therapies with other natural products or conventional medications could enhance treatment outcomes. Investigating pharmacokinetics and conducting studies in specific populations, such as children and the elderly, will provide valuable insights into their potential therapeutic use in respiratory diseases. Ultimately, this comprehensive research endeavour holds promise for the development of novel therapies for respiratory diseases.

9.7 Conclusion

In conclusion, the historical use of *D. nobile* in traditional medicine for treating respiratory ailments underscores its potential as a therapeutic remedy. Recent scientific enquiries have shed light on its multifaceted benefits in addressing conditions such as chronic bronchitis, asthma and COPD. These studies have revealed promising aspects of *D. nobile*, including its anti-inflammatory, antioxidant and immunomodulatory properties, which collectively contribute to its therapeutic effects. Despite a growing body of evidence supporting its efficacy, there remains a crucial need for more rigorous clinical trials and mechanistic studies. A deeper understanding of the precise mechanisms underlying *D. nobile* therapeutic actions is essential for researchers to confidently validate its use in respiratory disease management. Furthermore, comprehensive investigations are necessary to identify and characterize the active compounds responsible for its medicinal properties and to establish standardized preparation methods to ensure consistency and potency. Despite these ongoing research needs and challenges, the potential of *D. nobile* as a natural and complementary therapy for respiratory diseases remains promising. Its longstanding history of use, coupled with emerging scientific insights, warrants continued exploration and investigation to unlock its full therapeutic potential and facilitate integration into clinical practice for the benefit of respiratory health.

9.7.1 Summary of key points

D. nobile has served as a staple in traditional Chinese medicine, revered for its efficacy in alleviating respiratory ailments, including cough, asthma and bronchitis. Emerging research has underscored its multifaceted therapeutic potential, revealing its noteworthy anti-inflammatory, antioxidant and immunomodulatory properties. *D. nobile* can be administered orally, topically or through inhalation, and is available in various forms such as teas, extracts and capsules. Clinical studies have shown that *D. nobile* can improve cough, sputum and lung function in patients with respiratory diseases, with minimal adverse effects. However, there are limitations in the development of *D. nobile* as a therapeutic agent, including a lack of standardization in preparation, limited knowledge on its mechanisms of action, and insufficient clinical data in different populations. Further research is needed to identify the active compounds, establish optimal dosages and treatment regimens, investigate their potential in combination with other therapies and assess their safety and toxicity. Overall, *D. nobile* shows promising potential as a natural treatment option for respiratory diseases, but further research is necessary to fully evaluate its efficacy and safety.

9.7.2 Implications for clinical practice

The use of *D. nobile* in the treatment of respiratory diseases has potential implications for clinical practice. Based on the available evidence, *D. nobile*-containing herbal formulas may be considered as an adjunct therapy for respiratory diseases, particularly in cases where conventional therapies have failed or caused adverse effects. It is important to note that the quality and efficacy of *D. nobile*-based therapies can be influenced by factors such as the preparation method and quality of the raw materials. Therefore, it is important for clinicians to obtain high-quality preparations from reputable suppliers. Clinicians should also be aware of potential adverse effects and drug interactions, especially in patients who are taking other medications or who have pre-existing medical conditions. Patients should also be advised to disclose their use of *D. nobile*-containing herbal formulas to their healthcare providers. Overall, while further research is needed to fully establish the efficacy and safety of *D. nobile* in respiratory diseases, it has the potential to provide a valuable addition to the range of therapeutic options available to clinicians.

Acknowledgements

The authors express their gratitude for the financial support received through the Distinguished High-Level Talents Research Grant from the Guizhou Science and Technology Corporation Platform Talents Fund (grant nos. [2017]5733-001 and CK-1130-002), the National Natural Science Foundation of China (U1812403, 82373981 and 82060750), 2011 Collaborative Innovation Centre of Traditional Chinese Medicine in Guizhou Province (no. [2022]026) and the support provided by the Zunyi Medical University, China. Special appreciation is extended to all laboratory colleagues and research staff members for their valuable insights, constructive guidance and assistance with the contents of this chapter.

References

Abrami, M., Biasin, A., Tescione, F., Tierno, D., Dapas, B. *et al.* (2024) Mucus structure, viscoelastic properties, and composition in chronic respiratory diseases. *International Journal of Molecular Sciences* 25(3), 1–44.

Adeloye, D., Song, P., Zhu, Y., Campbell, H., Sheikh, A. *et al.* (2022) Global, regional, and national prevalence of, and risk factors for, chronic obstructive pulmonary disease (COPD) in 2019: a systematic review and modelling analysis. *The Lancet Respiratory Medicine* 10(5), 447–458.

Alamgir, A.N.M. (2017) Pharmacognostical botany: classification of medicinal and aromatic plants (maps), botanical taxonomy, morphology, and anatomy of drug plants. In: Alamgir, A.N.M. (ed.) *Therapeutic Use of Medicinal Plants and Their Extracts*, vol. 1. Springer Cham, 177–293.

Bălă, G.-P., Râjnoveanu, R.-M., Tudorache, E., Motişan, R. and Oancea, C. (2021) Air pollution exposure-the (in)visible risk factor for respiratory diseases. *Environmental Science and Pollution Research International* 28(16), 19615–19628.

Campbell, J.K. (2020) Interactive health literacy and self-management of lung cancer symptoms. Thesis, University of Wisconsin Milwaukee.

Chen, D.M., Yu, W.C., Hung, H.F., Tsai, J.C., Wu, H.Y. and Chiou, A.F. (2018) The effects of Baduanjin exercise on fatigue and quality of life in patients with heart failure: a randomized controlled trial. *European Journal of Cardiovascular Nursing* 17(5), 456–466.

Dar, R.A., Shahnawaz, M., Ahanger, M.A. and Majid, I. ul (2023) Exploring the diverse bioactive compounds from medicinal plants: a review. *The Journal of Phytopharmacology* 12(3), 189–195.

Debnath, S.K., Debnath, M. and Srivastava, R. (2022) Opportunistic etiological agents causing lung infections: emerging need to transform lung-targeted delivery. *Heliyon* 8(12), e12620.

Dong, J., Li, Z., Luo, L. and Xie, H. (2020) Efficacy of pulmonary rehabilitation in improving the quality of life for patients with chronic obstructive pulmonary disease: evidence based on nineteen randomized controlled trials. *International Journal of Surgery* 73, 78–86.

Duan, F.P., Li, Y.S., Hu, T.Y., Pan, X.Q., Ma, F. *et al.* (2022) *Dendrobium nobile* protects against ovalbumin-induced allergic rhinitis by regulating intestinal flora and suppressing lung inflammation. *Chinese Journal of Natural Medicines* 20(6), 443–457.

Esme, M., Topeli, A., Yavuz, B.B. and Akova, M. (2019) Infections in the elderly critically-ill patients. *Frontiers in Medicine* 6, 118.

Fan, C., Sun, X., Wang, X. and Yu, H. (2023) Therapeutic potential of the chemical composition of *Dendrobium nobile* Lindl. *Frontiers in Pharmacology* 14, 1–5.

Fan, Y., Yu, Q., Wang, G., Tan, J., Liu, S. *et al.* (2020) Effects of non-thermal plasma treatment on the polysaccharide from *Dendrobium nobile* Lindl. and its immune activities *in vitro*. *International Journal of Biological Macromolecules* 153, 942–950.

Gogoi, P. and Sen, S. (2023) Orchids of Dibru-Saikhowa: a systematic review on their traditional use, pharmacological activity and phytochemistry. *Current Topics in Medicinal Chemistry* 23(24), 2277–2299.

Gong, D., Li, B., Wu, B., Fu, D., Li, Z. *et al.* (2023) The integration of the metabolome and transcriptome for *Dendrobium nobile* Lindl. in response to methyl Jasmonate. *Molecules* 28(23), 1–27.

Gopallawa, I., Dehinwal, R., Bhatia, V., Gujar, V. and Chirmule, N. (2023) A four-part guide to lung immunology: invasion, inflammation, immunity, and intervention. *Frontiers in Immunology* 14, 1119564.

Haque, M., Islam, T., Rahman, N.A.A., McKimm, J., Abdullah, A. *et al.* (2020) Strengthening primary health -c are services to help prevent and control long-term (chronic) non-communicable diseases in low- and middle-income countries. *Risk Management and Healthcare Policy* 13, 409–426.

Heo, J.Y., Seo, Y.B., Choi, W.S., Kim, E.J., Jeong, H.W. *et al.* (2022) Effectiveness of pneumococcal vaccination against pneumococcal pneumonia hospitalization in older adults: a prospective, test-negative study. *The Journal of Infectious Diseases* 225(5), 836–845.

Hong, S., Kim, E.Y., Lim, S.E., Kim, J.H., Sohn, Y. *et al.* (2022) *Dendrobium nobile* Lindley administration attenuates atopic dermatitis-like lesions by modulating immune cells. *International Journal of Molecular Sciences* 23(8), 4470.

Hossain, M.M. (2011) Therapeutic orchids: traditional uses and recent advances — an overview. *Fitoterapia* 82(2), 102–140.

Jacobs, S.S., Krishnan, J.A., Lederer, D.J., Ghazipura, M., Hossain, T. *et al.* (2020) Home oxygen therapy for adults with chronic lung disease. An Official American Thoracic Society Clinical Practice Guideline. *American Journal of Respiratory and Critical Care Medicine* 202(10), e121–e141.

Jadhav, S.P., Singh, H., Hussain, S., Gilhotra, R., Mishra, A. *et al.* (2021) Introduction to lung diseases. In: Dua, K. (ed.) *Targeting Cellular Signalling Pathways in Lung Diseases*. Springer, Singapore, pp. 1–25.

Jansen, E.M., van de Hei, S.J., Dierick, B.J.H., Kerstjens, H.A.M., Kocks, J.W.H. *et al.* (2021) Global burden of medication non-adherence in chronic obstructive pulmonary disease (COPD) and asthma: a narrative review of the clinical and economic case for smart inhalers. *Journal of Thoracic Disease* 13(6), 3846–3864.

Jędrejko, K., Lazur, J. and Muszyńska, B. (2021) Risk associated with the use of selected ingredients in food supplements. *Chemistry & Biodiversity* 18(2), e2000686.

Jones, R., Gruffydd-Jones, K., Pinnock, H., Peffers, S.J., Lawrence, J. *et al.* (2010) Summary of the consultation on a strategy for services for chronic obstructive pulmonary disease (COPD) in England. *Primary Care Respiratory Journal* 19(2), S1–S17.

Lee, A., Hawcutt, D. and Sinha, I. (2023) What causes poverty and how does this contribute to respiratory inequality. *ERS Educational Publications* 99, 11–25.

Li, P.Y., Li, L. and Wang, Y.Z. (2023) Traditional uses, chemical compositions and pharmacological activities of *Dendrobium*: a review. *Journal of Ethnopharmacology* 310, 116382.

Liu, L., Li, M., Yu, M., Shen, M., Wang, Q. *et al.* (2019) Natural polysaccharides exhibit anti-tumor activity by targeting gut microbiota. *International Journal of Biological Macromolecules* 121, 743–751.

Long, R., Divangahi, M. and Schwartzman, K. (2022) Transmission and pathogenesis of tuberculosis. *Canadian Journal of Respiratory, Critical Care, and Sleep Medicine* 6(1), 22–32.

Long, Y., Wang, W., Zhang, Y., Zhang, S., Li, Z. *et al.* (2023) *Dendrobium nobile* Lindl polysaccharides attenuate UVB-induced photodamage by regulating oxidative stress, inflammation and MMPs expression in mice model. *Photochemistry and Photobiology* 99(5), 1269–1281.

Matera, M.G., Page, C.P., Calzetta, L., Rogliani, P. and Cazzola, M. (2020) Pharmacology and therapeutics of bronchodilators revisited. *Pharmacological Reviews* 72(1), 218–252.

McAllister, D.A., Liu, L., Shi, T., Chu, Y., Reed, C. *et al.* (2019) Global, regional, and national estimates of pneumonia morbidity and mortality in children younger than 5 years between 2000 and 2015: a systematic analysis. *The Lancet Global Health* 7(1), e47–e57.

Mishra, V., Seyedzenouzi, G., Almohtadi, A., Chowdhury, T., Khashkhusha, A. *et al.* (2021) Health inequalities during COVID-19 and their effects on morbidity and mortality. *Journal of Healthcare Leadership* 13, 19–26.

Momtazmanesh, S., Moghaddam, S.S., Ghamari, S.H., Rad, E.M., Rezaei, N. *et al.* (2023) Global burden of chronic respiratory diseases and risk factors, 1990–2019: an update from the global burden of disease study 2019. *E Clinical Medicine* 59, 1–22.

Mou, Z., Zhao, Y., Ye, F., Shi, Y., Kennelly, E.J. *et al.* (2021) Identification, biological activities and biosynthetic pathway of *Dendrobium* alkaloids. *Frontiers in Pharmacology* 12, 605994.

Nie, X., Chen, Y., Li, W. and Lu, Y. (2020) Anti-aging properties of *Dendrobium nobile* Lindl.: from molecular mechanisms to potential treatments. *Journal of Ethnopharmacology* 257, 112839.

Ninawe, A.S. and Swapna, T.S. (2019) Orchid diversity of northeast India–traditional knowledge and strategic plan for conservation. *Journal of Orchid Society of India* 31, 41–56.

Papadopoulos, N.G., Miligkos, M. and Xepapadaki, P. (2021) A current perspective of allergic asthma: from mechanisms to management. *Allergic Diseases–From Basic Mechanisms to Comprehensive Management and Prevention* 268, 69–93.

Patel, V.H., Thannir, S., Dhanani, M., Augustine, I., Sandeep, S.L. *et al.* (2023) Current limitations and recent advances in the management of asthma. *Disease-a-Month* 69(7), 101483.

Ramesh, T., Koperuncholan, M., Praveena, R., Ganeshkumari, K., Vanithamani, J. *et al.* (2019) Medicinal properties of some *Dendrobium* orchids – a review. *Journal of Applied and Advanced Research* 4(4), 119–128.

Rao, D., Hu, Y., Zhao, R., Li, H., Chun, Z. *et al.* (2022) Quantitative identification of antioxidant basis for *Dendrobium nobile* flower by high performance liquid chromatography-tandem mass spectrometry. *International Journal of Analytical Chemistry* 2022, 9510598.

Rao, D., Zhao, R., Hu, Y., Li, H., Chun, Z. *et al.* (2023) Revealing of intracellular antioxidants in *Dendrobium nobile* by high performance liquid chromatography-tandem mass spectrometry. *Metabolites* 13(6), 1–24.

Rossaki, F.M., Hurst, J.R., van Gemert, F., Kirenga, B.J., Williams, S. *et al.* (2021) Strategies for the prevention, diagnosis and treatment of COPD in low- and middle-income countries: the importance of primary care. *Expert Review of Respiratory Medicine* 15(12), 1563–1577.

Ruvuna, L. and Sood, A. (2020) Epidemiology of chronic obstructive pulmonary disease. *Clinics in Chest Medicine* 41(3), 315–327.

Sharma, A.K., Sharma, A.K., Sharma, M. and Sharma, M. (2021) Medicinal and aromatic plants: a potential source of novel bioactive compounds showing antiviral efficacy against coronavirus (SARS-cov-2). *Indian Journal of Ecology* 48(14), 7–16.

Su, X.L., Wang, J.W., Che, H., Wang, C.F., Jiang, H. *et al.* (2020) Clinical application and mechanism of traditional Chinese medicine in treatment of lung cancer. *Chinese Medical Journal* 133(24), 2987–2997.

Ti, H., Zhuang, Z., Yu, Q. and Wang, S. (2021) Progress of plant medicine derived extracts and alkaloids on modulating viral infections and inflammation. *Drug Design, Development and Therapy* 15, 1385–1408.

Wahab, S., Ahmad, I., Irfan, S., Siddiqua, A., Usmani, S. *et al.* (2022) Pharmacological efficacy and safety of *Glycyrrhiza glabra* in the treatment of respiratory tract infections. *Mini Reviews in Medicinal Chemistry* 22(11), 1476–1494.

Wang, G., Wang, J., Deng, Y., Qin, L., He, Y. *et al.* (2022) Chemical constituents and nutritional health functions of *Dendrobium nobile*: a review. *Food Science and Technology* 42, 1–19.

Wang, R.X., Zhou, M., Ma, H.L., Qiao, Y.B. and Li, Q.S. (2021) The role of chronic inflammation in various diseases and anti-inflammatory therapies containing natural products. *ChemMedChem* 16(10), 1576–1592.

Wang, W., Xu, J., Fang, H., Li, Z. and Li, M. (2020) Advances and challenges in medicinal plant breeding. *Plant Science* 298, 110573.

Wang, X., Li, J., Chen, R., Li, T. and Chen, M. (2023) Active ingredients from Chinese medicine for combination cancer therapy. *International Journal of Biological Sciences* 19(11), 3499–3525.

Wu, C., Duan, X., Wang, X. and Wang, L. (2023) Advances in the role of epigenetics in homocysteine-related diseases. *Epigenomics* 15(15), 769–795.

Xie, Z., Jiang, N., Lin, M., He, X., Li, B. *et al.* (2023) The mechanisms of polysaccharides from tonic chinese herbal medicine on the enhancement immune function: a review. *Molecules* 28(21), 7355.

Xu, H.X. and Lin, Z.X. (2021) Overview of research trends in precious Chinese medicines. *Chinese Medicine and Culture* 4(4), 197–200.

Yan, Y., Li, X., Zhang, C., Lv, L., Gao, B. *et al.* (2021) Research progress on antibacterial activities and mechanisms of natural alkaloids: a review. *Antibiotics* 10(3), 318.

Zhang, J., Xu, H.X., Zhao, Z.L., Xian, Y.F. and Lin, Z.X. (2021) *Dendrobium nobile* Lindl: a review on its chemical constituents and pharmacological effects. *Chinese Medicine and Culture* 4(4), 235–242.

Zhang, M.S., Linghu, L., Wang, G., He, Y.Q., Sun, C.X. *et al.* (2022) Dendrobine type alkaloids from *Dendrobium nobile*. *Natural Product Research* 36(21), 5393–5399.

Zhang, Q., Fu, X., Liu, H., Chen, Y., Chen, S. *et al.* (2023a) A systematical review on ethnobotanical, phytochemical and pharmacological aspects of *Dendrobium nobile* Lindl. *Phytochemistry Reviews* 22(3), 743–780.

Zhang, Y.W., Shi, Y.C. and Zhang, S.B. (2023b) Metabolic and transcriptomic analyses elucidate a novel insight into the network for biosynthesis of carbohydrate and secondary metabolites in the stems of a medicinal orchid *Dendrobium nobile Plant Diversity* 45(3), 326–336.

Zhang, Z.-J., Morris-Natschke, S.L., Cheng, Y.-Y., Lee, K.-H. and Li, R.-T. (2020) Development of anti-influenza agents from natural products. *Medicinal Research Reviews* 40(6), 2290–2338.

10 *Dendrobium nobile* in the Management of Gastrointestinal Disorders

Surendra Sarsaiya, Archana Jain, Qihai Gong, Qin Wu and Jingshan Shi*
Key Laboratory of Basic Pharmacology and Joint International Research Laboratory of Ethnomedicine of Ministry of Education, Zunyi Medical University, Zunyi, Guizhou, China

Abstract
This chapter explores the potential of *Dendrobium nobile* for managing gastrointestinal disorders. Traditional uses of *D. nobile* include treatments for nausea, vomiting, diarrhoea, gastritis, peptic ulcers and inflammatory bowel disease. Scientific research has identified its diverse phytochemical composition, including alkaloids, phenanthrenes, polysaccharides and flavonoids, which contributes to its anti-inflammatory, antioxidant, antiemetic, gastroprotective and immunomodulatory properties. Studies have demonstrated significant improvements in various gastrointestinal disorders such as chronic constipation, gastro-oesophageal reflux disease and peptic ulcers. However, current evidence is limited and often inconclusive, highlighting the need for large-scale clinical trials to validate its efficacy and safety. Future research should focus on standardization of extracts, identification of active compounds and detailed mechanistic studies. This chapter provides a comprehensive review of the phytochemical properties, traditional uses and scientific evidence supporting the use of *D. nobile* in gastrointestinal health, along with recommendations for clinical practice.

10.1 Introduction

The traditional uses of *Dendrobium nobile* Lindl. in gastrointestinal disorders range from treating nausea, vomiting and diarrhoea to more chronic conditions such as gastritis, peptic ulcers and inflammatory bowel disease (IBD) (Gong *et al.*, 2024). Whereas empirical evidence from centuries of traditional use provides a foundation for exploring its therapeutic potential, modern scientific research has sought to elucidate the mechanisms underlying its effects and evaluate its efficacy (Andreae *et al.*, 2018). Gastrointestinal disorders encompass a broad spectrum of ailments affecting the digestive system, ranging from mild discomfort to severe conditions that significantly impair quality of life and may even be life-threatening (Pabolu *et al.*, 2021). Despite advancements in medical science, the management of gastrointestinal disorders often presents challenges owing to the complexity of the digestive system and limitations of current treatment modalities (Drewes *et al.*, 2020). In recent years, there has been growing interest in exploring natural remedies derived from traditional medicine systems as potential alternatives or adjuncts to conventional therapies (Wang *et al.*, 2021).

*Corresponding author: shijs@zmu.edu.cn

© Jingshan Shi, Archana Jain, Surendra Sarsaiya, Qin Wu and Qihai Gong 2025. Dendrobium nobile: *Research and Therapeutic Applications* (J. Shi, A. Jain, S. Sarsaiya, Q. Wu and Q. Gong)
DOI: 10.1079/9781800627987.0010

Among the myriad natural substances with therapeutic potential, *D. nobile*, a species of orchid native to Asia, has emerged as a subject of scientific enquiry owing to its purported benefits to gastrointestinal health (Wang, 2021). With a rich history of use in traditional Chinese medicine and other traditional healing practices across Asia, *D. nobile* has been revered for its multifaceted pharmacological properties and wide-ranging therapeutic applications (De, 2023). The appeal of *D. nobile* lies not only in its historical usage but also in its diverse phytochemical composition, which includes alkaloids, phenanthrenes, polysaccharides, flavonoids and other bioactive compounds (Fu *et al.*, 2023). These constituents are believed to contribute to its medicinal properties, which include anti-inflammatory, antioxidant, antiemetic, gastroprotective and immunomodulatory effects (Ketsa and Warrington, 2023). Despite the growing interest in *D. nobile* as a natural remedy for gastrointestinal disorders, several questions remain unanswered. These include the specific bioactive compounds responsible for its therapeutic effects, optimal dosage and formulation for different gastrointestinal conditions and safety profile for long-term use. Moreover, rigorous clinical trials are needed to validate its efficacy and establish evidence-based guidelines for its use in clinical practice. Against this backdrop, this chapter aims to provide a comprehensive review of the current understanding of *D. nobile* in the management of gastrointestinal disorders. It will delve into the phytochemical composition of *D. nobile*, its traditional uses in gastrointestinal health, the scientific evidence supporting its efficacy and the mechanisms underlying its pharmacological actions. Furthermore, it will explore safety considerations, formulation strategies and future research directions to harness the full therapeutic potential of *D. nobile* in promoting gastrointestinal wellness and combating digestive ailments (Table 10.1).

10.2 Gastrointestinal Disorders: An Overview

Gastrointestinal disorders encompass a wide array of conditions that affect the digestive system, ranging from mild discomfort to severe illness. These disorders can arise from various factors, such as infections, inflammation, dietary habits, genetic predispositions and lifestyle choices. Understanding common gastrointestinal disorders, their underlying causes and the challenges associated with their management is crucial for developing effective treatment strategies.

10.2.1 Common gastrointestinal disorders

Gastrointestinal disorders may manifest as acute episodes or develop into chronic conditions, varying in severity from mild discomfort to severe symptoms that profoundly affect quality of life. Each disorder presents unique challenges and may require tailored management strategies to alleviate symptoms and improve overall well-being. Gastro-oesophageal reflux disease (GORD) is a chronic condition characterized by backward flow of stomach acid into the oesophagus, resulting in symptoms such as heartburn, regurgitation and chest pain (Chapelle *et al.*, 2021). The condition often leads to uncomfortable symptoms, such as heartburn, regurgitation and pain in the chest. Potential causes include a weakened lower oesophageal sphincter or hiatal hernia, among other factors contributing to compromised oesophageal function (Clarrett and Hachem, 2018). Irritable bowel syndrome is a functional disorder affecting the gastrointestinal tract. It manifests with symptoms such as abdominal discomfort, bloating and changes in bowel movements, but typically lacks any identifiable structural or biochemical irregularities (Camilleri, 2021). It may manifest as constipation, diarrhoea or a blend of both, with the precise aetiology remaining unclear, likely involving intricate interactions between gut motility, sensitivity and brain–gut axis function (Pan *et al.*, 2022).

Coeliac disease is an autoimmune disorder triggered by the ingestion of gluten, which is commonly found in wheat, barley and rye. Gluten consumption affects the small intestine (Sharma *et al.*, 2020), leading to damage to the villi of the small intestine, resulting in nutrient malabsorption and diverse symptoms, such as diarrhoea, abdominal discomfort, bloating, fatigue and weight loss. Effective treatment primarily involves adherence to a gluten-free diet (Orjiekwe, 2023). Chronic constipation refers to persistent

Table 10.1. Common gastrointestinal disorders (authors' own table).

Disorder	Description	Symptoms	Causes and risk factors	Treatment	Reference
GORD	Chronic condition where stomach acid flows back into the oesophagus, causing symptoms like heartburn and regurgitation	Heartburn, regurgitation, chest pain	Weakened lower oesophageal sphincter, hiatal hernia, dietary factors, obesity	Lifestyle changes (e.g. diet modification, weight loss), medications to reduce acid production, surgery in severe cases	Chapelle *et al.* (2021)
IBS	Functional disorder of the GI tract characterized by abdominal discomfort, bloating and changes in bowel movements	Abdominal discomfort, bloating, altered bowel habits (constipation, diarrhoea or both)	Unknown, but involves gut motility, sensitivity and brain–gut axis function	Dietary modifications, stress management, medications to alleviate symptoms (e.g. anti-spasmodics, laxatives, anti-diarrhoeals), psychotherapy	Camilleri (2021)
Coeliac disease	Autoimmune disorder triggered by gluten ingestion, leading to small intestine damage and nutrient malabsorption	Diarrhoea, abdominal discomfort, bloating, fatigue, weight loss	Ingestion of gluten (found in wheat, barley, rye)	Gluten-free diet	Sharma *et al.* (2020)
Chronic constipation	Persistent difficulty passing stool, characterized by infrequent bowel movements and straining during defecation	Infrequent bowel movements, straining during defecation, feeling of incomplete evacuation	Dietary habits, dehydration, lack of physical activity, medications, underlying medical conditions	Lifestyle modifications (e.g. dietary changes, increased fluid intake, fibre supplementation), medications, medical procedures if necessary	Schiller (2019)
IBD	Chronic inflammatory conditions affecting the GI tract, including Crohn's disease and ulcerative colitis	Abdominal pain, diarrhoea, rectal bleeding, weight loss, fatigue	Multifactorial: genetic, environmental, immune system elements	Medications (anti-inflammatory drugs, immunosuppressants), dietary changes, surgery in severe cases	Deshmukh *et al.* (2020)
Diverticulitis	Inflammation or infection of diverticula (small pouches) along the colon wall	Left-sided abdominal pain, fever, nausea, vomiting, changes in bowel habits	Diverticulosis, dietary factors, age, obesity, lack of exercise	Antibiotics, pain management, dietary changes, surgery in severe cases	Gunby and Strate (2024)

Continued

Table 10.1. Continued.

Disorder	Description	Symptoms	Causes and risk factors	Treatment	Reference
Colorectal cancer	Malignant tumour in the colon or rectum, often requiring multidisciplinary treatment	Varies depending on location and stage of cancer	Genetic predisposition, lifestyle factors (diet, smoking, alcohol consumption), age, IBD	Surgery, chemotherapy, radiation therapy, targeted therapy, immunotherapy, supportive care	Xi and Xu (2021)
Gastritis	Inflammation of the stomach lining, acute or chronic, often due to various factors	Abdominal discomfort, nausea, vomiting, indigestion, bloating, appetite loss	*Helicobacter pylori* infection, excessive alcohol intake, prolonged NSAID use, autoimmune conditions	Treatment of underlying cause, medications to reduce stomach acid, dietary and lifestyle changes	Borah *et al.* (2023)
Peptic ulcers	Open sores in the stomach, oesophagus or small intestine linings, often triggered by various factors	Abdominal pain, bloating, nausea, vomiting, dark or bloody stools	*H. pylori* infection, prolonged NSAID use, excessive alcohol consumption, smoking, stress	Antibiotics (to eradicate *H. pylori*), acid-reducing medications, lifestyle modifications, endoscopic therapy, surgery in severe cases	Bereda (2022)

GI, gastrointestinal; GORD, gastro-oesophageal reflux disease; IBS, irritable bowel syndrome; NSAID, non-steroidal anti-inflammatory drug.

difficulty in passing stool, characterized by infrequent bowel movements, straining during defaecation and a sense of incomplete evacuation (Schiller, 2019). This can result from factors such as dietary habits, dehydration, lack of physical activity, medications or underlying medical conditions. Management typically involves lifestyle modifications, dietary changes and pharmacological interventions if necessary (Raucci *et al.*, 2021). Management typically involves dietary and lifestyle modifications, increased fluid intake, fibre supplementation and, if necessary, pharmacological interventions or medical procedures aimed at addressing underlying causes and relieving symptoms (Włodarczyk *et al.*, 2021). IBD encompasses chronic inflammatory conditions that affect the gastrointestinal tract, including Crohn's disease and ulcerative colitis. Symptoms include abdominal pain, diarrhoea, rectal bleeding, weight loss and fatigue (Deshmukh *et al.*, 2020). Its exact cause is multifactorial, involving genetic, environmental and immune system elements, with treatments aimed at mitigating inflammation, managing symptoms and preventing complications (Ahlawat *et al.*, 2021). Treatment of IBD typically includes medications to reduce inflammation, control symptoms and manage complications, as well as lifestyle modifications and, in some cases, surgery (Cai *et al.*, 2021). Management of IBD aims to achieve symptom remission, prevent flare-ups and improve the overall quality of life of affected individuals (Byron *et al.*, 2020).

Diverticulitis is the inflammation or infection of small pouches called diverticula, which form along the colonic wall. The symptoms of diverticulitis often include left-sided abdominal pain, fever, nausea, vomiting and alterations in bowel habits (Gunby and Strate, 2024). Treatment typically includes antibiotics to clear the infection, pain management, dietary changes to prevent further irritation of the colon and surgery to remove the affected portion of the colon in severe cases (Puyuelo *et al.*, 2018). Management of diverticulitis also involves lifestyle modifications, such as increasing fibre intake, staying hydrated and regular exercise to prevent future flare-ups and promote overall digestive health (You *et al.*, 2019). Colorectal cancer is a malignant tumour arising in the colon or rectum and is the third most prevalent cancer worldwide (Xi and Xu, 2021). Early detection through screening tests such as colonoscopy is vital for prognosis improvement, with treatment options including surgery, chemotherapy, radiation therapy, targeted therapy and immunotherapy, often necessitating multidisciplinary approaches (Dey *et al.*, 2023). Gastritis denotes stomach lining inflammation, acute or chronic, stemming from factors such as *Helicobacter pylori* infection, excessive alcohol intake, prolonged non-steroidal anti-inflammatory drug (NSAID) use or autoimmune conditions (Borah *et al.*, 2023). Symptoms may encompass abdominal discomfort, nausea, vomiting, indigestion, bloating and appetite loss, with treatments addressing the underlying cause, such as reducing stomach acid with medications and implementing dietary and lifestyle changes (Singh *et al.*, 2023). Peptic ulcers are open sores that develop in the stomach, oesophagus or small intestine linings and are often triggered by *H. pylori* infection, prolonged NSAID use, excessive alcohol consumption, smoking or stress (Bereda, 2022). Symptoms may include abdominal pain, bloating, nausea, vomiting and dark or bloody stools, necessitating comprehensive management of the root cause and symptom relief (Obeagu and Obeagu, 2024). Treatment aims to relieve symptoms, promote ulcer healing and prevent complications, typically involving antibiotics to eradicate *H. pylori* infection, acid-reducing medications, lifestyle modifications and endoscopic therapy or surgery in severe cases.

10.2.2 Current treatment modalities and limitations

Managing gastrointestinal disorders often entails a multifaceted approach that integrates lifestyle adjustments, dietary modifications, pharmacotherapy and, occasionally, surgical procedures (El-Salhy, 2019). Lifestyle modifications, including avoiding trigger foods, mitigating stress and achieving a healthy weight, play pivotal roles in alleviating symptoms associated with conditions such as GORD and irritable bowel syndrome. Pharmacological interventions, such as proton pump inhibitors, H_2-receptor antagonists, antacids and prokinetic agents, are commonly prescribed to alleviate symptoms and manage underlying gastrointestinal conditions (Targownik *et al.*, 2022). In cases where conservative measures and medications are ineffective, surgical interventions may be considered to address anatomical abnormalities or severe complications of gastrointestinal disorders (Lee *et al.*, 2021). This comprehensive approach aims to optimize patient outcomes and improve overall gastrointestinal health, and agents are commonly used to manage symptoms of GORD and peptic ulcers by reducing gastric acid production or promoting gastric emptying (Chapelle *et al.*, 2021).

IBDs, such as Crohn's disease and ulcerative colitis, are managed with medications, such as corticosteroids, immunomodulators, biologic agents and aminosalicylates, to reduce inflammation and control symptoms. Individuals diagnosed with coeliac disease are required to strictly follow a gluten-free diet to mitigate symptom recurrence and potential complications (Le Berre *et al.*, 2020). In instances of severe manifestation or complication occurrence, surgical interventions such as bowel resection, stricture plasty or ostomy might be warranted (Wang and Shen, 2018). Despite the availability of various treatment modalities, managing gastrointestinal disorders can be challenging owing to several limitations. Many medications used to treat these disorders have side effects, and long-term use may be associated with adverse effects such as osteoporosis, vitamin deficiencies and increased risk of infections. Additionally, some patients may not respond adequately to conventional treatments or experience disease flares despite optimal therapy. Moreover, surgical interventions carry risks such as complications related to anaesthesia, infection, bleeding and bowel obstruction.

10.2.3 Need for alternative therapies

Alternative therapies for gastrointestinal disorders encompass a wide range of approaches, including herbal medicine, acupuncture, probiotics, prebiotics, dietary supplements and mind/body therapies. Individuals diagnosed with coeliac disease are required to strictly follow a gluten-free diet to mitigate symptom recurrence and potential complications. In instances of severe manifestation or complication occurrence, surgical interventions such as bowel resection, strictureplasty or ostomy might be warranted (Ghimire *et al.*, 2019). Given the limitations and challenges associated with current treatment modalities, there is growing recognition of the need for alternative therapies for gastrointestinal disorders. Natural remedies derived from plants and traditional medicine systems offer potential advantages, such as fewer side effects, greater accessibility and cultural acceptance. Moreover, botanicals and herbal supplements may exert beneficial effects through multiple mechanisms of action, including anti-inflammatory, antioxidant, antimicrobial and analgesic effects. In conclusion, gastrointestinal disorders pose significant challenges for patients and healthcare providers due to their diverse manifestations, complex aetiology and limitations of current treatment modalities. Understanding common gastrointestinal disorders, their treatment options and the need for alternative therapies is essential for providing comprehensive care and improving patient outcomes. Alternative therapies such as *D. nobile* offer promising avenues for further exploration and may complement existing treatments in the management of gastrointestinal disorders.

10.3 Traditional Use of *D. nobile* for Gastrointestinal Disorders

The traditional use of *D. nobile* in the management of gastrointestinal disorders dates back centuries in various traditional medicine systems, particularly in East Asia. In traditional Chinese medicine, *D. nobile*, known as 'shi hu' or 'shi hu gen', has been valued for its ability to promote digestive health and alleviate symptoms associated with gastrointestinal ailments (Wu *et al.*, 2023). Practitioners of traditional Chinese medicine have utilized *D. nobile* to treat a wide range of gastrointestinal complaints, including nausea, vomiting, diarrhoea, indigestion, gastritis and peptic ulcers (Teng *et al.*, 2018, Zhang *et al.*, 2024). Its effectiveness in relieving these symptoms is attributed to its purported properties, such as soothing irritated mucous membranes, regulating gastrointestinal motility and promoting gastric mucosal integrity (Herdiana, 2023). Furthermore, *D. nobile* has been traditionally prescribed as part of herbal formulations or decoctions tailored to address specific gastrointestinal conditions. These traditional formulations often combine *D. nobile* with other herbs believed to complement its therapeutic effects and enhance overall efficacy (Cheng *et al.*, 2018).

The traditional use of *D. nobile* in gastrointestinal disorders is grounded in the principles of traditional medicine, which emphasize restoring balance and harmony in the body to promote health and well-being (Wang, 2021). By addressing the underlying imbalances and supporting the body's innate healing mechanisms, *D. nobile* has been valued as a natural remedy for gastrointestinal ailments in traditional medicine systems (Duan *et al.*, 2022). Although traditional usage over centuries offers valuable insights into the potential therapeutic benefits of *D. nobile*, contemporary scientific investigation is essential to confirm its effectiveness and understand the mechanisms driving its effects. Preclinical studies and clinical trials are essential for corroborating traditional knowledge with scientific evidence and establishing evidence-based guidelines for the management of gastrointestinal disorders.

10.4 Phytochemical Compounds of *D. nobile*

D. nobile contains a diverse array of phytochemical compounds that contribute to its pharmacological properties and potential therapeutic effects. These compounds include alkaloids, flavonoids, phenanthrenes, bibenzyls, polysaccharides and organic acids (Liu *et al.*, 2023). Alkaloids found in *D. nobile*, such as dendrobine and nobiline, have been studied for their potential properties. Flavonoids, including quercetin and kaempferol derivatives, exhibit antioxidant and anti-inflammatory activities, which may contribute to the protective effects of the plant against oxidative stress and inflammation-related gastrointestinal disorders (Qiu *et al.*, 2023). Polysaccharides found in *D. nobile* possess immunomodulatory properties and may play a role in enhancing gut health by promoting the growth of beneficial gut bacteria (Li *et al.*, 2019). Additionally, organic acids, such as malic acid and citric acid, contribute to the plant's acidic taste and may have implications for gastrointestinal physiology and function. Together, these phytochemical compounds present in *D. nobile* underscore its potential as a natural remedy for gastrointestinal disorders and warrant further investigation into their specific mechanisms of action and therapeutic applications. Recent studies have investigated the phytochemistry of *D. nobile* and its potential health benefits. One study found that the alkaloids in *D. nobile* exhibited anti-inflammatory and analgesic effects in mice. Another study found that the flavonoids in *D. nobile* had antioxidant properties and could protect against oxidative stress (Bose *et al.*, 2017). In addition to its alkaloids and flavonoids, *D. nobile* contains polysaccharides, which have been shown to have immunomodulatory effects. One study found that *D. nobile* polysaccharides could enhance the immune response in mice (Fan *et al.*, 2020). While these studies suggest that *D. nobile* may have potential health benefits, more research is needed to determine its safety and effectiveness for human use. Nonetheless, its long history of use in traditional Chinese medicine and other traditional medicine systems, as well as its promising phytochemistry, make it an interesting candidate for further investigation as a natural remedy (Fig. 10.1).

10.4.1 Identification of active compounds and properties of *D. nobile*

D. nobile encompasses a diverse array of bioactive compounds. These phytochemicals contribute to the therapeutic properties of plants and have been the subject of scientific investigation to elucidate their pharmacological activities. Dendrobine, dendroxine, nobiline and dendrobine-*N*-oxide are among the alkaloids identified in *D. nobile*. These alkaloids exhibit various pharmacological effects, including analgesic, anti-inflammatory and anti-spasmodic activities. Dendrobine, dendrocandins and moscatilin are phenanthrene derivatives isolated from *D. nobile*. These compounds have demonstrated antioxidant, anti-inflammatory and anti-cancer properties in preclinical studies (Zhai *et al.*, 2022). *D. nobile* polysaccharides are complex carbohydrates with immunomodulatory and antioxidant activities. These polysaccharides have been shown to enhance immune function, scavenge free radicals and exert protective effects against oxidative stress (Li *et al.*, 2019).

D. nobile contains flavonoids such as luteolin, quercetin and kaempferol, which possess

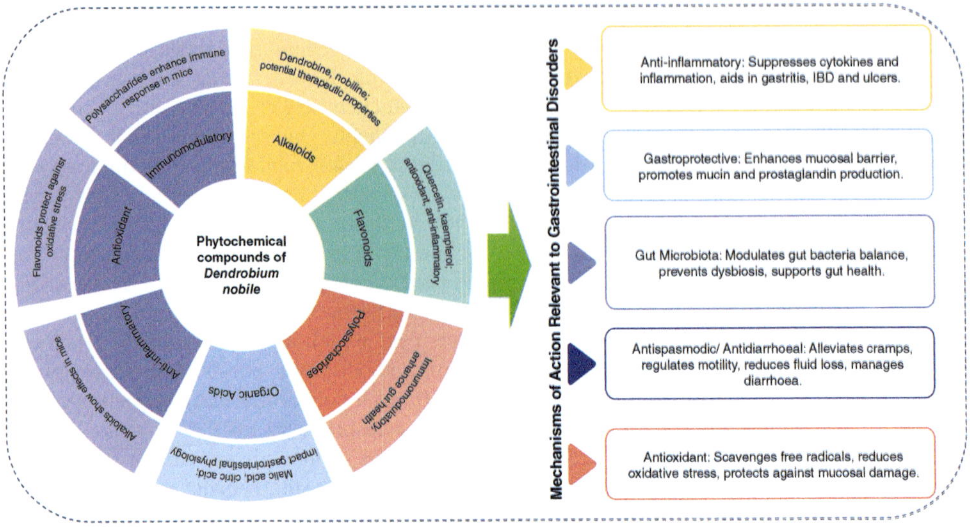

Fig. 10.1. Phytochemical compounds and mechanisms of action relevant to gastrointestinal disorders (authors' own figure).

antioxidant, anti-inflammatory and anti-cancer properties. These flavonoids contribute to the plant's pharmacological effects by scavenging free radicals, modulating inflammatory pathways and inhibiting cancer cell proliferation (Qiu *et al.*, 2023). Chlorogenic acid, caffeic acid and vanillic acid are phenolic acids identified in *D. nobile*. These compounds exhibit antioxidant, anti-inflammatory and antimicrobial activities, which may contribute to the therapeutic effects of the plant (Raj *et al.*, 2022). *D. nobile* contains various terpenoids, including monoterpenes, sesquiterpenes and diterpenes, with diverse pharmacological activities. These terpenoids have been reported to possess anti-inflammatory, antimicrobial and neuroprotective properties (Xu *et al.*, 2022). *D. nobile* is rich in amino acids, such as proline, glutamine and arginine, which play essential roles in protein synthesis and metabolism. These amino acids may contribute to the nutritional value and potential health benefits of plants (Rao *et al.*, 2023). Overall, the phytochemical profile of *D. nobile* is characterized by a complex mixture of bioactive compounds with diverse pharmacological properties.

10.4.2 Mechanisms of action relevant to gastrointestinal disorders

Gastrointestinal disorders can cause significant discomfort and negatively impact quality of life. Many traditional medicines and natural remedies have been used to treat these conditions, and recent research has shed light on the mechanisms of action of these treatments (Table 10.2). *D. nobile* exhibits potent anti-inflammatory properties attributed to its ability to suppress pro-inflammatory cytokines and signalling pathways involved in inflammation. *D. nobile* helps mitigate inflammation in the gastrointestinal tract (Wu *et al.*, 2023). This anti-inflammatory activity is particularly beneficial in gastritis, IBD and peptic ulcers, where excessive inflammation contributes to tissue damage and disease progression (Wu *et al.*, 2023). *D. nobile* exerts gastroprotective effects by enhancing the integrity of the gastric mucosal barrier and promoting mucin secretion. Mucins form a protective layer on the gastric mucosa, preventing damage from gastric acid and other harmful substances (Zhang *et al.*, 2018). Additionally, *D. nobile* may stimulate the production of prostaglandins, which help maintain mucosal blood flow and enhance mucosal defense mechanisms. These gastroprotective effects are valuable for the prevention and treatment of gastric ulcers, gastritis and other mucosal injuries in the gastrointestinal tract (Pei *et al.*, 2021).

D. nobile may exert beneficial effects on gut microbiota composition and diversity, thereby promoting gastrointestinal health. By modulating the balance of beneficial and pathogenic bacteria in the gut, *D. nobile* helps maintain microbial homeostasis and prevents dysbiosis,

Table 10.2. Mechanisms of action of *D. nobile* in the treatment of gastrointestinal disorders (authors' own table).

Effect	Mechanism	Relevant disorders	References
Anti-inflammatory effects	Suppression of pro-inflammatory cytokines and signalling pathways	Gastritis, IBD, peptic ulcers	Wu *et al.* (2023)
Gastroprotective effects	Enhancing gastric mucosal barrier integrity, promoting mucin secretion, stimulating prostaglandin production	Gastric ulcers, gastritis	Zhang *et al.* (2018), Pei *et al.* (2021)
Modulation of gut microbiota	Promoting GI health by balancing gut microbiota composition and diversity	Various GI disorders	Duan *et al.* (2022)
Anti-spasmodic and anti-diarrhoeal actions	Relaxation of smooth muscle cells, regulation of intestinal motility and fluid secretion	IBS	Lu *et al.* (2022), Fan *et al.* (2023)
Antioxidant activity	Scavenging free radicals, reducing oxidative damage in the GI mucosa	Gastritis, peptic ulcers, IBD	Cakova *et al.* (2017), Wu *et al.* (2023)

GI, gastrointestinal; IBS, irritable bowel syndrome.

which is associated with various gastrointestinal disorders (Duan *et al.*, 2022). *D. nobile* exhibits anti-spasmodic effects on smooth muscle cells in the gastrointestinal tract, which can help alleviate abdominal cramps and spasms associated with conditions such as irritable bowel syndrome (Lu *et al.*, 2022). Additionally, *D. nobile* may regulate intestinal motility and fluid secretion, leading to its anti-diarrhoeal properties. By modulating gut motility and reducing excessive fluid loss, *D. nobile* can effectively manage diarrhoea and improve stool consistency in patients with gastrointestinal disorders (Fan *et al.*, 2023). Oxidative stress plays an important role in the pathogenesis of gastrointestinal disorders by inducing tissue damage and inflammation. *D. nobile* exhibits potent antioxidant activity, scavenging free radicals and reducing oxidative damage in the gastrointestinal mucosa (Cakova *et al.*, 2017). This antioxidant activity protects against oxidative stress-related injury and may contribute to the prevention and treatment of gastrointestinal disorders including gastritis, peptic ulcers and IBD (Wu *et al.*, 2023). Overall, the mechanisms of action of *D. nobile* relevant to gastrointestinal disorders involve a multifaceted approach, including anti-inflammatory, gastroprotective, modulation of gut microbiota, anti-spasmodic, anti-diarrhoeal and antioxidant effects. These mechanisms contribute to its therapeutic efficacy in alleviating symptoms and promoting gastrointestinal health, and may offer promising interventions for gastrointestinal disorders.

10.5 Studies on *D. nobile* and Gastrointestinal Disorders

Research on the effects of *D. nobile* on gastrointestinal disorders is limited, but some preclinical studies have investigated its potential benefits for gut health. One clinical study focused on evaluating the impact of *D. nobile* supplementation on bowel movement frequency and stool consistency in individuals with chronic constipation, aiming to assess its effectiveness as a natural remedy for promoting regularity and relieving constipation symptoms (Choudhary *et al.*, 2023). Additionally, a case study documented significant improvements in bowel movements and quality of life in a patient with chronic constipation after 6 months of *D. nobile* intake, suggesting its potential as a treatment option, although further research is warranted to validate these findings (Duan *et al.*, 2022). Furthermore, a pilot study explored the potential of *D. nobile* in alleviating symptoms of GORD, such as heartburn and acid reflux, with the aim of determining its efficacy in reducing oesophageal irritation and improving quality of life in GORD patients (Voon *et al.*, 2020).

A study conducted by Chapelle *et al.* (2021) examined the use of *D. nobile* in a patient with GORD. The patient experienced notable improvements in GORD symptoms after 6 weeks of *D. nobile* supplementation, suggesting a potential therapeutic effect. However, the authors emphasized the need for further studies to validate these findings. Additionally, Zhang *et al.* (2018) conducted an observational study investigating the traditional use of *D. nobile* preparations in patients with peptic ulcer disease. This study aimed to evaluate the effectiveness of *D. nobile* in promoting ulcer healing, reducing gastric acidity and alleviating ulcer-related symptoms. Furthermore, Zhang *et al.* (2018) conducted a study on the effects of *D. nobile* on peptic ulcer disease in rats, revealing significant anti-ulcer effects and reduced inflammation in the gastric mucosa. Fan *et al.* (2023) explored the effects of *D. nobile* on colitis in mice, demonstrating substantial anti-inflammatory effects and decreased oxidative stress in the colon tissue.

Limited research exists on the direct effects of *D. nobile* on coeliac disease. However, preclinical studies have suggested its potential as an adjunct therapy for managing symptoms associated with gluten intolerance and promoting intestinal healing (Deva *et al.*, 2023). Furthermore, Wu *et al.* (2023) conducted a trial evaluating the efficacy of *D. nobile* in improving gastric mucosal integrity, reducing ulcer size and alleviating ulcer-related symptoms such as pain and dyspepsia. These findings contribute to the understanding of the potential therapeutic benefits of *D. nobile* in managing gastritis and peptic ulcers.

Additionally, Nie *et al.* (2020) investigated the effects of *D. nobile* extract on gastrointestinal motility in rats, revealing a prokinetic effect that may benefit individuals with conditions such as constipation or gastroparesis. Lastly, Pei *et al.* (2021) explored the protective effects of *D. nobile* alkaloids on gastric mucosal injury in rats, demonstrating their efficacy in reducing oxidative stress and inflammation.

Oskouei *et al.* (2023) investigated the effects of *D. nobile* polysaccharides in mice with acetic acid-induced ulcerative colitis, revealing significant reductions in inflammation and improved colon histology, suggesting potential therapeutic effects for ulcerative colitis. In a study by Xiong *et al.* (2021), the effects of *D. nobile* on exercise performance in healthy adults were explored, indicating improved exercise performance and reduced markers of muscle damage. However, the limitations of this study include a small sample size and the absence of a control group. Pi *et al.* (2022) conducted a review encompassing various studies on the safety and efficacy of *D. nobile* in humans, focusing on its impacts on exercise performance, cognitive function, and other health outcomes. Additionally, Lu *et al.* (2022) described a patient with irritable bowel syndrome who experienced significant symptom improvement after 2 weeks of taking a *D. nobile*-containing herbal formula, reporting reduced frequency and severity of abdominal pain, bloating and diarrhoea.

Figure 10.2 outlines the effects of *D. nobile* on gastrointestinal disorders and Table 10.3 summarizes the studies that have been conducted.

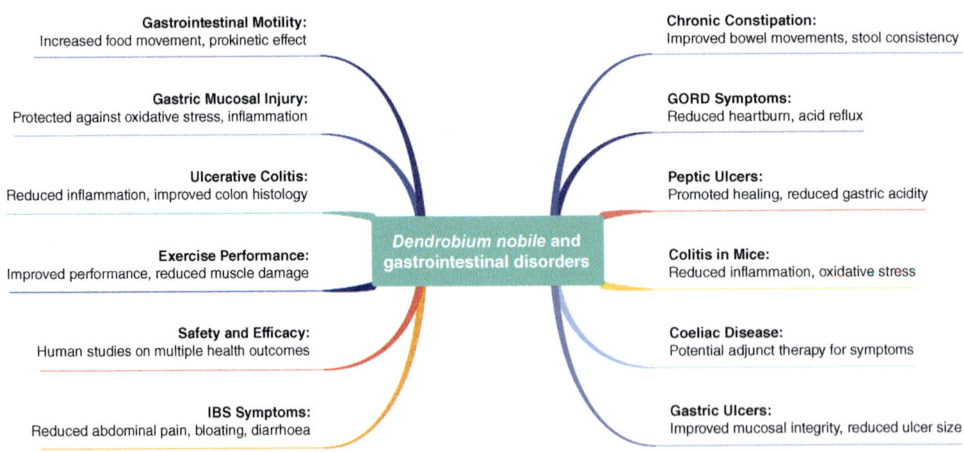

Fig. 10.2. *D. nobile* and gastrointestinal disorders (authors' own figure). IBS, irritable bowel syndrome.

Table 10.3. Studies on *D. nobile* and gastrointestinal disorders (authors' own table).

Subject of study and disorder	Study design and main findings	Participants/ subjects	Intervention/ methodology	Key findings	Reference
Effects of *D. nobile* supplementation on chronic constipation	Clinical trial evaluating the efficacy of *D. nobile* supplementation in improving bowel movement frequency and stool consistency in chronic constipation; significant improvements were observed	Individuals with chronic constipation	*D. nobile* supplementation	Improved bowel movement frequency and stool consistency	Choudhary *et al.* (2023)
Potential of *D. nobile* in chronic constipation	Case study reporting significant improvements in bowel movements and quality of life in a patient with chronic constipation after 6 months of *D. nobile* supplementation	Patient with chronic constipation	*D. nobile* supplementation	Significant improvements in bowel movements and quality of life	Duan *et al.* (2022)
Pilot study on *D. nobile* for GORD	Pilot study assessing the potential of *D. nobile* in reducing symptoms of GORD; significant improvements were observed in symptom severity	Patients with GORD	*D. nobile* supplementation	Reduction in symptoms severity of GORD	Voon *et al.* (2020)
D. nobile treatment for GORD	Study reporting significant improvements in GORD symptoms after 6 weeks of *D. nobile* treatment; suggests potential therapeutic effects on GORD symptoms	Patients with GORD	*D. nobile* treatment	Significant improvements in GORD symptoms	Chapelle *et al.* (2021)
Traditional use of *D. nobile* in peptic ulcer disease	Observational study investigating the traditional use of *D. nobile* in peptic ulcer disease; assessed its effectiveness in promoting ulcer healing and relieving symptoms	Patients with peptic ulcer disease	Traditional use of *D. nobile*	Promotion of ulcer healing and symptom relief	Zhang *et al.* (2018)
Anti-ulcer effects of *D. nobile* in rats	Preclinical study demonstrating significant anti-ulcer effects of *D. nobile* in rats with peptic ulcers, along with reduced inflammation in the gastric mucosa	Animal model: rats with peptic ulcers	*D. nobile* treatment	Anti-ulcer effects and reduced inflammation in gastric mucosa	Zhang *et al.* (2018)
Effects of *D. nobile* on colitis in mice	Preclinical study showing significant anti-inflammatory effects of *D. nobile* in mice with colitis, along with reduced oxidative stress in colon tissue	Animal model: mice with colitis	*D. nobile* treatment	Anti-inflammatory effects and reduced oxidative stress in colon tissue	Fan *et al.* (2023)
Potential of *D. nobile* in coeliac disease	Limited preclinical research suggesting potential as an adjunct therapy for managing symptoms and promoting intestinal healing in coeliac disease	Animal model/ studies	*D. nobile* treatment	Potential as adjunct therapy for symptom management and intestinal healing	Deva *et al.* (2023)
Therapeutic benefits of *D. nobile* in gastritis and peptic ulcers	Clinical trial demonstrating the efficacy of *D. nobile* in improving gastric mucosal integrity, reducing ulcer size and alleviating ulcer-related symptoms	Patients with gastritis and peptic ulcers	*D. nobile* treatment	Improved gastric mucosal integrity, reduced ulcer size and symptom relief	Wu *et al.* (2023)
Prokinetic effects of *D. nobile* extract	Study on rats showing prokinetic effects of *D. nobile* extract, indicating potential benefits for conditions such as constipation or gastroparesis	Animal model: rats	*D. nobile* extract	Prokinetic effects, potential benefits for constipation or gastroparesis	Nie *et al.* (2020)

Continued

Table 10.3. Continued.

Subject of study and disorder	Study design and main findings	Participants/ subjects	Intervention/ methodology	Key findings	Reference
Protective effects of *D. nobile* alkaloids on gastric mucosal injury	Preclinical study demonstrating the protective effects of *D. nobile* alkaloids against gastric mucosal injury by reducing oxidative stress and inflammation	Animal model: rats with gastric mucosal injury	*D. nobile* alkaloids treatment	Protective effects against gastric mucosal injury	Pei *et al.* (2021)
Therapeutic effects of *D. nobile* polysaccharides in ulcerative colitis	Preclinical study showing significant anti-inflammatory effects and histological improvement in mice with ulcerative colitis treated with *D. nobile* polysaccharides	Animal model: mice with ulcerative colitis	*D. nobile* polysaccharides treatment	Anti-inflammatory effects and histological improvement in ulcerative colitis	Oskouei *et al.* (2023)
Effects of *D. nobile* on exercise performance	Study on healthy adults showing improved exercise performance and reduced markers of muscle damage with *D. nobile* supplementation	Healthy adult participants	*D. nobile* supplementation	Improved exercise performance and reduced markers of muscle damage	Xiong *et al.* (2021)
Safety and efficacy of *D. nobile* in humans	Review of several studies evaluating the safety and efficacy of *D. nobile* in humans, covering various health outcomes including GI disorders	Human participants	*D. nobile* supplementation or treatment	Safety and efficacy in various health outcomes including GI disorders	Pi *et al.* (2022)
Therapeutic potential of *D. nobile* in GORD	Case report describing potential therapeutic effects of *D. nobile* in a patient with GORD who did not respond to conventional treatments	Patient with GORD	*D. nobile* treatment	Potential therapeutic effects on GORD symptoms	Ebrahimi *et al.* (2024)
D. nobile-containing herbal formula for IBS	Case report demonstrating significant symptom improvement in a patient with IBS after taking a *D. nobile*-containing herbal formula	Patient with IBS	*D. nobile*-containing herbal formula	Significant symptom improvement in IBS	Lu *et al.* (2022)

GI, gastrointestinal; IBS, irritable bowel syndrome.

10.6 Future Research Directions for *D. nobile* in Gastrointestinal Disorders

Future research directions for *D. nobile* in gastrointestinal disorders should focus on addressing the current limitations and challenges in the existing research. To thoroughly assess the safety and efficacy of *D. nobile* in managing gastrointestinal disorders, meticulously planned clinical trials with larger sample sizes and extended follow-up durations are necessary. Standardizing methods for preparing and administering *D. nobile* extracts is crucial for accurate comparison of results across studies and determining optimal dosages for therapeutic use. Investigating the mechanisms of action of *D. nobile* in treating gastrointestinal disorders can provide insights into its therapeutic potential and guide the development of new treatments. Evaluating the long-term safety of *D. nobile*, especially at high doses or with prolonged use, is essential to fully understand its safety profile and address potential concerns. Studies exploring the synergistic effects of *D. nobile* in combination with other natural products or conventional therapies will help to improve therapeutic outcomes. Comprehensive assessments of the potential benefits and risks of *D. nobile* in treating gastrointestinal disorders, considering factors such as efficacy, safety, tolerability and patient preferences, are necessary. Further research is needed to elucidate optimal therapeutic strategies involving *D. nobile*, ultimately aiming to develop more effective and safer treatments for patients with gastrointestinal disorders. These research objectives aimed to advance our understanding of the role of *D. nobile* in gastrointestinal health and enhance patient care through evidence-based therapeutic approaches.

10.6.1 Development of standardized extracts and dosages

One important area for future research on *D. nobile* and gastrointestinal disorders is the development of standardized extracts and dosages. As mentioned earlier, the lack of standardized methods for preparing and administering *D. nobile* extracts has made it difficult to compare results across studies and determine optimal dosages for therapeutic use. Therefore, future studies should focus on developing standardized methods for the extraction and dosing of *D. nobile*. This would allow for a more accurate comparison of results across studies and could help to establish optimal dosages for therapeutic use. One potential approach could be to use validated extraction methods and quality control procedures to ensure consistency in the composition and purity of extracts. Furthermore, the development of standardized extracts could facilitate the development of *D. nobile*-based pharmaceuticals for the treatment of gastrointestinal disorders. This could involve the identification and isolation of specific bioactive compounds from *D. nobile* extracts, which could then be formulated into pharmaceutical products with known dosages and standardized quality. Overall, the development of standardized extracts and dosages for *D. nobile* could lead to more effective and consistent therapeutic use in the treatment of gastrointestinal disorders. This could also facilitate the development of *D. nobile*-based pharmaceuticals, which could potentially offer a safer and more targeted approach to treatment.

10.6.2 Large-scale clinical trials

Large-scale clinical trials are necessary to determine the safety and efficacy of *D. nobile* in treating gastrointestinal disorders, but most of them have been small and short term, limiting the strength of the conclusions that can be drawn. These trials could also compare the effectiveness of *D. nobile* with other conventional treatments for gastrointestinal disorders. Mechanistic studies are needed to elucidate the underlying mechanism of action of *D. nobile* in treating gastrointestinal disorders. These studies could investigate the specific compounds within the plant responsible for its therapeutic effects, as well as their molecular targets. Understanding the mechanisms of action of *D. nobile* could provide insights into new therapeutic strategies for gastrointestinal disorders. To fully understand the long-term safety profile of *D. nobile*, more studies are needed to investigate the potential adverse effects of prolonged use or high doses of the plant. This could include the investigation of the potential for drug interactions or toxicity with other medications or supplements. Long-term safety studies are essential to ensure that the use of *D. nobile* in treating gastrointestinal disorders is safe. Standardization of the preparation and dosages of *D. nobile* extracts is crucial for accurate comparison of results across studies and for determining optimal dosages for therapeutic use. This could involve developing a set of standardized extraction and preparation methods, as well as guidelines for dosing based on patient factors such as age, weight and severity of the gastrointestinal disorder. Standardization of the preparation and dosages would improve the reproducibility and reliability of the results obtained from clinical trials.

10.7 Implications for Clinical Practice

The current evidence on *D. nobile* suggests that it may have potential therapeutic effects in treating gastrointestinal disorders. However, the limited number and small size of existing studies, as well as the lack of standardized preparations and dosages, make it difficult to draw definitive conclusions. Therefore, it is important for clinicians to be cautious in recommending *D. nobile* as a treatment option for patients with gastrointestinal disorders until more robust evidence is available. Clinicians should also consider the potential risks associated with the use of *D. nobile*, especially in high doses or with prolonged use, and be vigilant for any adverse effects. Patients who are considering using *D. nobile* as a complementary or alternative treatment for their gastrointestinal disorder should also be advised to speak to their healthcare provider first to ensure its safe and appropriate use. Overall, while *D. nobile* shows promise as a potential therapeutic option for gastrointestinal disorders, more rigorous studies are needed to establish its safety and efficacy. Clinicians should be aware of the current evidence and limitations and should approach its use with caution until more definitive evidence is available.

10.7.1 Recommendations for safe and effective use of *D. nobile* in gastrointestinal disorders

While *D. nobile* shows promise in treating gastrointestinal disorders, caution should be exercised until further research is conducted. To safely and effectively use *D. nobile* for these conditions, it is advisable for patients to consult a healthcare professional before starting supplementation, especially for those with underlying medical conditions or who are taking other medications or supplements. Patients should adhere to the recommended dosage guidelines provided by the manufacturer or healthcare provider and refrain from exceeding the suggested dosage. They should also monitor themselves for any potential adverse effects while taking *D. nobile* and discontinue use if any adverse reactions occur, and opt for standardized products to ensure consistency in dosage and potency. Women who are pregnant or breastfeeding should avoid *D. nobile* due to limited safety data. Patients should be advised to be mindful of potential interactions with other medications or supplements and seek advice before starting supplementation. Importantly, *D. nobile* should not replace conventional treatment for gastrointestinal disorders but rather complement it under the guidance of a healthcare professional.

10.8 Conclusion

In conclusion, *D. nobile* shows potential as a natural remedy for the management of gastrointestinal disorders. Several studies have reported the ability of plants to alleviate the symptoms of these disorders, such as abdominal pain, diarrhoea and vomiting. The long-term safety and efficacy of *D. nobile* needs to be investigated, particularly through large-scale clinical trials encompassing extended follow-up periods and larger participant cohorts. Despite promising initial findings suggesting potential benefits for gastrointestinal disorders, the current body of evidence remains limited and inconclusive. Small-scale and short-term studies provide only a glimpse into the potential therapeutic effects of *D. nobile* on gastrointestinal symptoms. However, comprehensive research involving rigorous methodologies and robust data collection is essential to establish its true value as a treatment option for such conditions. Through meticulously designed large-scale clinical trials, we can attain a deeper understanding of the safety profile and effectiveness of *D. nobile* over prolonged periods, offering invaluable insights to guide clinical decision-making and enhance patient care in the realm of gastrointestinal health. Mechanistic studies are required to understand the underlying mechanisms of action of the plant in treating gastrointestinal disorders.

In clinical practice, it is important to note that *D. nobile* is not yet approved by regulatory bodies for the treatment of gastrointestinal disorders. Patients should consult their healthcare providers before using *D. nobile* or any other natural remedy, particularly if they are taking other medications or supplements. The preparation and dosages of *D. nobile* extracts should be standardized to improve the reproducibility and reliability of results obtained from clinical trials. Overall, *D. nobile* has the potential to serve as a complementary or alternative therapy in the management of gastrointestinal disorders. However, further research is required to determine its optimal use and safety profile.

10.8.1 Summary of current evidence and future directions

Current research on the therapeutic potential of *D. nobile* for gastrointestinal disorders remains limited and inconclusive. Although some preliminary studies have suggested possible benefits for gastrointestinal symptoms, the existing evidence primarily comprises small-scale trials of short duration. These studies, while suggestive, lack the robustness and duration necessary to draw definitive conclusions about the safety and effectiveness of *D. nobile* in treating gastrointestinal disorders. Therefore, larger-scale and longer-term clinical trials are imperative to comprehensively evaluate the therapeutic potential, safety profile and efficacy of *D. nobile* in managing various gastrointestinal conditions. Only through rigorous and extensive investigation can we ascertain the true value of *D. nobile* as a treatment option for gastrointestinal disorders and guide clinical practice effectively.

Mechanistic studies are also needed to understand the underlying mechanisms of action of *D. nobile* in treating gastrointestinal disorders. Long-term safety studies are essential to ensure that the use of *D. nobile* in treating gastrointestinal disorders is safe, and standardization of preparation and dosages is crucial for accurate comparison of results across studies and for determining optimal dosages for therapeutic use. Future research directions could include conducting large-scale clinical trials with longer follow-up periods and larger sample sizes, investigating specific compounds in the plant responsible for its therapeutic effects, and developing a set of standardized extraction and preparation methods, as well as guidelines for dosing based on patient factors such as age, weight and the severity of the gastrointestinal disorder. Overall, while *D. nobile* shows potential as a therapeutic option for gastrointestinal disorders, further research is necessary to determine its safety and efficacy and to fully understand its mechanisms of action.

Acknowledgements

The authors express their gratitude for the financial support received through the Distinguished High-Level Talents Research Grant from the Guizhou Science and Technology Corporation Platform Talents Fund (grant nos. [2017]5733-001 and CK-1130-002), the National Natural Science Foundation of China (U1812403, 82373981 and 82060750), 2011 Collaborative Innovation Centre of Traditional Chinese Medicine in Guizhou Province (no. [2022]026) and the support provided by the Zunyi Medical University, China. Special appreciation is extended to all laboratory colleagues and research staff members for their valuable insights, constructive guidance and assistance with the contents of this chapter.

References

Ahlawat, S., Kumar, P., Mohan, H., Goyal, S. and Sharma, K.K. (2021) Inflammatory bowel disease: Tri-directional relationship between microbiota, immune system and intestinal epithelium. *Critical Reviews in Microbiology* 47(2), 254–273.

Andreae, M.H., Rhodes, E., Bourgoise, T., Carter, G.M., White, R.S. *et al.* (2016) An ethical exploration of barriers to research on controlled drugs. *The American Journal of Bioethics* 16(4), 36–47.

Bereda, G. (2022) Peptic ulcer disease: definition, pathophysiology, and treatment. *Journal of Biomedical and Biological Sciences* 1(2), 1–10.

Borah, M., Biswas, R., Saha, A. and Pegu, A.K. (2023) Gastritis in young adults, the evil in disguise: a review of literature. *Mathews Journal of Case Reports* 8(3), 1–7.

Bose, B., Choudhury, H., Tandon, P. and Kumaria, S. (2017) Studies on secondary metabolite profiling, anti-inflammatory potential, in vitro photoprotective and skin-aging related enzyme inhibitory activities of *Malaxis acuminata*, a threatened orchid of nutraceutical importance. *Journal of Photochemistry and Photobiology B: Biology* 173, 686–695.

Byron, C., Cornally, N., Burton, A. and Savage, E. (2020) Challenges of living with and managing inflammatory bowel disease: a meta-synthesis of patients' experiences. *Journal of Clinical Nursing* 29(3–4), 305–319.

Cai, Z., Wang, S. and Li, J. (2021) Treatment of inflammatory bowel disease: a comprehensive review. *Frontiers in Medicine* 8, 765474.

Cakova, V., Bonte, F. and Lobstein, A. (2017) *Dendrobium:* Sources of active ingredients to treat age-related pathologies. *Aging and Disease* 8(6), 827–849.

Camilleri, M. (2021) Diagnosis and treatment of irritable bowel syndrome: a review. *JAMA* 325(9), 865–877.

Chapelle, N., Ben Ghezala, I., Barkun, A. and Bardou, M. (2021) The pharmacotherapeutic management of gastroesophageal reflux disease (GERD). *Expert Opinion on Pharmacotherapy* 22(2), 219–227.

Cheng, Y.Y., Hsieh, C.H. and Tsai, T.H. (2018) Concurrent administration of anticancer chemotherapy drug and herbal medicine on the perspective of pharmacokinetics. *Journal of Food and Drug Analysis* 26(2S), S88–S95.

Choudhary, D., Mashkey, V.K., Goutam, E., Shrivastava, M., Rawat, M. *et al.* (2023) Medicinal orchids: Traditional uses and recent advances. *Annals of Phytomedicine* 12(1), 1–9.

Clarrett, D.M. and Hachem, C. (2018) Gastroesophageal reflux disease (GERD). *Missouri Medicine* 115(3), 214–218.

De, L.C. (2023) Pharmaceutical properties of orchids. *The Pharma Innovation Journal* 12(6), 1454–1460.

Deshmukh, R., Kumari, S. and Harwansh, R.K. (2020) Inflammatory bowel disease: a snapshot of current knowledge. *Research Journal of Pharmacy and Technology* 13(2), 956–962.

Deva, K.D., Bose, B.V. and Basavan, D. (2023) A review on plant-derived immunomodulatory agents: Hopes as an alternative medicine in the management of immune-related disorders. *Traditional and Integrative Medicine* 8(2), 1–22.

Dey, A., Mitra, A., Pathak, S., Prasad, S., Zhang, A.S. *et al.* (2023) Recent advancements, limitations, and future perspectives of the use of personalized medicine in treatment of colon cancer. *Technology in Cancer Research & Treatment* 22, 15330338231178403.

Drewes, A.M., Olesen, A.E., Farmer, A.D., Szigethy, E., Rebours, V. *et al.* (2020) Gastrointestinal pain. *Nature Reviews Disease Primers* 6(1), 1.

Duan, F.P., Li, Y.S., Hu, T.Y., Pan, X.Q., Ma, F. *et al.* (2022) *Dendrobium nobile* protects against ovalbumin-induced allergic rhinitis by regulating intestinal flora and suppressing lung inflammation. *Chinese Journal of Natural Medicines* 20(6), 443–457.

Ebrahimi, N., Abarghooee, E.F., Vazin, A., Ansari, R. and Zarshenas, M.M. (2024) The effectiveness of a poly-herbal formulation from traditional Persian medicine (TPM) in gastroesophageal reflux disease (GERD), a double-blinded randomized clinical trial. *Current Traditional Medicine* 10(7), 1–8.

El-Salhy, M. (2019) Nutritional management of gastrointestinal diseases and disorders. *Nutrients* 11(12), 3013.

Fan, Y., Yu, Q., Wang, G., Tan, J., Liu, S. *et al.* (2020) Effects of non-thermal plasma treatment on the polysaccharide from *Dendrobium nobile* Lindl. and its immune activities in vitro. *International Journal of Biological Macromolecules* 153, 942–950.

Fan, C., Sun, X., Wang, X. and Yu, H. (2023) Therapeutic potential of the chemical composition of *Dendrobium nobile* Lindl. *Frontiers in Pharmacology* 14, 1163830.

Fu, X., Chen, S., Xian, S., Wu, Q., Shi, J. *et al.* (2023) *Dendrobium* and its active ingredients: emerging role in liver protection. *Biomedicine & Pharmacotherapy* 157, 114043.

Ghimire, S., Yang, T., Shah, H., Sarker, A. and Srivatana, U. (2019) 2716 A rare case of gastric glomus tumor. *American Journal of Gastroenterology* 114(1), S1499–S1500.

Gong, H., Zhao, N., Zhu, C., Luo, L. and Liu, S. (2024) Treatment of gastric ulcer, traditional chinese medicine may be a better choice. *Journal of Ethnopharmacology* 324, 117793.

Gunby, S.A. and Strate, L.L. (2024) Acute colonic diverticulitis. *Annals of Internal Medicine* 177(3), ITC33–ITC48.

Herdiana, Y. (2023) Functional food in relation to gastroesophageal reflux disease (GERD). *Nutrients* 15(16), 1–17.

Ketsa, S. and Warrington, I.J. (2023) The *Dendrobium* orchid: botany, horticulture, and utilization. *Crop Science* 63(4), 1829–1888.

Le Berre, C., Ananthakrishnan, A.N., Danese, S., Singh, S. and Peyrin-Biroulet, L. (2020) Ulcerative colitis and crohn's disease have similar burden and goals for treatment. *Clinical Gastroenterology and Hepatology* 18(1), 14–23.

Lee, J.H., Kedia, P., Stavropoulos, S.N. and Carr-Locke, D. (2021) AGA clinical practice update on endoscopic management of perforations in gastrointestinal tract: expert review. *Clinical Gastroenterology and Hepatology* 19(11), 2252–2261.

Li, Z., Shi, J., Hu, D. and Song, B. (2019) A polysaccharide found in *Dendrobium nobile* Lindl stimulates calcium signaling pathway and enhances tobacco defense against TMV. *International Journal of Biological Macromolecules* 137, 1286–1297.

Liu, J., Li, Z., Zeng, M., Liang, Y., Liang, B., Ge, Q., Zhang, X., Zhang, S., and Zhou, W. (2023) Rapid comparison of antitumor chemical constituents and mechanisms between *Dendrobium nobile* and *Dendrobium officinale* by UPLC-IT-TOF, network pharmacology and experimental verification. *Acta Chromatographica* 35(2), 123–138.

Lu, Q., Tan, D., Luo, J., Ye, Y., Zuo, M., Wang, S., and Li, C. (2022) Potential of natural products in the treatment of irritable bowel syndrome. *Phytomedicine* 106, 154419.

Nie, X., Chen, Y., Li, W., and Lu, Y. (2020) Anti-aging properties of *Dendrobium nobile* Lindl.: from molecular mechanisms to potential treatments. *Journal of Ethnopharmacology* 257, 112839.

Obeagu, E.I., and Obeagu, G.U. (2024) Exploring the intersection: peptic ulcers and hemolysis—Unraveling the complex relationship. *Medicine* 103(11), 1–6.

Orjiekwe, O.A. (2023) Nutritional management of celiac disease. *Journal of Clinical and Experimental Immunology* 8(2), 561–572.

Oskouei, Z., Ghasemzadeh Rahbardar, M. and Hosseinzadeh, H. (2023) The effects of *Dendrobium* species on the metabolic syndrome: a review. *Iranian Journal of Basic Medical Sciences* 26(7), 738–752.

Pabolu, S., Dudekula, A. and Pitchumoni, C.S. (2021) Gastrointestinal manifestations of non-GI disorders. *Geriatric Gastroenterology* 2, 2117–2166.

Pan, R., Wang, L., Xu, X., Chen, Y., Wang, H. *et al.* (2022) Crosstalk between the gut microbiome and colonic motility in chronic constipation: potential mechanisms and microbiota modulation. *Nutrients* 14(18), 1–28.

Pei, Z., Lou, Z., Zhang, B., Wang, H. and Li, Y. (2021) Development of a compound oral liquid containing herbal extracts and its effect on immunity and gastric mucosa. *Journal of Food Science* 86(6), 2684–2699.

Pi, T., Lang, G., Liu, B. and Shi, J. (2022) Protective effects of *Dendrobium nobile* Lindl. alkaloids on Alzheimer's disease-like symptoms induced by high-methionine diet. *Current Neuropharmacology* 20(5), 983–997.

Puyuelo, C.J.G., Sopeña, F. and Arbeloa, A.L. (2018) Colonic diverticular disease. Treatment and prevention. *Gastroenterología y Hepatología* 38(10), 590–599.

Qiu, Y., Cai, C., Mo, X., Zhao, X., Wu, L. *et al.* (2023) Transcriptome and metabolome analysis reveals the effect of flavonoids on flower color variation in *Dendrobium nobile* Lindl. *Frontiers in Plant Science* 14, 1220507.

Raj, D., Shah, M.S. and Sharma, D. (2022) A review: extraction, isolation and determination of secondary metabolite (phenols) from medicinal plants and its benefits. *Journal of Medicinal Plants Studies* 10(4), 181–193.

Rao, D., Zhao, R., Hu, Y., Li, H., Chun, Z. *et al.* (2023) Revealing of intracellular antioxidants in *Dendrobium nobile* by high performance liquid chromatography-tandem mass spectrometry. *Metabolites* 13(6), 1–24.

Raucci, U., Boni, A., Evangelisti, M., Della Vecchia, N., Velardi, M. *et al.* (2021) Lifestyle modifications to help prevent headache at a developmental age. *Frontiers in Neurology* 11, 618375.

Schiller, L.R. (2019) Chronic constipation: New insights, better outcomes. *The Lancet Gastroenterology & Hepatology* 4(11), 873–882.

Sharma, N., Bhatia, S., Chunduri, V., Kaur, S., Sharma, S. *et al.* (2020) Pathogenesis of celiac disease and other gluten related disorders in wheat and strategies for mitigating them. *Frontiers in Nutrition* 7, 6.

Singh, S., Chakole, S., Agrawal, S., Shetty, N., Prasad, R. *et al.* (2023) A comprehensive review of upper gastrointestinal symptom management in autoimmune gastritis: current insights and future directions. *Cureus* 15(8), 1–11.

Targownik, L.E., Fisher, D.A. and Saini, S.D. (2022) AGA clinical practice update on de-prescribing of proton pump inhibitors: expert review. *Gastroenterology* 162(4), 1334–1342.

Teng, L., Shaw, D. and Barnes, J. (2018) Characteristics and practices of traditional Chinese medicine retail shops in London, UK: a cross-sectional study using an observational approach. *Journal of Ethnopharmacology* 173, 318–329.

Voon, H.C., Bhat, R. and Rusul, G. (2020) Flower extracts and their essential oils as potential antimicrobial agents for food uses and pharmaceutical applications. *Comprehensive Reviews in Food Science and Food Safety* 11(1), 34–55.

Wang, Y.H. (2021) Traditional uses, chemical constituents, pharmacological activities, and toxicological effects of *Dendrobium* leaves: a review. *Journal of Ethnopharmacology* 270, 113851.

Wang, X. and Shen, B. (2018) Management of crohn's disease and complications in patients with ostomies. *Inflammatory Bowel Diseases* 24(6), 1167–1184.

Wang, Y., Chen, S., Du, K., Liang, C., Wang, S. *et al.* (2021) Traditional herbal medicine: therapeutic potential in rheumatoid arthritis. *Journal of Ethnopharmacology* 279, 114368.

Włodarczyk, J., Waśniewska, A., Fichna, J., Dziki, A., Dziki, Ł. *et al.* (2021) Current overview on clinical management of chronic constipation. *Journal of Clinical Medicine* 10(8), 1–17.

Wu, W., Lin, Y., Farag, M.A., Li, Z. and Shao, P. (2023) *Dendrobium* as a new natural source of bioactive for the prevention and treatment of digestive tract diseases: a comprehensive review with future perspectives. *Phytomedicine* 114, 154784.

Xi, Y. and Xu, P. (2021) Global colorectal cancer burden in 2020 and projections to 2040. *Translational Oncology* 14(10), 101174.

Xiong, T.W., Liu, B., Wu, Q., Xu, Y.Y., Liu, P. *et al.* (2021) Beneficial effects of *Dendrobium nobile* Lindl. Alkaloids (DNLA) on anxiety and depression induced by chronic unpredictable stress in rats. *Brain Research* 1771, 147647.

Xu, Q., Niu, S.C., Li, K.L., Zheng, P.J., Zhang, X.J. *et al.* (2022) Chromosome-scale assembly of the *Dendrobium nobile* genome provides insights into the molecular mechanism of the biosynthesis of the medicinal active ingredient of *Dendrobium*. *Frontiers in Genetics* 13, 1–15.

You, H., Sweeny, A., Cooper, M.L., Papen, M. and Innes, J. (2019) The management of diverticulitis: a review of the guidelines. *Medical Journal of Australia* 211(9), 421–427.

Zhai, D., Lv, X., Chen, J., Peng, M. and Cai, J. (2022) Recent research progress on natural stilbenes in *Dendrobium* species. *Molecules* 27(21), 1–38.

Zhang, Y., Wang, H., Mei, N., Ma, C., Lou, Z. *et al.* (2018) Protective effects of polysaccharide from *Dendrobium nobile* against ethanol-induced gastric damage in rats. *International Journal of Biological Macromolecules* 107, 230–235.

Zhang, Y., Yang, Q., Song, B., Tang, W., Yu, F. *et al.* (2024) Efficacy and safety of Piwei Peiyuan Prescription in the treatment of chronic atrophic gastritis: a multicenter, double-blind, double-simulated, randomized, controlled clinical trial. *Medicine* 103(19), 1–8.

11 *Dendrobium nobile* in the Treatment of Renal Diseases

Surendra Sarsaiya, Archana Jain, Qihai Gong, Qin Wu and Jingshan Shi*

Key Laboratory of Basic Pharmacology and Joint International Research Laboratory of Ethnomedicine of Ministry of Education, Zunyi Medical University, Zunyi, Guizhou, China

Abstract

This chapter explores the therapeutic potential of *Dendrobium nobile* in the treatment of renal diseases. *D. nobile* exhibits notable antioxidant, anti-inflammatory and diuretic properties that could protect kidneys from oxidative stress and inflammation, which are crucial factors in renal damage progression. It may also enhance renal blood flow and the glomerular filtration rate, offering a multifaceted approach to renal health. Preclinical studies have shown promising results using models of chronic kidney disease, acute kidney injury, nephrolithiasis and diabetic nephropathy. However, the clinical evidence remains limited, necessitating further research to confirm its efficacy and safety in humans. While the bioactive compounds of *D. nobile* have potential benefits, understanding their mechanisms, optimal dosages, long-term safety and potential drug interactions is essential. Future research should focus on well-designed clinical trials and pharmacokinetic studies to establish *D. nobile* as a viable adjunctive therapy for renal diseases, integrating it into comprehensive treatment plans under professional guidance.

11.1 Introduction

Research into *Dendrobium nobile* Lindl. has unveiled compelling findings for renal diseases. The bioactive compounds in this orchid species exhibit remarkable antioxidant capabilities, which are believed to protect the kidneys from oxidative stress, a pivotal factor in the progression of renal damage (Fu *et al.*, 2023). Moreover, the presence of specific constituents in *D. nobile* has shown promise in mitigating inflammation within the renal system, thereby fostering an environment conducive to improving kidney function (Rabidas *et al.*, 2023). Renal diseases can lead to the progressive loss of renal function, impaired quality of life and increased morbidity and mortality. Despite advances in medical treatment, there is a need for alternative therapeutic approaches to effectively manage renal diseases and improve patient outcomes (Kalariya *et al.*, 2023). The diuretic properties of *D. nobile* offer another dimension of potential benefits for renal health (Ramesh *et al.*, 2019). By facilitating the expulsion of waste products and excess fluids from the body, these properties could play a crucial role in supporting the overall kidney function (Jafari and Rohn, 2022). Additionally, the ability of *D. nobile* to influence renal blood flow and enhance glomerular filtration rate suggests a multifaceted approach to bolster renal health (Liu *et al.*, 2024a).

*Corresponding author: shijs@zmu.edu.cn

© Jingshan Shi, Archana Jain, Surendra Sarsaiya, Qin Wu and Qihai Gong 2025. Dendrobium nobile: *Research and Therapeutic Applications* (J. Shi, A. Jain, S. Sarsaiya, Q. Wu and Q. Gong)
DOI: 10.1079/9781800627987.0011

While initial investigations underscore the potential of *D. nobile* in the management of renal diseases, further exploration is indispensable to unravel its intricate mechanisms of action and ascertain its clinical efficacy (Li *et al.*, 2022a). Equally imperative is a comprehensive assessment of the safety profile of *D. nobile*, particularly concerning its long-term use and potential interactions with concurrent medications (Zhang *et al.*, 2023a). The use of *D. nobile* in the treatment of renal diseases has gained attention because of its renoprotective, anti-inflammatory, antioxidant and antifibrotic properties (Wu *et al.*, 2021). Preclinical studies conducted on animal models and *in vitro* cell cultures have shown promising results, indicating the potential of *D. nobile* as a therapeutic agent for renal diseases (Rao *et al.*, 2023). However, despite the growing interest in *D. nobile*, comprehensive research is needed to evaluate its effectiveness, safety and underlying mechanisms of action in the context of renal disease. Clinical studies examining the efficacy of *D. nobile* in patients with chronic kidney disease (CKD), acute kidney injury (AKI) and nephrolithiasis are limited, and existing evidence is largely based on preclinical investigations.

11.2 Overview of Renal Diseases

Renal disease refers to a wide range of conditions that affect the kidneys, impairing their ability to function properly. These diseases can lead to significant health challenges and considerably impact a person's quality of life (Table 11.1).

11.2.1 Chronic kidney disease

CKD is a chronic and progressive disorder characterized by a gradual decline in kidney function. It commonly arises from underlying ailments, such as diabetes, hypertension and specific kidney disorders. CKD can result in complications, such as disruptions in fluid and electrolyte balance, anaemia, bone issues and cardiovascular complications (Kang *et al.*, 2021). Symptoms may include fatigue, swelling in the legs, ankles or around the eyes (oedema), difficulty concentrating, decreased appetite and changes in urine output (either increased or decreased frequency) (Fletcher *et al.*, 2022). Treatment focuses on managing underlying conditions, controlling blood pressure and blood sugar levels, and addressing complications, such as anaemia and bone disease (Hao *et al.*, 2023).

11.2.2 Nephrolithiasis (kidney stones)

Nephrolithiasis, also known as kidney stones, occurs when hard deposits form in the kidneys. These stones vary in size and composition, and can lead to symptoms such as intense back or side pain, haematuria (blood in the urine) and urinary tract infections (UTIs). The factors contributing to kidney stone formation include dehydration, certain dietary habits, genetics and metabolic disorders (Gaur *et al.* 2022). Treatment typically involves pain management, increased fluid intake to facilitate stone passage, muscle-relaxing medications for the ureter and, occasionally, surgical interventions to address larger stones (Tamang *et al.*, 2023).

11.2.3 Urinary tract infections (UTIs)

UTIs are bacterial infections that affect various parts of the urinary tract including the kidneys, bladder, ureters and urethra (Mancuso *et al.*, 2023). Symptoms commonly include pain or burning sensations during urination, increased frequency of urination and cloudy or bloody urine. In severe cases, UTIs can progress to kidney infections (pyelonephritis), necessitating immediate medical attention. Additional symptoms of UTIs may include pelvic pain. UTIs can sometimes lead to kidney infections (pyelonephritis), which are more severe and require prompt treatment (Kaur and Kaur, 2021). Treatment typically involves antibiotics to clear the infection, increased fluid intake and measures to prevent recurrent UTIs such as proper hygiene practices (Sosland and Stewart, 2021).

11.2.4 Acute kidney injury

AKI manifests as a sudden and abrupt deterioration of kidney function. This condition may arise due to multiple factors such as dehydration,

Table 11.1. Overview of renal diseases (authors' own table).

Renal disease	Description	Symptoms	Treatment	References
Chronic kidney disease (CKD)	A chronic, progressive disorder marked by a gradual decline in kidney function	Fatigue, oedema, difficulty concentrating, decreased appetite, changes in urine output	Managing underlying conditions, controlling blood pressure and blood sugar levels, addressing complications such as anaemia and bone disease	Kang *et al.* (2021); Fletcher *et al.* (2022); Hao *et al.* (2023)
Nephrolithiasis (kidney stones)	Formation of hard deposits in the kidneys	Intense back or side pain, haematuria, urinary tract infections	Pain management, increased fluid intake, muscle-relaxing medications, surgical interventions for larger stones	Gaur *et al.* (2022); Tamang *et al.* (2023);
Urinary tract infections (UTIs)	Bacterial infections affecting various parts of the urinary tract	Pain during urination, increased frequency of urination, cloudy or bloody urine	Antibiotics, increased fluid intake, proper hygiene practices	Mancuso *et al.* (2023); Kaur and Kaur (2021); Sosland and Stewart (2021)
Acute kidney injury (AKI)	Sudden deterioration in kidney function	Reduced urine output, fluid retention, fatigue, confusion, nausea	Addressing underlying cause, intravenous fluids, medications to mitigate complications	Gaut and Liapis (2021); Hao *et al.* (2021); Kellum *et al.* (2021)
Glomerulonephritis	Inflammation affecting the glomeruli, the kidneys' blood-filtering units	Proteinuria, haematuria, swelling, hypertension	Symptom management, controlling blood pressure and proteinuria, immunosuppressive medications	Pesce *et al.* (2021); Sethi *et al.* (2022); Zhang *et al.* (2023b)
Polycystic kidney disease (PKD)	Hereditary condition characterized by fluid-filled cysts in the kidneys	Back or flank discomfort, hypertension, recurrent infections, kidney stones	Symptom management, blood pressure control, prevention of complications, renal replacement therapy (dialysis or transplantation)	Cho *et al.* (2021); Zhang *et al.* (2021); Lakhia *et al.* (2022)
Renal failure (end-stage renal disease)	Kidneys lose their ability to sustain life	Fatigue, nausea, vomiting, swelling, changes in urine output, difficulty concentrating	Dialysis, kidney transplantation	Hei *et al.* (2021); Olsen and Galen (2022); Liu *et al.* (2024a)

kidney infections, certain medications and diminished blood flow to the kidneys. AKI can lead to a rapid build up of waste products and fluids in the body, requiring immediate medical intervention (Kellum *et al.*, 2021). Symptoms of AKI can vary, encompassing reduced urine output and fluid retention, resulting in swelling, fatigue, confusion, nausea and chest discomfort (Gaut and Liapis, 2021). Treatment primarily revolves around pinpointing and rectifying the underlying cause, providing supportive measures such as intravenous fluids to sustain hydration and administering medications to mitigate complications such as electrolyte imbalances (Hao *et al.*, 2021).

11.2.5 Glomerulonephritis

Glomerulonephritis comprises a cluster of conditions affecting the glomeruli and intricate blood vessels in the kidneys, which are tasked with filtering waste and surplus fluids from the bloodstream. Inflammation of the glomeruli can compromise their function, resulting in proteinuria (excessive protein in urine), haematuria and diminished kidney function (Sethi *et al.*, 2022). Symptoms may include both the aforesaid proteinuria and haematuria, and also swelling of the face, abdomen or legs, and hypertension (Pesce *et al.*, 2021). Treatment aims to control symptoms and manage blood pressure and proteinuria, and may involve immunosuppressive medications to reduce inflammation and preserve kidney function (Zhang *et al.*, 2023b).

11.2.6 Polycystic kidney disease

Polycystic kidney disease (PKD), is a hereditary condition characterized by the proliferation of multiple fluid-filled cysts in the kidneys. As these cysts expand over time, they prompt kidney enlargement. PKD can cause progressive impairment of kidney function and, in certain instances, culminate in end-stage renal disease, necessitating interventions such as dialysis or kidney transplantation (Lakhia *et al.*, 2022). Symptoms associated with PKD may include back or flank discomfort, hypertension, recurrent tract infections and kidney stone formation (Cho *et al.*, 2021). Treatment focuses on managing symptoms, controlling blood pressure, preventing complications, such as infections and kidney stones, and, in advanced cases, renal replacement therapy with dialysis or kidney transplantation (Zhang *et al.*, 2021).

11.2.7 Renal failure

Renal failure, also referred to as end-stage renal disease, is a condition in which the kidneys lose their ability to sustain life adequately. It frequently represents the ultimate stage of CKD or arises from severe AKI. Renal failure requires dialysis or kidney transplantation to replace lost kidney function (Olsen and Galen, 2022). Symptoms include fatigue, nausea, vomiting, swelling of the legs and abdomen, changes in urine output and difficulty concentrating (Liu *et al.*, 2024b). Treatment involves renal replacement therapy, such as dialysis to remove waste products and excess fluids from the body, or kidney transplantation to permanently replace lost kidney function (Hei *et al.*, 2021).

Management and treatment strategies for renal diseases are contingent on a particular condition and its root cause. Approaches to treatment may encompass lifestyle adjustments, pharmaceutical interventions, dialysis (for advanced kidney failure) and, in some cases, kidney transplantation. Early detection, regular monitoring and appropriate medical care are crucial for managing renal disease and improving patient outcomes.

11.3 Therapeutic Potential of *D. nobile* in Renal Diseases

D. nobile, an orchid species native to China, has garnered significant interest in recent years for its potential therapeutic applications in the management of renal diseases. Traditional medicine systems have long recognized their medicinal properties, and modern research has begun to unveil their potential benefits in treating various renal conditions. Traditional medicine systems such as traditional Chinese medicine and Ayurveda have extensively acknowledged the therapeutic attributes of *D. nobile*. These systems have been utilized for various health purposes, including promoting kidney health and treating renal disorders. *D. nobile* is a prime example of the rich botanical heritage of Asia and its potential contributions to modern medicine. With its traditional roots firmly grounded in centuries of cultural wisdom and its therapeutic potential increasingly validated by scientific research, *D. nobile* offers a promising avenue for the management of renal diseases. Further exploration and validation of its efficacy and safety profile are warranted to fully harness its therapeutic benefits in individuals suffering from renal conditions. Healthcare professionals continue to explore the therapeutic potential of *D. nobile* as a complementary or alternative treatment option for renal diseases (Fig. 11.1).

11.3.1 Renoprotective effects of *D. nobile*

D. nobile, which is used in traditional medicine, has emerged as a promising agent for safeguarding renal health. It can shield kidneys from injury and dysfunction (Rao *et al.*, 2023). By preserving renal function and impeding the progression of renal diseases, *D. nobile* showcases its renoprotective effects. Through intricate interactions with cellular pathways, it orchestrates a defence mechanism against detrimental factors, advocating renal resilience and vitality (Hei *et al.*, 2021).

11.3.2 Anti-inflammatory properties

Within the intricate landscape of renal disease, inflammation plays a pivotal role in exacerbating renal injury and dysfunction. *D. nobile*, with its reported anti-inflammatory properties, has emerged as a beacon of hope in this domain. By modulating inflammatory signalling pathways within the kidneys it can alleviate renal inflammation and its consequent complications (Li *et al.*, 2023a). Through its multifaceted action, *D. nobile* presents itself as a potential therapeutic agent against renal inflammation (Ti *et al.*, 2021).

11.3.3 Antioxidant activity

The journey towards renal health is often hindered by oxidative stress. Here, *D. nobile* has been shown to have potent antioxidant activity in the intricate dance between free radicals and cellular defences

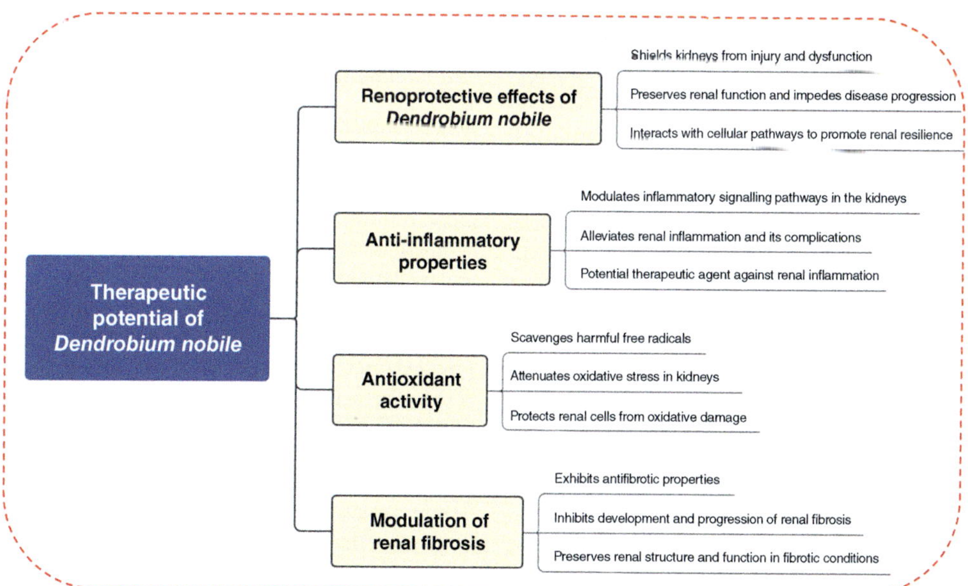

Fig. 11.1. Therapeutic potential of *D. nobile* in renal disease (authors' own figure).

(Chen, 2021). *D. nobile* has emerged as a champion that scavenges harmful free radicals and attenuates oxidative stress. It fortifies the foundation of renal health and vitality by shielding renal cells from oxidative damage (Li *et al.*, 2022a).

11.3.4 Modulation of renal fibrosis

Renal fibrosis is a hallmark of progressive renal disease and casts a shadow over the renal health landscape. However, *D. nobile* is a beacon of hope (Zhou *et al.*, 2020). Through its reported antifibrotic properties, its molecular interactions inhibit the development and progression of renal fibrosis (Tian *et al.*, 2023). By navigating the intricate pathways of fibrotic cascades, *D. nobile* has emerged as a potential guardian of renal structure and function (He *et al.*, 2023). *D. nobile* has shown potential in modulating renal fibrosis, which is a common pathological process in various renal diseases (Zhou *et al.*, 2020).

11.4 Clinical Studies on *D. nobile* in Renal Diseases

Currently, there is limited clinical evidence regarding the use of *D. nobile* in the treatment of renal diseases. Most research conducted on *D. nobile* and its effects on renal health has been in the form of preclinical studies, primarily using animal models and *in vitro* cell cultures. However, few studies have explored the potential clinical applications of *D. nobile* in renal diseases. The study of the efficacy and safety of *D. nobile* in CKD patients is an area of ongoing research. While there is limited clinical evidence specifically focused on the effects of *D. nobile* in CKD, certain properties of *D. nobile* and its constituents suggest potential benefits and safety considerations (Hao *et al.*, 2023). The use of *D. nobile* as an adjunctive therapy in AKI is an area of emerging interest, although clinical studies specifically on its use in AKI are limited (Hao *et al.*, 2021). Additionally, the potential of *D. nobile* in preventing the recurrence of nephrolithiasis, or kidney stones, is another area that warrants further investigation (Tamang *et al.*, 2023). A study exploring the effects of *D. nobile* in a rat model of kidney ischaemia/reperfusion injury found that treatment with *D. nobile* alleviated kidney injury, reduced oxidative stress and enhanced antioxidant enzyme activity, thereby exhibiting renoprotective effects (Luo *et al.*, 2024). Furthermore, research on *D. nobile* alkaloids in a rat model of early diabetic nephropathy demonstrated that these alkaloids can attenuate renal damage, improve renal function and reduce oxidative stress and inflammation, indicating their potential as therapeutic interventions for diabetic nephropathy (Oskouei *et al.*, 2023). Another study examined the renoprotective effects of *D. nobile* aqueous extract in a rat model of five sixths nephrectomy-induced renal injury, where approximately five sixths of the kidney tissue is removed or rendered nonfunctional. The findings revealed that *D. nobile* treatment ameliorated renal fibrosis, reduced oxidative stress and improved renal function, suggesting its therapeutic value in the prevention and treatment of renal injury (Feng *et al.*, 2018).

A study explored the effects of *D. nobile* alkaloids in a rat model of diabetic nephropathy, revealing that these alkaloids reduced renal inflammation and improved renal function, suggesting their potential anti-inflammatory effects in the context of renal diseases (Li *et al.*, 2018). Similarly, research on *D. nobile* alkaloids in a rat model of aristolochic acid-induced AKI indicated that these alkaloids attenuated renal inflammation and fibrosis, highlighting their potential as anti-inflammatory agents in kidney injury (Wu *et al.*, 2023). In another study, the impact of *D. nobile* alkaloids on metabolism-related gene expression was examined in the livers of mice. These results suggest that these alkaloids exert anti-inflammatory effects by modulating the expression of genes involved in inflammation and oxidative stress (Liu *et al.*, 2020). Additionally, research on the protective effects of *D. nobile* alkaloids against cisplatin-induced AKI in mice revealed that these alkaloids exert antioxidant activity by activating the nuclear factor erythroid 2-related factor 2 (Nrf2)-antioxidant response element (ARE) pathway, leading to the upregulation of antioxidant enzymes and reduction of oxidative stress in the kidneys (Dai *et al.*, 2024). Furthermore, a study on *D. nobile* alkaloids in a rat model of diabetic nephropathy demonstrated that these alkaloids alleviated renal oxidative stress and inflammation, indicating their antioxidant properties as a potential mechanism for protection against diabetic nephropathy (Li *et al.*, 2023b).

A study examined the effects of *D. nobile* alkaloids on renal fibrosis in a rat model of unilateral ureteral obstruction. These findings suggest that these alkaloids ameliorate renal fibrosis by reducing oxidative stress and inflammation, indicating their antioxidant activity in renal diseases (Li *et al.*, 2022b). Similarly, another study using the same rat model showed that *D. nobile* alkaloids significantly reduced renal fibrosis by suppressing oxidative stress and inflammation, suggesting that their antifibrotic effects may be mediated through the regulation of these pathways (Wu *et al.*, 2021). Further research investigated the potential antifibrotic effects of *D. nobile* alkaloids in a rat model of unilateral ureteral obstruction. This study suggests that the modulation of oxidative stress and inflammation pathways by *D. nobile* alkaloids may contribute to their antifibrotic effects (Cakova *et al.*, 2017). Additionally, a study on cisplatin-induced ototoxicity provided insights into the antifibrotic effects of *D. nobile*. The results indicated that treatment with *D. nobile* alkaloids effectively suppressed renal fibrosis by modulating the AMP-activated protein kinase (AMPK)/mammalian target of rapamycin (mTOR) pathway, suggesting that the regulation of autophagy and the AMPK/mTOR pathway may play a role in their antifibrotic effects (Nie *et al.*, 2020).

One study assessed the clinical effectiveness and safety of *D. nobile* injection in individuals with CKD. The results demonstrated that *D. nobile* injection improved renal function and reduced proteinuria in patients with CKD. The treatment was well tolerated, with no significant adverse effects. However, it is essential to acknowledge the study's limited sample size, underscoring the necessity for further research with larger cohorts and robust study designs to validate these results definitively (Eslamifar *et al.*, 2021). Another study investigated the therapeutic effects of oral *D. nobile* solution combined with losartan potassium tablets in patients with early diabetic nephropathy. Combination therapy resulted in significant improvements in renal function, a reduction in urinary albumin excretion and decreased blood pressure compared to the control group receiving losartan potassium tablets alone. These findings suggest that *D. nobile* oral solution may have favourable effects in the management of early diabetic nephropathy. Nevertheless, further extensive and meticulously controlled studies are required to corroborate these observations (Table 11.2) (Zhang *et al.*, 2023c).

11.5 Mechanisms of Action of *D. nobile* in Renal Diseases

The mechanisms of action of *D. nobile* in renal disease are not fully understood and require further research. However, based on the available studies and preliminary evidence, several potential mechanisms have been proposed. It is important to note that these mechanisms are theoretical and may not have been conclusively proven. *D. nobile* is believed to possess anti-inflammatory, antioxidant and antifibrotic properties that may benefit kidney health. The anti-inflammatory characteristics of *D. nobile* can help mitigate renal inflammation by inhibiting pro-inflammatory pathways and the release of inflammatory mediators, thereby reducing kidney inflammation and the associated damage (Li *et al.*, 2023b). Additionally, the bioactive compounds in *D. nobile* exhibit antioxidant activity, counteract reactive oxygen species (ROS) and alleviate oxidative stress within the kidneys. These compounds may protect renal cells from injury and support renal function by reducing oxidative damage (Li *et al.*, 2022b). Furthermore, *D. nobile* has been suggested to have antifibrotic effects, potentially inhibiting the activation of fibrogenic pathways and the deposition of extracellular matrix proteins, which may help prevent or slow the progression of renal fibrosis, a common feature of advancing renal diseases (Li *et al.*, 2018; Panizo *et al.*, 2021).

D. nobile may offer therapeutic benefits for renal health through various mechanisms, including the regulation of key signalling pathways. Its alkaloids are thought to modulate the AMPK/mTOR pathway, which influences cellular processes, such as cell growth, proliferation and metabolism, potentially contributing to the inhibition of renal fibrosis (Dai *et al.*, 2024). Additionally, *D. nobile* alkaloids might mitigate renal fibrosis by modulating the transforming growth factor beta (TGF-β)1/Smad and mitogen-activated protein kinase (MAPK) signalling

Table 11.2. Clinical studies of *D. nobile* in renal diseases (authors' own table).

Study topic	Focus	Findings	Reference
Efficacy and safety of *D. nobile* in CKD patients	CKD	Limited evidence, potential benefits, safety considerations	Hao *et al.* (2023)
D. nobile as adjunctive therapy in AKI	AKI	Emerging interest, theoretical considerations, potential benefits	Hao *et al.* (2021)
Effectiveness of *D. nobile* in preventing nephrolithiasis recurrence	Nephrolithiasis	Requires further investigation	Tamang *et al.* (2023)
Renoprotective effects of *D. nobile* in kidney ischaemia/reperfusion injury	Kidney ischaemia/ reperfusion injury	Alleviates kidney injury, reduces oxidative stress, enhances antioxidant enzyme activity	Luo *et al.* (2024)
Renoprotective effects of *D. nobile* alkaloids in early diabetic nephropathy	Diabetic nephropathy	Attenuates renal damage, improves renal function, reduces oxidative stress and inflammation	Oskouei *et al.* (2023)
Renoprotective effects of *D. nobile* aqueous extract in nephrectomy-induced renal injury	Renal injury	Ameliorates renal fibrosis, reduces oxidative stress, improves renal function	Feng *et al.* (2018)
Renoprotective effects of *D. nobile* alkaloids in diabetic nephropathy	Diabetic nephropathy	Reduces renal inflammation, improves renal function	Li *et al.* (2018)
Renoprotective effects of *D. nobile* alkaloids in aristolochic acid-induced AKI	AKI	Attenuates renal inflammation and fibrosis	Wu *et al.* (2023)
Anti-inflammatory effects of *D. nobile* alkaloids on liver gene expression in mice	Liver gene expression	Modulates inflammation and oxidative stress	Liu *et al.* (2020)
Antioxidant activity of *D. nobile* alkaloids in cisplatin-induced AKI in mice	Cisplatin-induced AKI	Activates the Nrf2-ARE pathway, upregulates antioxidant enzymes, reduces oxidative stress	Dai *et al.* (2024)
Antioxidant properties of *D. nobile* alkaloids in diabetic nephropathy	Diabetic nephropathy	Alleviates renal oxidative stress and inflammation	Li *et al.* (2023a)
Antifibrotic effects of *D. nobile* alkaloids in unilateral ureteral obstruction	Renal fibrosis	Reduces renal fibrosis by suppressing oxidative stress and inflammation	Li *et al.* (2022b)
Antifibrotic effects of *D. nobile* alkaloids in unilateral ureteral obstruction	Renal fibrosis	Significant reduction in renal fibrosis through the regulation of oxidative stress and inflammation pathways	Wu *et al.* (2021)
Antifibrotic effects of *D. nobile* alkaloids in unilateral ureteral obstruction	Renal fibrosis	Modulation of pathways contributing to antifibrotic effects	Cakova *et al.* (2017)
Potential antifibrotic effects of *D. nobile* alkaloids in cisplatin-induced ototoxicity	Ototoxicity	Suppresses renal fibrosis by modulating the AMPK/ mTOR pathway	Nie *et al.* (2020)
Clinical effectiveness and safety of *D. nobile* injection in CKD	CKD	Improves renal function, reduces proteinuria, well-tolerated	Eslamifar *et al.* (2021)
Therapeutic effects of *D. nobile* oral solution in early diabetic nephropathy combined with losartan potassium tablets	Early diabetic nephropathy	Significant improvements in renal function, reduction in urinary albumin excretion, decreased blood pressure	Zhang *et al.* (2023c)

pathways, both crucial in the fibrotic cascade within renal tissues (Zhang *et al.*, 2023c). More broadly, *D. nobile* may affect multiple signalling pathways involved in renal injury and repair, including the nuclear factor kappa B (NF-κB), TGF-β and MAPK pathways, thereby exerting protective effects on renal cells and tissues (Yang *et al.*, 2023). Furthermore, some studies have suggested that *D. nobile* may have vasodilatory effects, improving renal blood flow and maintaining adequate perfusion to the kidneys, which supports renal function and contributes to renal protection (Hao *et al.*, 2023). The mechanisms of action of *D. nobile* in renal diseases are still being explored, and further research is needed to validate these potential mechanisms and to understand their clinical significance. Clinical studies, including randomized controlled trials, are necessary to determine the efficacy and safety of *D. nobile* as a therapeutic agent for renal diseases and elucidate its precise mechanisms of action (Fig. 11.2).

11.5.1 Regulation of renal inflammation

The regulation of renal inflammation is critical in the management of renal diseases. *D. nobile* has been investigated for its role in the regulation of renal inflammation. Several studies have suggested that *D. nobile* modulates various pathways and processes that are involved in renal inflammation. Studies have demonstrated that *D. nobile* can hinder the release of pro-inflammatory mediators, including cytokines (such as tumour necrosis factor alpha (TNF-α) and interleukin 1 beta (IL-1β)) and chemokines (such as monocyte chemoattractant protein 1 (MCP-1)). These mediators are pivotal for initiating and intensifying the inflammatory response in the kidneys (Su *et al.*, 2023). *D. nobile* may regulate signalling pathways implicated in renal inflammation, notably the NF-κB pathway. NF-κB is a critical transcription factor that governs the expression of pro-inflammatory genes. Studies have indicated that *D. nobile* can impede NF-κB activation and the subsequent production of inflammatory molecules (Tseng *et al.*, 2016). *D. nobile* may regulate the function and activation of immune cells involved in renal inflammation, such as macrophages and T lymphocytes. It has been reported to inhibit macrophage infiltration and polarization towards pro-inflammatory phenotypes. Additionally, *D. nobile* has immunomodulatory effects on T lymphocytes, promoting a regulatory T cell phenotype that can suppress inflammation (Li *et al.*, 2022c). Oxidative stress plays an important role in renal inflammation. *D. nobile* contains bioactive compounds with antioxidant properties that can counteract the damaging effects of ROS. By reducing oxidative stress, *D. nobile* may attenuate inflammation in the kidneys

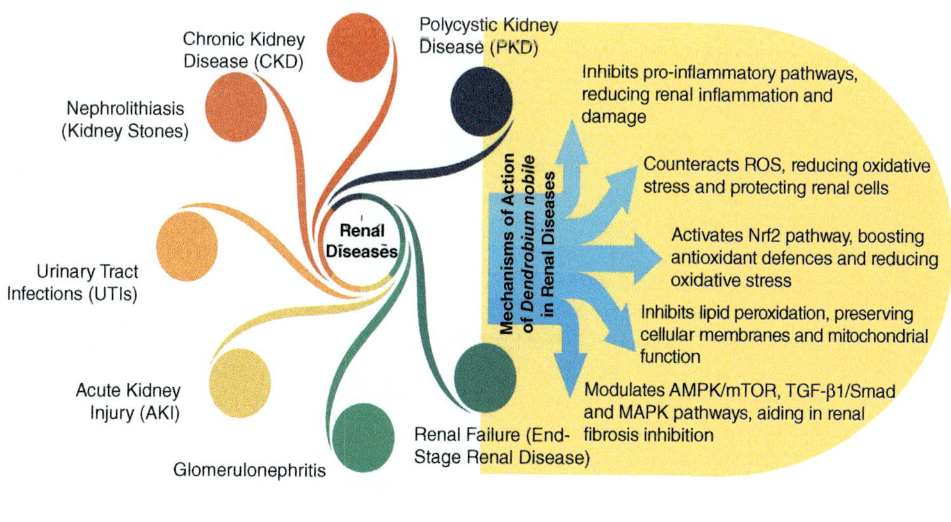

Fig. 11.2. Mechanisms of action of *D. nobile* in renal disease (authors' own figure).

(Hsu *et al.*, 2022). *D. nobile* has been reported to inhibit the adhesion and migration of inflammatory cells, such as monocytes and neutrophils, to kidney tissues. This can help reduce the infiltration of immune cells into the kidneys and limit inflammatory responses (Wu *et al.*, 2024).

11.5.2 Modulation of oxidative stress

D. nobile has been investigated for its potential to modulate oxidative stress, a factor implicated in the pathogenesis of renal diseases. Numerous studies have suggested that *D. nobile* possesses antioxidant properties and can modulate oxidative stress-related processes in the kidneys. *D. nobile* contains bioactive compounds such as phenolic compounds and flavonoids, which have been shown to possess antioxidant properties. These compounds have the capacity to function as scavengers of free radicals, effectively neutralizing ROS and diminishing oxidative stress within the kidneys (Natta *et al.*, 2022). *D. nobile* has the potential to boost the function of endogenous antioxidant enzymes such as superoxide dismutase (SOD), catalase (CAT) and glutathione peroxidase (GPX). These enzymes are pivotal in defence against oxidative stress, as they neutralize ROS and limit cellular damage (Li *et al.*, 2022c).

D. nobile may modulate signalling pathways associated with oxidative stress, such as the Nrf2 pathway, which represents crucial cellular mechanisms. Nrf2 serves as a transcription factor responsible for orchestrating the expression of genes involved in antioxidant defences and cellular protection. When activated, Nrf2 orchestrates the upregulation of various antioxidant enzymes and cytoprotective proteins, contributing to cellular defences against oxidative stress and maintaining overall cellular health. *D. nobile* has been shown to activate the Nrf2 pathway, leading to increased expression of antioxidant enzymes and reduced oxidative stress in the kidneys (Liu *et al.*, 2021; Mou *et al.*, 2021). Oxidative stress can induce lipid peroxidation, resulting in damage to the cellular membranes and exacerbation of renal injury. *D. nobile* has been reported to inhibit lipid peroxidation and preserve the integrity of cellular membranes, thereby protecting against oxidative damage (Long *et al.*, 2023). Mitochondrial dysfunction is closely associated with oxidative stress and renal disease. *D. nobile* has been shown to preserve mitochondrial function and attenuate mitochondrial oxidative stress, thereby protecting renal cells from oxidative damage (He *et al.*, 2022).

11.5.3 Inhibition of renal fibrosis pathways

D. nobile has shown potential in inhibiting renal fibrosis pathways, which play a critical role in the progression of renal diseases. Renal fibrosis is distinguished by the abnormal build up of extracellular matrix proteins, resulting in the disruption of the typical structure and function of the kidneys. *D. nobile* has been found to play a significant role in mitigating renal fibrosis through multiple mechanisms, including the regulation of the TGF-β pathway, inhibition of fibroblast activation and myofibroblast differentiation, and suppression of inflammation-associated fibrosis. By modulating the TGF-β pathway, *D. nobile* inhibits TGF-β signalling and reduces the expression of fibrotic markers such as collagen and fibronectin (Yang *et al.*, 2023). Additionally, it inhibits the activation of fibroblasts and their differentiation into myofibroblasts, thus decreasing the excessive production of extracellular matrix proteins associated with renal fibrosis (Lin *et al.*, 2022). Furthermore, the anti-inflammatory properties of *D. nobile* help attenuate the activation of fibrogenic pathways by suppressing inflammatory responses, thereby reducing renal fibrosis (Ge *et al.*, 2017). *D. nobile* may influence remodelling of the extracellular matrix by regulating matrix metalloproteinases (MMPs) and tissue inhibitors of metalloproteinases (TIMPs), which are instrumental in renal physiology. MMPs are responsible for degrading extracellular matrix components, whereas TIMPs inhibit their activity. *D. nobile* promotes the expression of MMPs and inhibits TIMP expression, promoting extracellular matrix remodelling (Huang *et al.*, 2022; Zhao *et al.*, 2023). Autophagy is a fundamental cellular process that degrades damaged proteins and organelles in cells. Perturbations in autophagy have been associated with the progression of renal fibrosis, highlighting their significance in maintaining renal health and function. *D. nobile* has been found to induce autophagy in renal cells, which may help remove damaged components and mitigate fibrotic processes (He *et al.*, 2020).

11.6 Safety and Side Effects of *D. nobile*

Limited clinical studies evaluating the safety of *D. nobile* in humans, particularly concerning renal diseases, have been conducted, with most research based on preclinical studies and traditional Chinese medicine. Despite its potential benefits, individuals should be aware of the possibility of allergic reactions to *D. nobile* or its components, which may manifest as skin rashes, itching, swelling or respiratory symptoms, requiring immediate discontinuation and medical attention if experienced. Additionally, the quality and purity of *D. nobile* products can vary, underscoring the importance of obtaining them from reputable sources to minimize the risk of contamination or adulteration. Furthermore, the appropriate dosage and duration of *D. nobile* use for renal diseases remain unclear, emphasizing the need to adhere to recommended guidelines and consult healthcare professionals for accurate dosing instructions. Limited information is available regarding the safety of using *D. nobile* during pregnancy and breastfeeding, necessitating consultation with a healthcare provider before considering its use during these phases. It is advisable for pregnant or breastfeeding women to avoid *D. nobile* unless specifically recommended by a healthcare professional, as its safety during these periods remains uncertain.

Although generally considered safe when used appropriately, there have been limited reports of adverse reactions to *D. nobile*, including gastrointestinal discomfort such as nausea, vomiting or diarrhoea, as well as rare allergic reactions. Additionally, although there is limited information available on its toxicity profile, studies on animals suggest that high doses of *D. nobile* extract may cause mild liver toxicity, emphasizing the importance of adhering to recommended dosages and exercising caution. Furthermore, due to the limited clinical data available, other potential side effects or long-term effects of *D. nobile* are not well characterized, highlighting the need to monitor adverse reactions during use.

11.6.1 Drug interactions and contraindications

D. nobile has the potential to interact with specific medications, including anticoagulants and antiplatelet drugs. Therefore, caution is warranted when using *D. nobile* concurrently with these medications, and consulting a healthcare professional is advisable to mitigate any potential risks. *D. nobile* presents potential interactions with various medications, warranting caution when used concurrently. Its anticoagulant properties may increase the risk of bleeding when used alongside anticoagulants and antiplatelet drugs, such as warfarin, heparin, aspirin and clopidogrel. Additionally, its immunomodulatory effects could interfere with the action of immunosuppressant drugs such as cyclosporine, tacrolimus and corticosteroids. Moreover, its vasodilatory effects may potentiate the blood pressure-lowering effect of antihypertensive medications, leading to excessively low blood pressure, including angiotensin-converting enzyme (ACE) inhibitors, beta blockers and calcium channel blockers. Furthermore, the influence of *D. nobile* on blood sugar levels suggests the potential for interactions with diabetes medications, such as metformin, insulin and sulfonylureas, possibly amplifying the hypoglycaemic effect and necessitating close monitoring of blood sugar levels. Therefore, consulting a healthcare professional is crucial before using *D. nobile* alongside these medications. While *D. nobile* is generally considered safe, certain individuals should exercise caution or avoid its use because of specific contraindications. Individuals with known allergies to *D. nobile* or its constituents should avoid its use. Additionally, due to its potential anticoagulant effects, it is advisable to discontinue *D. nobile* at least 2 weeks before scheduled surgery to minimize the risk of excessive bleeding.

11.6.2 Long-term safety considerations

Long-term safety considerations regarding the use of *D. nobile* are important, especially for individuals who are considering its prolonged or chronic use. Due to the scarcity of long-term clinical studies evaluating the safety of *D. nobile*, most research relies on short-term studies or traditional Chinese medicine practices. Consequently, the long-term adverse effects or potential risks associated with its use are not well understood, necessitating monitoring for adverse reactions and regular medical check-ups during prolonged use. Individuals with pre-existing renal conditions or at risk of renal impairment should exercise caution when using *D. nobile* long term, with regular monitoring of renal function recommended. Additionally, responses to *D. nobile* may vary among individuals, with some experiencing adverse effects or intolerance and others tolerating it well. Ensuring the quality and purity of *D. nobile* products is crucial for long-term safety, highlighting the importance of obtaining them from reputable sources and following proper storage instructions to minimize the risk of contamination or adulteration. As with any herbal or dietary supplement, a balanced approach is recommended, incorporating *D. nobile* as part of a comprehensive treatment plan under the guidance of a healthcare professional rather than relying solely on it for the management of renal conditions.

11.7 Future Directions and Research Perspectives

Future directions and research perspectives in the field of *D. nobile* and its potential therapeutic use in renal diseases are essential to advance our understanding and clinical applications. The implementation of meticulously planned and rigorous clinical trials is imperative for assessing the efficacy, safety and ideal dosing of *D. nobile* in managing renal diseases. Randomized controlled trials incorporating larger participant cohorts and extended follow-up durations would yield more definitive evidence. Therefore, prioritizing such trials is essential for elucidating the potential benefits of *D. nobile* in renal health. Further investigation into the underlying mechanisms of action of *D. nobile* in renal diseases is required. Understanding the molecular pathways and cellular targets involved can provide insights into their therapeutic potential and aid in the development of targeted interventions.

Studying the pharmacokinetics (absorption, distribution, metabolism and excretion) and pharmacodynamics (how a drug affects the body) of *D. nobile* in humans is crucial. This information can help determine the optimal dosing regimens, identify potential drug interactions and understand the overall effects on renal function. Long-term safety studies are warranted to assess the potential risks and adverse effects associated with prolonged use of *D. nobile*. This includes evaluating its effects on renal function, assessing cumulative toxicity and monitoring for potential long-term complications. Establishing standardized extraction methods and quality control measures for *D. nobile* products is important to ensure consistency in its composition and potency. This can enhance product reliability and reproducibility of results, and minimize the risk of contamination or adulteration. Exploring the potential synergistic effects of *D. nobile* in combination with conventional therapies or other herbal medicines could be an interesting area of investigation. Combination therapy approaches may offer enhanced therapeutic outcomes and improve the management of renal disease. Research focusing on the safety and efficacy of *D. nobile* in special populations, such as children, elderly individuals and individuals with comorbidities, is needed. These populations often have unique considerations and tailored studies are necessary to guide their safe and effective use. Identifying specific biomarkers that can predict the response to *D. nobile* treatment and monitor its therapeutic effects on renal diseases would be valuable. Biomarkers can aid patient selection, treatment monitoring and assessment of treatment outcomes.

11.7.1 Novel therapeutic approaches

Novel therapeutic approaches in the context of renal disease involve exploring innovative strategies to improve patient outcomes and address unmet medical needs. Although *D. nobile* is a potential therapeutic agent, there are other emerging approaches worth considering. Precision medicine strives to customize treatment approaches for individual patients by considering their unique characteristics including genetic, molecular and environmental factors. This approach enables healthcare providers to deliver personalized and targeted therapies, thereby optimizing patient outcomes. By understanding the unique molecular profile of renal disease, personalized therapies can be developed to target specific pathways or molecular targets implicated in the disease. Stem cell therapy offers potential benefits in renal diseases by fostering tissue regeneration, alleviating inflammation and modulating immune responses. Scientists are investigating various stem cell categories, including mesenchymal stem cells and induced pluripotent stem cells, for their potential therapeutic benefits in the treatment of AKI, CKD and kidney transplantation. This research avenue holds promise for improving the management and outcomes of various renal conditions.

Gene therapy entails the introduction of genetic material into cells to rectify or alter genes associated with diseases. In the context of renal diseases, gene therapy approaches may target specific genes involved in the pathogenesis of renal disorders to restore normal cellular function, inhibit disease progression or enhance tissue repair. Nanotechnology offers new possibilities for targeted drug delivery, imaging and diagnostics of renal diseases. Nanoparticles have the potential to be designed and tailored to deliver therapeutic agents directly to the kidneys, thereby enhancing drug effectiveness while minimizing systemic adverse effects. This targeted drug delivery approach holds promise for improving the treatment outcomes and patient safety in renal diseases. Additionally, nanomaterials can be used for targeted imaging and the early detection of renal diseases. Immunotherapy strategies, including immune checkpoint inhibitors and adoptive cell therapy, are currently under investigation for their potential applications in renal cancer and autoimmune kidney conditions. These therapies aim to modulate the immune system to recognize and eliminate cancer cells or aberrant immune responses that cause renal damage.

Biomaterials and tissue engineering strategies involve the development of scaffolds, grafts and artificial organs to repair or replace damaged renal tissues. These approaches aim to enhance tissue regeneration, improve renal function and provide alternatives to traditional transplantations. Gut microbiota has emerged as an important factor that influences renal health and disease. Modulation of the gut microbiota through interventions such as probiotics, prebiotics and faecal microbiota transplantation holds potential in the management of renal diseases, to modify the course of renal disease and improve the outcomes. The integration of artificial intelligence (AI) and machine learning algorithms can streamline the analysis of intricate datasets, bolster disease diagnosis and prognosis, and aid in treatment decision-making. AI-based tools can help identify patterns, predict disease progression and optimize personalized treatment strategies for renal disease. These novel therapeutic approaches represent exciting areas of ongoing research and development in the field of renal diseases. Continued advancements in these areas have the potential to revolutionize the management and treatment outcomes of patients with renal conditions.

11.7.2 Combination therapies

Combination therapies, also known as combination treatments or polytherapies, involve the use of multiple therapeutic agents or interventions simultaneously or sequentially to enhance treatment efficacy and improve patient outcomes. In the context of renal disease, combination therapies are being explored to target different pathological mechanisms, enhance drug synergy and overcome treatment resistance. Different classes of medications with complementary mechanisms of action can be combined to achieve additive or synergistic effects in the treatment of renal disease. For example, co-administration of ACE inhibitors or angiotensin receptor blockers (ARBs) with diuretics may result in enhanced control of blood pressure and a reduction in proteinuria among individuals with CKD. In immune-mediated renal diseases, combining immunomodulatory agents with different targets or modes of action can provide more comprehensive suppression of the immune response. For instance, combining glucocorticoids with other immunosuppressive drugs such as calcineurin inhibitors or mycophenolate mofetil may be effective in treating certain types of glomerulonephritis.

For example, a synergistic approach utilizing targeted therapies that individually inhibit distinct molecular pathways implicated in renal diseases can be employed to achieve a holistic and efficacious blockade. In advanced renal cell carcinoma treatment, the concomitant utilization of targeted agents antagonizing both the vascular endothelial growth factor (VEGF) and mTOR signalling pathways have shown heightened response rates and extended progression-free survival. In some cases, combining drug therapy with specialized medical devices or interventions can enhance the treatment outcomes. For instance, combining intravenous drug therapy with haemodialysis or peritoneal dialysis may be necessary in certain situations to optimize drug dosing and minimize drug accumulation in patients with impaired renal function. Complementary and alternative therapies, such as herbal medicines, acupuncture or mind/body interventions, may be used in combination with conventional medical treatments to augment their effects or address specific symptoms. However, it is essential to ensure the safety and compatibility of these combinations and consult healthcare professionals for guidance.

11.7.3 Pharmacokinetic and pharmacodynamic studies

Pharmacokinetic and pharmacodynamic studies play a crucial role in understanding the behaviour of drugs in the body and their effects on the target tissues or organs. In the context of renal diseases, these studies provide valuable insights into the absorption, distribution, metabolism and excretion of drugs as well as their pharmacological effects (Sun *et al.*, 2021). Pharmacokinetic studies have investigated how drugs are absorbed into the bloodstream following administration. They assess parameters such as bioavailability, which indicates the fraction of the administered dose that reaches the systemic circulation. Pharmacokinetic studies may explore different routes of drug administration, including oral, intravenous or transdermal, to determine the most effective and efficient method for drug delivery (Jain, 2020). Distribution studies have examined the distribution of drugs throughout the body, including in the kidneys. Jain (2020) investigated factors such as drug binding to plasma proteins, tissue penetration and drug accumulation in specific organs or compartments. In the case of renal diseases, pharmacokinetic studies help assess drug distribution in the kidneys and determine whether any alterations in drug distribution occur due to impaired renal function (Hei *et al.*, 2021).

Metabolism studies have investigated how drugs are metabolized in the body, primarily in the liver and kidneys. Zhang and Tang (2018) assessed the roles of enzymes and metabolic pathways involved in drug metabolism, including phase I and phase II reactions. Pharmacokinetic studies have also explored the impact of renal impairment on drug metabolism and clearance because impaired kidney function can affect the elimination of drugs from the body (Zhang and Tang, 2018). Excretion studies have focused on how drugs and their metabolites are eliminated from the body, primarily through renal excretion. They evaluated parameters such as renal clearance, which indicates the ability of the kidneys to remove drugs from the bloodstream. Pharmacokinetic studies of renal diseases help determine alterations in drug excretion due to impaired kidney function and guide dosage adjustments to prevent drug accumulation and potential toxicity (Wang, 2021).

Pharmacodynamic studies aim to understand the specific molecular and cellular mechanisms by which drugs exert their pharmacological effects on the kidneys. They explore drug–receptor interactions, signal transduction pathways and downstream physiological or biochemical changes induced by a drug. Pharmacodynamic studies help to elucidate the targets and pathways through which drugs modulate renal function and pathology (Wootten *et al.*, 2018). Dose–response studies have investigated the relationship between drug dosage and pharmacological effects. They determined the minimum effective dose, maximum tolerated dose and therapeutic range of the drugs for renal diseases. Pharmacodynamic studies also assess the duration of drug action and potential dose-dependent adverse effects (Guan *et al.*, 2023). Drug interaction studies have explored the interactions between different drugs, including their potential synergistic or antagonistic effects. They investigated how the concomitant use of multiple drugs could affect their pharmacological action in the kidneys. Pharmacodynamic studies help identify potential drug–drug interactions and guide the selection and dosing of combination therapies (Papotti *et al.*, 2021). Pharmacokinetic and pharmacodynamic studies provide essential information for optimizing drug therapies for renal diseases. They contribute to the development of personalized treatment regimens, dosage adjustments based on renal function and identification of potential drug–drug interactions. These studies also help ensure the efficacy and safety of drug interventions in patients with renal diseases.

11.8 Conclusion

In conclusion, *D. nobile* shows promising potential for the treatment of renal diseases. Its various bioactive compounds and pharmacological properties have been investigated in preclinical studies, suggesting its beneficial effects in modulating renal inflammation, oxidative stress, fibrosis and other pathways involved in renal pathology. However, it is important to note that clinical evidence for the efficacy of *D. nobile* in renal diseases is limited, and further research is needed to establish its therapeutic benefits. Safety considerations are essential when using *D. nobile*, as limited clinical studies and the variability in product quality and purity pose potential risks. Allergic reactions and potential drug interactions, particularly with anticoagulants and antiplatelet drugs, should be monitored carefully. The appropriate dosage and duration of *D. nobile* use for renal diseases have not been well established, emphasizing the need for professional guidance and adherence to recommended guidelines. The long-term safety and toxicity profiles of *D. nobile*, as well as its use during pregnancy and breastfeeding, require further investigation. Monitoring for adverse reactions and conducting comprehensive studies are necessary to better understand the potential side effects and long-term effects of *D. nobile* in the context of renal diseases. Future research should focus on conducting well-designed clinical trials to evaluate the efficacy and safety of *D. nobile* in treating renal diseases. This includes investigating the optimal dosages, treatment durations and identifying patient populations that may benefit the most from their therapeutic effects. Further in-depth pharmacokinetic and pharmacodynamic studies are warranted to elucidate the precise mechanisms of action and potential drug interactions associated with *D. nobile*. While *D. nobile* holds promise as a potential therapeutic option for renal diseases, it is crucial to approach its use with caution and consult healthcare professionals for personalized guidance. Integrating *D. nobile* into comprehensive treatment approaches and combination therapies may provide synergistic benefits, but careful evaluation of potential interactions and monitoring of treatment outcomes are necessary.

11.8.1 Summary of findings

The use of *D. nobile* in the treatment of renal diseases holds promise based on preclinical studies and traditional Chinese medicine. However, current evidence is limited, and more research is needed to establish its efficacy and safety in clinical settings. *D. nobile* may exert its effects in renal diseases through anti-inflammatory, antioxidant and antifibrotic mechanisms, and modulation of signalling pathways, as proposed by preclinical studies that require further validation. There is a scarcity of clinical studies evaluating its safety and efficacy in renal diseases, with most research being based on preclinical studies and traditional use. Adverse reactions and potential drug interactions, particularly with anticoagulants and antiplatelet drugs, should be monitored, and the quality and purity of *D. nobile* products can vary, underscoring the importance of obtaining them from reputable sources. The appropriate dosage and duration of use for renal diseases have not been well established, making it crucial to follow the recommended guidelines and consult healthcare professionals. Well-designed clinical trials are needed to evaluate the efficacy and safety of *D. nobile* in renal diseases, focusing on optimal dosages and treatment durations, and identifying patient populations that may benefit the most. Pharmacokinetic and pharmacodynamic studies are necessary to understand the mechanisms of action and potential drug interactions. In conclusion, whereas *D. nobile* shows promise as a potential therapeutic option for renal diseases, more research is needed to establish its efficacy, safety, optimal usage and long-term effects. It is important to approach its use with caution, consult healthcare professionals and integrate it into a comprehensive treatment approach.

11.8.2 Implications for clinical practice

The existing evidence on the utilization of *D. nobile* in renal disease management is limited, necessitating further research to ascertain its clinical effectiveness and safety profile. However, based on the available information, healthcare professionals should be aware of *D. nobile* as a potential therapeutic option for renal diseases, but caution is necessary due to the limited clinical data, making it crucial to consider the potential advantages and drawbacks of each patient's unique characteristics and medical background. Given the lack of established dosages and treatment durations, an individualized approach is essential, with healthcare professionals considering the specific needs and characteristics of each patient and carefully monitoring their responses to treatment. *D. nobile* should be integrated into comprehensive care, including other established therapeutic approaches, such as medication, lifestyle modifications and regular monitoring of renal function, rather than being considered a standalone treatment. Patients should be educated about the limited evidence regarding *D. nobile* in renal diseases and the need for caution, understanding the potential risks and benefits, and encouraged to communicate openly with their healthcare professionals about any changes or concerns during treatment. Continued research is needed to establish the efficacy, safety and optimal use of *D. nobile* in renal diseases, with healthcare professionals collaborating with researchers and participating in well-designed clinical trials to contribute to the body of knowledge in this area. Overall, while *D. nobile* shows promise as a potential adjunctive therapy in renal diseases, its integration into clinical practice should be guided by the available evidence, individual patient characteristics and a comprehensive approach to patient care.

Acknowledgements

The authors express their gratitude for the financial support received through the Distinguished High-Level Talents Research Grant from the Guizhou Science and Technology Corporation Platform Talents Fund (grant nos. [2017]5733-001 and CK-1130-002), the National Natural Science Foundation of China (U1812403, 82373981 and 82060750), 2011 Collaborative Innovation Centre of Traditional Chinese Medicine in Guizhou Province (no. [2022]026) and the support provided by the Zunyi Medical University, China. Special appreciation is extended to all laboratory colleagues and research staff members for their valuable insights, constructive guidance and assistance with the contents of this chapter.

References

Cakova, V., Bonte, F. and Lobstein, A. (2017) *Dendrobium:* Sources of active ingredients to treat age-related pathologies. *Aging and Disease* 8(6), 827–849.

Chen, J.T. (2021) Orchid biochemistry. *International Journal of Molecular Sciences* 21(7), 2338.

Cho, Y., Tong, A., Craig, J.C., Mustafa, R.A., Chapman, A. *et al.* (2021) Establishing a core outcome set for autosomal dominant polycystic kidney disease: report of the standardized outcomes in nephrology-polycystic kidney disease (SONG-PKD) consensus workshop. *American Journal of Kidney Diseases* 77(2), 255–263.

Dai, Y.-P., Duan, Y., Lu, Y.-T., Ni, X.-T., Zhang, Y.-K. *et al.* (2024) Nourishing yin traditional chinese medicine: potential role in the prevention and treatment of type 2 diabetes. *American Journal of Translational Research* 16(1), 234–254.

Eslamifar, Z., Moridnia, A., Sabbagh, S., Ghaffaripour, R., Jafaripour, L. *et al.* (2021) Ameliorative effects of gallic acid on cisplatin-induced nephrotoxicity in rat variations of biochemistry, histopathology, and gene expression. *BioMed Research International* 2021, 2195238.

Feng, T., Hui, J., Zhong, Y.G., Lei, Z.Y. and Ping, T.Y. (2018) Effects of the aqueous extract of dendrobium nobile lindl on the nonenzymatic glycation and oxidative stress in diabetic rat kidney. *Journal of Shandong University (Health Sciences)* 50(10), 11–15.

Fletcher, B.R., Damery, S., Aiyegbusi, O.L., Anderson, N., Calvert, M. *et al.* (2022) Symptom burden and health-related quality of life in chronic kidney disease: a global systematic review and meta-analysis. *PLoS Medicine* 19(4), e1003954.

Fu, X., Chen, S., Xian, S., Wu, Q., Shi, J. *et al.* (2023) Dendrobium and its active ingredients: Emerging role in liver protection. *Biomedicine & Pharmacotherapy* 157, 114043.

Gaur, A.S., Mandal, S., Pandey, A., Das, M.K. and Nayak, P. (2022) Efficacy of PCNL in the resolution of symptoms of nephrolithiasis. *Urolithiasis* 50(4), 487–491.

Gaut, J.P. and Liapis, H. (2021) Acute kidney injury pathology and pathophysiology: a retrospective review. *Clinical Kidney Journal* 14(2), 526–536.

Ge, M.X., He, H.W., Shao, R.G. and Liu, H. (2017) Recent progression in the utilization of autophagy-regulating nature compound as anti-liver fibrosis agents. *Journal of Asian Natural Products Research* 19(2), 109–113.

Guan, Y., Li, X., Li, H., Ren, J., Tang, K. *et al.* (2023) Sacubitril/valsartan in heart failure with hypertension patients: real-world experiences on different ages, drug doses, and renal functions. *High Blood Pressure & Cardiovascular Prevention* 30(6), 561–572.

Hao, Y., Lao, S., Liu, H., Chen, X., Ye, G. *et al.* (2023) Isolation and characterization of a nephroprotective polysaccharide from *Dendrobium chrysotoxum* Lindl against LPS-induced acute kidney injury mice. *International Journal of Biological Macromolecules* 253, 126614.

He, L., Su, Q., Bai, L., Li, M., Liu, J. *et al.* (2020) Recent research progress on natural small molecule bibenzyls and its derivatives in dendrobium species. *European Journal of Medicinal Chemistry* 204, 112530.

He, Q., Xiao, L., Shi, Y., Li, W. and Xin, X. (2023) Natural products: Protective effects against ischemia-induced retinal injury. *Frontiers in Pharmacology* 14, 1149708.

Hsu, W.-H., Chung, C.-P., Wang, Y.-Y., Kuo, Y.-H., Yeh, C.-H. *et al.* (2022) *Dendrobium nobile* protects retinal cells from UV-induced oxidative stress damage via nrf2/HO-1 and MAPK pathways. *Journal of Ethnopharmacology* 288, 114886.

Huang, Q., Yan, Y.A.N.G., Rui, Z.E.N.G., Menglin, Y.A.O. and Qin, S.U.N. (2022) Regulation of liver fibrosis by matrix metalloproteinase/tissue inhibitor of metalloproteinase and research advances in related therapeutic drugs. *Journal of Clinical Hepatobiliary Diseases* 38(6), 1420–1425.

Jafari, N.V. and Rohn, J.L. (2022) The urothelium: a multi-faceted barrier against a harsh environment. *Mucosal Immunology* 15(6), 1127–1142.

Jain, K.K. (2020) An overview of drug delivery systems. *Drug Delivery* Systems15, 1–54.

Kalariya, Y., Kumar, A., Ullah, A., Umair, A., Neha, F. *et al.* (2023) Integrative medicine approaches: bridging the gap between conventional and renal complementary therapies. *Cureus* 15(9), e46033.

Kang, H., Lee, J.P. and Choi, K. (2021) Exposure to phthalates and environmental phenols in association with chronic kidney disease (CKD) among the general US population participating in multi-cycle NHANES (2005–2016). *Science of The Total Environment* 791, 148343.

Kaur, R. and Kaur, R. (2021) Symptoms, risk factors, diagnosis and treatment of urinary tract infections. *Postgraduate Medical Journal* 97(1154), 803–812.

Kellum, J.A., Romagnani, P., Ashuntantang, G., Ronco, C., Zarbock, A. *et al.* (2021) Acute kidney Injury. *Nature Reviews Disease Primers* 7(1), 1–17.

Lakhia, R., Ramalingam, H., Chang, C.-M., Cobo-Stark, P., Biggers, L. *et al.* (2022) PKD1 and PKD2 mRNA Cis-inhibition drives polycystic kidney disease progression. *Nature Communications* 13(1), 4765.

Li, D.-D., Fan, H.-X., Yang, R., Li, Y.-Y., Zhang, F. *et al.* (2022a) *Dendrobium nobile* Lindl. alkaloid suppresses NLRP3-mediated pyroptosis to alleviate LPS-induced neurotoxicity. *Frontiers in Pharmacology* 13, 846541.

Li, M., Trapika, I.G.S.C., Tang, S.Y.S., Cho, J.-L., Qi, Y. *et al.* (2022b) Mechanisms and active compounds polysaccharides and bibenzyls of medicinal dendrobiums for diabetes management. *Frontiers in Nutrition* 8, 811870.

Li, W., Mu, X., Wu, X., He, W., Liu, Y. *et al.* (2022c) *Dendrobium nobile* Lindl. polysaccharides protect fibroblasts against UVA-induced photoaging via JNK/c-Jun/MMPs pathway. *Journal of Ethnopharmacology* 298, 115590.

Li, P.Y., Li, L. and Wang, Y.Z. (2023a) Traditional uses, chemical compositions and pharmacological activities of dendrobium: a review. *Journal of Ethnopharmacology* 310, 116382.

Li, Z., Zeng, M., Geng, K., Lai, D., Xu, Z. *et al.* (2023b) Chemical constituents and hypoglycemic mechanisms of *Dendrobium nobile* in treatment of type 2 diabetic rats by UPLC-ESI-Q-orbitrap, network pharmacology and in vivo experimental verification. *Molecules* 28(6), 2683.

Li, S., Tan, H.Y., Wang, N., Cheung, F., Hong, M. *et al.* (2018) The potential and action mechanism of polyphenols in the treatment of liver diseases. *Oxidative Medicine and Cellular Longevity* 2018, 1–26.

Lin, Z.H.U. and Yun-sheng, X.U. (2022) Research progress on inflammatory mechanism and traditional Chinese medicine in the treatment of diabetic retinopathy. *Journal of Hainan Medical College* 28(18), 75.

Liu, J., Zhu, T., Niu, Q., Yang, X., Suo, H. *et al.* (2020) *Dendrobium nobile* alkaloids protects against H2O2-induced neuronal injury by suppressing JAK–stats pathway activation in N2A cells. *Biological and Pharmaceutical Bulletin* 43(4), 716–724.

Liu, Y., Pi, T., Yang, X. and Shi, J. (2021) Protective effects and mechanisms of *Dendrobium nobile* Lindl. alkaloids on PC12 cell damage induced by Aβ25-35. *Behavioural Neurology* 2021, 1–11.

Liu, X.M., Shi, H. and Li, W. (2024a) Review on the potential roles of traditional Chinese medicines in the prevention, treatment, and postoperative recovery of age-related cataract. *Journal of Ethnopharmacology* 324, 117786.

Liu, Y.H., Chen, Y.H. and Chen, C.M. (2024b) Fulminant leptospirosis presenting with rapidly developing acute renal failure and multiorgan failure. *Biomedicines* 12(2), 435.

Long, Y., Wang, W., Zhang, Y., Zhang, S., Li, Z. *et al.* (2023) *Dendrobium nobile* Lindl polysaccharides attenuate UVB-induced photodamage by regulating oxidative stress. Inflammation and MMPs expression in mice model. *Photochemistry and Photobiology* 99(5), 1269–1281.

Luo, Y., Yang, D., Xu, Y., Wu, D., Tan, D. *et al.* (2024) Hypoglycemic effects and quality marker screening of *Dendrobium nobile* Lindl. at different growth years. *Molecules* 29(3), 699.

Mancuso, G., Midiri, A., Gerace, E., Marra, M., Zummo, S. *et al.* (2023) Urinary tract infections: the current scenario and future prospects. *Pathogens* 12(4), 623.

Mou, Z., Zhao, Y., Ye, F., Shi, Y., Kennelly, E.J. *et al.* (2021) Identification, biological activities and biosynthetic pathway of dendrobium alkaloids. *Frontiers in Pharmacology* 12, 605994.

Natta, S., Mondol, M.S.A., Pal, K., Mandal, S., Sahana, N. *et al.* (2022) Chemical composition, antioxidant activity and bioactive constituents of six native endangered medicinal orchid species from north-eastern Himalayan region of India. *South African Journal of Botany* 150, 248–259.

Nie, X., Chen, Y., Li, W. and Lu, Y. (2020) Anti-aging properties of *Dendrobium nobile* Lindl.: from molecular mechanisms to potential treatments. *Journal of Ethnopharmacology* 257, 112839.

Olsen, E. and van Galen, G. (2022) Chronic renal failure-causes, clinical findings, treatments and prognosis. *Veterinary Clinics: Equine Practice* 38(1), 25–46.

Oskouei, Z., Ghasemzadeh Rahbardar, M. and Hosseinzadeh, H. (2023) The effects of *Dendrobium* species on the metabolic syndrome: a review. *Iranian Journal of Basic Medical Sciences* 26(7), 738–752.

Panizo, S., Martínez-Arias, L., Alonso-Montes, C., Cannata, P., Martín-Carro, B. *et al.* (2021) Fibrosis in chronic kidney disease: pathogenesis and consequences. *International Journal of Molecular Sciences* 22(1), 408.

Papotti, B., Marchi, C., Adorni, M.P. and Potì, F. (2021) Drug-drug interactions in polypharmacy patients: the impact of renal impairment. *Current Research in Pharmacology and Drug Discovery* 2, 100020.

Pesce, F., Stea, E.D., Rossini, M., Fiorentino, M., Piancone, F. *et al.* (2020) Glomerulonephritis in AKI: from pathogenesis to therapeutic intervention. *Frontiers in Medicine* 7, 582272.

Rabidas, S.S., Prakash, C., Tyagi, J., Suryavanshi, J., Kumar, P. *et al.* (2023) A comprehensive review on anti-inflammatory response of flavonoids in experimentally-induced epileptic seizures. *Brain Sciences* 13(1), 102.

Ramesh, T., Koperuncholan, M., Praveena, R., Ganeshkumari, K., Vanithamani, J. *et al.* (2019) Medicinal properties of some dendrobium orchids – a review. *Journal of Applied and Advanced Research* 4(4), 119–128.

Rao, D., Zhao, R., Hu, Y., Li, H., Chun, Z. *et al.* (2023) Revealing of intracellular antioxidants in *dendrobium nobile* by high performance liquid chromatography-tandem mass spectrometry. *Metabolites* 13(6), 702.

Sethi, S., De Vriese, A.S. and Fervenza, F.C. (2022) Acute glomerulonephritis. *The Lancet* 399(10335), 1646–1663.

Sosland, R. and Stewart, J.N. (2021) Management of recurrent urinary tract infections in women: how providers can improve the patient experience. *Urology* 151, 8–12.

Su, Z.W., Yan, T.Y., Feng, J., Zhang, M.Y. and Han, L. (2023) Protective effects and mechanism of polysaccharides from edible medicinal plants in alcoholic liver injury: a review. *International Journal of Molecular Sciences* 24(22), 16530.

Sun, J., Ren, J., Hu, X., Hou, Y. and Yang, Y. (2021) Therapeutic effects of Chinese herbal medicines and their extracts on diabetes. *Biomedicine & Pharmacotherapy* 142, 111977.

Tamang, S., Singh, A., Bussmann, R.W., Shukla, V. and Nautiyal, M.C. (2023) Ethno-medicinal plants of tribal people: a case study in Pakyong subdivision of east Sikkim, India. *Acta Ecologica Sinica* 43(1), 34–46.

Ti, H., Zhuang, Z., Yu, Q. and Wang, S. (2021) Progress of plant medicine derived extracts and alkaloids on modulating viral infections and inflammation. *Drug Design, Development and Therapy* 15, 1385–1408.
Tian, J., Fu, D., Liu, Y., Guan, Y., Miao, S. *et al.* (2023) Rectifying disorder of extracellular matrix to suppress urethral stricture by protein nanofilm-controlled drug delivery from urinary catheter. *Nature Communications* 14(1), 2816.
Tseng, T.H., Lin, W.L., Chen, Z.H., Lee, Y.J., Shie, M.S. *et al.* (2016) Monili for mediquin one as a potential therapeutic agent, inactivation of hepatic stellate cell and inhibition of liver fibrosis in vivo. *Journal of Translational Medicine* 14, 1–10.
Wan Hei, J., Hung, W., Yanbo, Z. and Zhang, Z.-J. (2021) Dendrobium in diabetes: a comprehensive review of its phytochemistry, pharmacology, and safety. *Journal of Translational Science* 7(6), 1–10.
Wang, Y.H. (2021) Traditional uses, chemical constituents, pharmacological activities, and toxicological effects of dendrobium leaves: a review. *Journal of Ethnopharmacology* 270, 113851.
Wootten, D., Christopoulos, A., Marti-Solano, M., Babu, M.M. and Sexton, P.M. (2018) Mechanisms of signalling and biased agonism in G protein-coupled receptors. *Nature Reviews Molecular Cell Biology* 19(10), 638–653.
Wu, D., Qin, L., Feng, C., Cao, L., Ye, J. *et al.* (2024) Sesquiterpenoids of *Dendrobium nobile* Lindl. aqueous extract for inhibition alcoholic liver injury through RTK/elf4 regulated inflammation in mouse hepatic macrophages. *Arabian Journal of Chemistry* 17(2), 1–12.
Wu, W., Lin, Y., Farag, M.A., Li, Z. and Shao, P. (2023) Dendrobium as A new natural source of bioactive for the prevention and treatment of digestive tract diseases: a comprehensive review with future perspectives. *Phytomedicine* 114, 154784.
Wu, X., Huang, J., Wang, J., Xu, Y., Yang, X. *et al.* (2021) Multi-pharmaceutical activities of chinese herbal polysaccharides in the treatment of pulmonary fibrosis: concept and future prospects. *Frontiers in Pharmacology* 12, 707491.
Yang, W., Chen, D., Ji, Q., Zheng, J., Ma, Y. *et al.* (2023) Molecular mechanisms underlying the anticancer property of dendrobium in various systems of the human body: a review. *Biomedicine & Pharmacotherapy* 165, 115223.
Zhang, P., Zhang, X., Zhu, X. and Hua, Y. (2023a) Chemical constituents, bioactivities, and pharmacological mechanisms of *Dendrobium officinale*: a review of the past decade. *Journal of Agricultural and Food Chemistry* 71(41), 14870–14889.
Zhang, Q., Fu, X., Liu, H., Chen, Y., Chen, S. *et al.* (2023b) A systematical review on ethnobotanical, phytochemical and pharmacological aspects of *Dendrobium nobile* Lindl. *Phytochemistry Reviews* 22(3), 743–780.
Zhang, Y., Lu, J., Li, H. and Song, H. (2023c) Advances in dietary polysaccharides as hypoglycemic agents: mechanisms, structural characteristics, and innovative applications. *Critical Reviews in Food Science and Nutrition* 64(18), 1–21.
Zhang, X., Wang, M., Zhang, C., Liu, Z. and Zhou, S. (2021) Clinical study of *Dendrobium nobile* Lindl intervention on patients with metabolic syndrome. *Medicine* 100(12), e24574.
Zhang, Z. and Tang, W. (2018) Drug metabolism in drug discovery and development. *Acta Pharmaceutica Sinica. B* 8(5), 721–732.
Zhao, J., You, S., Wang, D., Su, N., Yang, L. *et al.* (2023) Attenuating UVA-induced oxidative stress of human skin fibroblasts by enhancing bioactive components of *Armillaria luteo-virens* by *Lactobacillus delbrueckii* subsp. bulgaricus fermentation. *Food Bioscience* 54, 102899.
Zhou, J., Zhang, Y., Li, S., Zhou, Q., Lu, Y. *et al.* (2020) *Dendrobium nobile* Lindl. alkaloids-mediated protection against CCl4-induced liver mitochondrial oxidative damage is dependent on the activation of Nrf2 signaling pathway. *Biomedicine & Pharmacotherapy* 129, 110351.

12 Neuroprotective Effects of *Dendrobium nobile*

Surendra Sarsaiya, Archana Jain, Qihai Gong, Qin Wu and Jingshan Shi*
Key Laboratory of Basic Pharmacology and Joint International Research Laboratory of Ethnomedicine of Ministry of Education, Zunyi Medical University, Zunyi, Guizhou, China

Abstract
Dendrobium nobile, a renowned traditional Chinese medicinal orchid, demonstrates significant neuroprotective potential owing to its bioactive compounds such as alkaloids, flavonoids and polysaccharides. This chapter explores the mechanisms underlying the mitigation of neuronal damage, oxidative stress and neuroinflammation while enhancing neuronal resilience. The neuroprotective properties of *D. nobile* show promise in treating neurodegenerative diseases, such as Alzheimer's and Parkinson's diseases. Experimental studies have highlighted its ability to reduce oxidative stress, inhibit apoptosis and promote neurotrophic factor expression. However, further clinical trials are necessary to validate its efficacy and safety in humans. This chapter also addresses the challenges in standardizing extraction protocols, identifying active compounds and determining optimal dosing. This comprehensive review underscores the potential of *D. nobile* as a neuroprotective agent and calls for further research to translate preclinical findings into clinical applications, ultimately aiming to develop novel therapeutic strategies for neurodegenerative conditions.

12.1 Introduction

Dendrobium nobile Lindl., revered in traditional Chinese medicine as 'shi hu' or 'shi hu gen', exhibits promising neuroprotective potential because of its rich array of bioactive compounds, including alkaloids, flavonoids and polysaccharides. Additionally, its ability to impede beta-amyloid aggregation, enhance cognitive function and mitigate ischaemic brain injury underscores its therapeutic potential (Ganapathy *et al.*, 2023). However, while these findings are encouraging, further clinical exploration is indispensable to fully elucidate its mechanisms and validate its efficacy and safety in neurological interventions (Davis *et al.*, 2020). *D. nobile* can mitigate neuronal damage, curb oxidative stress, quell neuroinflammation and foster neuronal resilience, presenting a compelling avenue for combating neurodegenerative conditions such as Alzheimer's and Parkinson's diseases (Li *et al.*, 2022a). Neuroprotection refers to the ability of a substance or intervention to protect neurons from damage or degeneration and to promote their survival and function (Lee *et al.*, 2020). Understanding the neuroprotective effects of *D. nobile* is of significant interest, as it may offer potential therapeutic options for managing these debilitating conditions (Jain *et al.*, 2021).

The neuroprotective effects of *D. nobile* are attributed to its diverse bioactive compounds, including alkaloids, flavonoids, phenols,

*Corresponding author: shijs@zmu.edu.cn

© Jingshan Shi, Archana Jain, Surendra Sarsaiya, Qin Wu and Qihai Gong 2025. Dendrobium nobile: *Research and Therapeutic Applications* (J. Shi, A. Jain, S. Sarsaiya, Q. Wu and Q. Gong)
DOI: 10.1079/9781800627987.0012

polysaccharides and other phytochemicals (Wang *et al.*, 2022). These compounds possess antioxidant, anti-inflammatory, anti-apoptotic and neurotrophic properties that contribute to their potential neuroprotective actions (Wang *et al.*, 2023). Studies have demonstrated that *D. nobile* extract or its isolated compounds exert neuroprotective effects through various mechanisms. Additionally, *D. nobile* inhibits neuroinflammation by suppressing the production of pro-inflammatory cytokines and modulating inflammatory signalling pathways (Liu *et al.*, 2020).

Furthermore, *D. nobile* has been shown to enhance neuronal survival and promote neuronal regeneration. It stimulates neurotrophic factors, which play crucial roles in neuronal growth, differentiation and synaptic plasticity (Fan *et al.*, 2023). These neurotrophic factors are essential for maintaining neuronal health and for promoting neurogenesis. Animal studies and *in vitro* experiments have provided evidence for the neuroprotective effects of *D. nobile* (Zhang *et al.*, 2022). These studies demonstrated improved cognitive function, enhanced neuronal viability and reduced neuronal damage in various neurodegenerative disease models and experimental paradigms.

12.1.1 Importance of neuroprotection

Neuroprotection is crucial in neuroscience and neurology, as it preserves and defends the health and functionality of neurons in the central nervous system. Multiple research endeavours have underscored its importance (Shohag *et al.*, 2022). Neuroprotection plays a vital role in the management of neurodegenerative diseases. It aims to slow down or halt degenerative processes, preserve neuronal function, alleviate symptoms and delay disease progression (Sindhu *et al.*, 2022). Maintaining cognitive function is another critical aspect of neuroprotection, and is essential for memory, learning, attention and executive function (Cammisuli *et al.*, 2022). Preserving the health and integrity of neurons in critical brain regions involved in cognition is of the utmost importance for optimal cognitive performance throughout life (Cabeza *et al.*, 2018). Neuroprotection also plays a significant role in stroke prevention and rehabilitation. Neuroprotection can contribute to reducing the severity and long-term consequences of strokes by protecting neurons from ischaemic injury and promoting their survival and recovery (Zhao *et al.*, 2022).

In cases of traumatic brain injury, neuroprotective strategies aim to minimize neuronal damage and promote recovery (Crupi *et al.*, 2020). Neuroprotection can lead to improved outcomes and functional recovery following traumatic brain injury by minimizing secondary injury processes and supporting neuronal survival (Jarrahi *et al.*, 2020). Moreover, neuroprotection is important to promote optimal brain ageing. As individuals age, the risk of neurodegenerative diseases and cognitive decline increases (Kim *et al.*, 2022). Neuroprotective strategies can help optimize brain ageing by preserving neuronal health, reducing oxidative stress and inflammation, and supporting neuroplasticity and neurogenesis (Maharjan *et al.*, 2020). The concept of neuroprotection also has potential for the development of novel therapeutics. Understanding the mechanisms of neuroprotection and identifying neuroprotective agents can pave the way for the development of targeted interventions (Villoslada and Steinman, 2020). These interventions enhance neuronal survival, prevent degeneration and promote functional recovery. Neuroprotection is of significant importance for preserving neuronal health, preventing neurodegenerative diseases, maintaining cognitive function and improving outcomes in neurological conditions. Extensive research conducted by various authors has highlighted the crucial role of neuroprotection in multiple contexts, emphasizing the need for continued efforts to develop effective neuroprotective strategies.

12.2 Mechanisms of Neuroprotection

Neuroprotection involves a range of mechanisms aimed at preserving and safeguarding the health and function of neurons in the central nervous system. Understanding these mechanisms is crucial for developing effective strategies to protect neurons against damage and degeneration. One important mechanism of neuroprotection is modulation of oxidative stress and scavenging of reactive oxygen species (ROS). Oxidative stress contributes to neuronal damage in various neurodegenerative disorders (Singh and Devasahayam,

2020). Neuroprotective agents can counteract oxidative stress by enhancing antioxidant defences and reducing ROS production (Singh *et al.*, 2019). Another key mechanism is inhibition of neuroinflammation. Inflammatory processes involving the activation of microglia and release of pro-inflammatory cytokines contribute to neuronal damage in neurodegenerative diseases (Voet *et al.*, 2019). Neuroprotective compounds can suppress neuroinflammation, thereby reducing neuronal injury and promoting neuronal survival (Zhu *et al.*, 2022).

Neurotrophic support is crucial to neuroprotection. Neuroprotective interventions can enhance the production or activity of neurotrophic factors, thereby providing support to vulnerable neurons (Cobianchi *et al.*, 2017). Modulation of apoptotic pathways is another mechanism involved in neuroprotection. Excessive apoptosis contributes to neuronal loss under various neurological conditions. Neuroprotective agents can inhibit apoptotic pathways, promote cell survival and prevent neuronal death (Chen *et al.*, 2022a). Moreover, neuroprotection may encompass the regulation of excitotoxicity, denoting overstimulation of glutamate receptors, resulting in neuronal damage. Neuroprotective strategies can target glutamate receptors and their downstream signalling pathways to prevent excitotoxic neuronal injury (Marcelli, 2018). Neuroprotection refers to the strategies or mechanisms aimed at preserving the structure, integrity and/or function of neurons in the brain. Several mechanisms have been identified and explored for their potential neuroprotective effects (Saragovi *et al.*, 2019; Lee *et al.*, 2020). Among these, neurotrophic factors play essential roles in fostering the survival, growth and maintenance of neurons, thereby enhancing their resilience to pathological insults (Zarneshan *et al.*, 2022).

Inflammation plays a significant role in neurodegenerative diseases, and agents that modulate the inflammatory response, such as nonsteroidal anti-inflammatory drugs (NSAIDs) or cytokine inhibitors, may offer neuroprotection by reducing inflammation-mediated neuronal damage (Degan *et al.*, 2018). Oxidative stress is also implicated in various neurodegenerative disorders, including antioxidants, such as vitamins C and E, glutathione and coenzyme Q10, which neutralize harmful free radicals and protect neurons from oxidative damage (Teleanu *et al.*, 2022). Additionally, neurotrophic factors, which are crucial for the growth, survival and differentiation of neurons, are integral components of neuroprotective strategies. Administration of these factors or compounds that enhance their expression can promote neuronal survival and function (Colucci-D'Amato *et al.*, 2020).

Mitochondrial dysfunction is implicated in neurodegenerative diseases, and compounds that support mitochondrial function, such as coenzyme Q10, creatine and nicotinamide adenine dinucleotide (NAD+) precursors, may offer neuroprotection by enhancing cellular energy production and reducing oxidative stress (Jurcau, 2021). Excitotoxicity, caused by excessive activation of glutamate receptors, contributes to neuronal damage in conditions such as stroke and traumatic brain injury. Agents that modulate glutamate signalling, such as N-methyl-d-aspartate (NMDA) receptor antagonists (e.g. memantine) and inhibitors of glutamate release (e.g. riluzole), can protect neurons from excitotoxic damage (Carvajal *et al.*, 2016). Additionally, apoptosis or programmed cell death is a significant mechanism of neuronal loss in neurodegenerative diseases, and compounds that inhibit apoptotic pathways or promote cell survival signalling, such as caspase inhibitors or Bcl-2 family modulators, may confer neuroprotection (Li *et al.*, 2022b). Chaperone proteins assist in protein folding and prevent the aggregation of misfolded proteins, which are characteristic features of many neurodegenerative disorders, including Alzheimer's and Parkinson's diseases. Enhancing the activity of chaperone proteins such as heat shock proteins (HSPs), may mitigate protein aggregation and protect neurons from damage (Wankhede *et al.*, 2022). Stimulating neurogenesis, the process of generating new neurons, offers a potential avenue for neuroprotection and repair. Agents that promote neurogenesis, such as certain drugs (e.g. antidepressants) or environmental enrichment, may enhance brain plasticity and cognitive function (Marchetti *et al.*, 2020). These mechanisms represent diverse strategies for preserving neuronal integrity and function, and ongoing research aims to further elucidate their potential for therapeutic intervention in neurodegenerative diseases and other neurological disorders (Table 12.1 and Fig. 12.1).

Table 12.1. Mechanisms of neuroprotection (authors' own table).

Mechanism	Description	Key processes/targets	Neuroprotective agents	Current research/key references
Oxidative stress	Counteracting oxidative stress by enhancing antioxidant defences and reducing ROS production	Enhancement of antioxidant defences, reduction of ROS production	Vitamins C and E, glutathione, coenzyme Q10	Singh *et al.* (2019); Singh and Devasahayam (2020)
Neuroinflammation	Suppressing neuroinflammation to reduce neuronal injury and promote survival	Inhibition of microglial activation, reduction of pro-inflammatory cytokines	NSAIDs, cytokine inhibitors	Voet *et al.* (2019); Zhu *et al.* (2022)
Neurotrophic support	Enhancing the production or activity of neurotrophic factors to support vulnerable neurons	Promotion of neuronal growth, survival and differentiation	Nerve growth factor (NGF), brain-derived neurotrophic factor (BDNF)	Cobianchi *et al.* (2017); Zarneshan *et al.* (2022)
Apoptotic pathways	Inhibiting apoptotic pathways to promote cell survival and prevent neuronal death	Inhibition of caspases, promotion of cell survival signalling	Caspase inhibitors, Bcl-2 family modulators	Chen *et al.* (2022a); Li *et al.* (2022)
Excitotoxicity	Targeting glutamate receptors and downstream signalling pathways to prevent excitotoxic injury	Modulation of glutamate signalling, prevention of neuronal overstimulation	NMDA receptor antagonists (e.g. memantine), glutamate release inhibitors (e.g. riluzole)	Carvajal *et al.* (2016); Marcelli (2018)
Mitochondrial support	Supporting mitochondrial function to enhance cellular energy production and reduce oxidative stress	Enhancement of ATP production, reduction of ROS	Coenzyme Q10, creatine, NAD+ precursors	Jurcau (2021)
Anti-inflammatory	Modulating inflammatory response to reduce inflammation-mediated neuronal damage	Suppression of inflammatory mediators, attenuation of microglial activation	NSAIDs, cytokine inhibitors	Degan *et al.* (2018)
Antioxidants	Neutralizing harmful free radicals and protecting neurons from oxidative damage	Scavenging of free radicals, enhancement of antioxidant defences	Vitamins C and E, glutathione, coenzyme Q10	Teleanu *et al.* (2022)
Chaperone proteins	Assisting in protein folding and preventing the aggregation of misfolded proteins	Facilitation of proper protein folding, prevention of protein aggregation	Heat shock proteins	Wankhede *et al.* (2022)
Neurogenesis stimulation	Promoting the generation of new neurons to enhance brain plasticity and cognitive function	Stimulation of neurogenesis, enhancement of synaptic connectivity	Antidepressants, environmental enrichment	Marchetti *et al.* (2020)

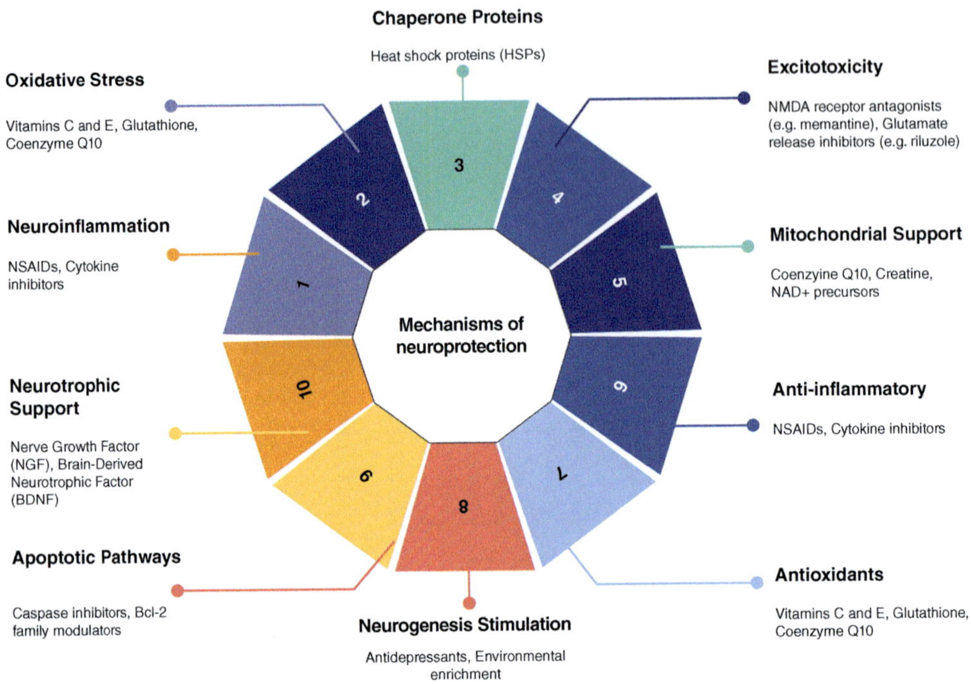

Fig. 12.1. Mechanisms of neuroprotection (authors' own figure).

12.3 Clinical Studies

Clinical studies play a crucial role in translating the neuroprotective effects observed in preclinical models into potential therapeutic strategies for human subjects. Several clinical studies have evaluated the efficacy and safety of neuroprotective interventions for various neurological conditions. In a randomized controlled trial, the authors investigated the neuroprotective effects of a novel drug in Parkinson's disease patients. The study showed that the drug was able to slow the progression of motor symptoms and preserve dopaminergic function, suggesting its potential as a neuroprotective therapy for Parkinson's (Carrera and Cacabelos, 2019). Another clinical study examined the neuroprotective effects of dietary intervention in individuals at risk of developing Alzheimer's disease. The intervention involved a combination of diet, exercise, cognitive training and vascular risk monitoring. The study demonstrated improvements in cognitive function and brain imaging measures, providing evidence for the potential neuroprotective effects of multimodal intervention in Alzheimer's disease (Perus *et al.*, 2022; Liu *et al.*, 2023). Furthermore, a clinical trial was conducted to investigate the neuroprotective effects of a pharmacological agent in patients with acute ischaemic stroke. The study showed that the agent, when administered within a specific time window after stroke onset, improved functional outcomes and reduced neurological deficits, highlighting its potential as a neuroprotective treatment for acute ischaemic stroke (Goenka *et al.*, 2019; Paul and Jalil, 2021). Another study utilized a randomized controlled design, in which rats were randomly assigned to different treatment groups and assessed for neurobehavioural outcomes, histopathological changes, and biochemical markers of neuroprotection (Hashim *et al.*, 2024).

Rigorous study designs such as randomized controlled trials are commonly employed to evaluate the neuroprotective effects of interventions in human subjects (Kong *et al.*, 2020). For example, a multicentre, double-blind, placebo-controlled trial investigated the neuroprotective effects of pharmacological interventions in patients with multiple sclerosis. The trial utilized a randomized controlled design, with patients randomly assigned to receive either the experimental drug or the placebo. Neurological assessments, imaging techniques and biomarker analyses were used to evaluate the neuroprotective effects of the intervention (Chataway *et al.*, 2020). In addition to the randomized controlled trials design mentioned above, various methodologies have been used to assess neuroprotection. This study specifically investigated the neuroprotective effects of cognitive interventions in individuals at risk of developing dementia. This study employed a longitudinal design and assessed changes in brain activity, connectivity and metabolic patterns as measures of neuroprotection (Schoeps *et al.*, 2022). In clinical studies, neurobehavioural assessments are commonly used to evaluate functional outcomes of interventions. For example, a mouse model of Parkinson's disease has been used to assess the neuroprotective effects of a natural compound. The study utilized a battery of behavioural tests, including the rotarod test and pole test, to evaluate motor function and coordination as indicators of neuroprotection (Dos *et al.*, 2022). Comprehensive neurological assessments have been conducted in clinical trials to evaluate the effects of interventions on neurological function (Pena *et al.*, 2022). For example, one study closely monitored participants for any adverse events and recorded them using standardized protocols. The findings of this trial provide valuable insights into the safety profile of interventions in the target population (Tang *et al.*, 2020). Another factor is the safety of neuroprotective herbal supplements in patients with mild cognitive impairment. The trial included rigorous safety monitoring such as regular assessment of vital signs, laboratory tests and patient-reported adverse events. The study findings shed light on the safety aspects of herbal supplements and help determine their suitability for further clinical development (Table 12.2) (Chen *et al.*, 2022b).

12.4 Neuroprotective Applications

D. nobile, also known as the noble dendrobium, has potential neuroprotective properties. Several studies have explored its neuroprotective application, highlighting its promising effects in various contexts. The rich phytochemical composition of *D. nobile* includes alkaloids, flavonoids, phenols, polysaccharides and other bioactive compounds that possess antioxidant, anti-inflammatory, anti-apoptotic and neurotrophic properties. *D. nobile* exhibits antioxidant properties by scavenging reactive oxygen species (ROS) and reducing oxidative stress, which are key contributors to neuronal damage and neurodegeneration. In addition, *D. nobile* has demonstrated significant anti-neuroinflammatory effects. *D. nobile* has demonstrated potential in mitigating neuroinflammation by reducing the production of inflammatory cytokines and influencing inflammatory pathways. Additionally, it exhibits the capacity to enhance neurotrophic factors, which are pivotal for neuronal development, differentiation and synaptic adaptability. Further investigation is imperative to comprehensively understand its mode of operation, refine dosages and treatment durations, and to confirm its effectiveness and safety in human trials (Fig. 12.2).

12.4.1 Neurodegenerative diseases

Neurodegenerative conditions refer to a broad spectrum of disorders in which neurons in the central nervous system gradually deteriorate over time. These disorders include Alzheimer's disease, Parkinson's disease, Huntington's disease and amyotrophic lateral sclerosis (ALS), profoundly impacting cognitive function and the overall well-being of affected individuals. Understanding the underlying mechanisms and developing effective interventions for neurodegenerative diseases are areas of active research. Alzheimer's is particularly devastating, causing a gradual decline in cognitive abilities and memory. The build up of beta-amyloid plaques and tau protein tangles in the brain leads to neuronal damage and disrupts synaptic function, contributing to memory loss, confusion and eventual loss of independence (Perumal and Pichaivel, 2024). A significant focus has been on targeting the tau protein for neuroprotection. The crucial role of tau pathology in Alzheimer's disease progression emphasizes the necessity for interventions aimed at mitigating tau aggregation and propagation (Guo *et al.*, 2022). Understanding and addressing tau-related mechanisms offer promising avenues for the development of effective treatments to combat Alzheimer's disease and its devastating effects on cognitive function and quality of life (Guha *et al.*, 2020).

Table 12.2. Clinical studies on neuroprotective effects of *D. nobile* (authors' own table).

Reference	Condition studied	Intervention	Study design	Key outcomes and approaches
Carrera and Cacabelos (2019)	Parkinson's disease	Novel drug	Randomized controlled trial	Slowed progression of motor symptoms, preserved dopaminergic function
Perus *et al.* (2022)	Alzheimer's disease	Dietary intervention	Clinical study	Improved cognitive function and brain imaging measures
Goenka *et al.* (2019)	Acute ischaemic stroke	Pharmacological agent	Clinical trial	Improved functional outcomes, reduced neurological deficits
Liu *et al.* (2023)	Alzheimer's disease	Dietary intervention	Clinical trial	Demonstrated improvements in cognitive function and brain imaging measures
Paul and Jalil (2021)	Acute ischaemic stroke	Pharmacological agent	Clinical trial	Improved functional outcomes, reduced neurological deficits
Hashim *et al.* (2024)	Neurotoxicity	Pharmacological agent	Experimental study	Assessed neurobehavioural outcomes, histopathological changes, biochemical markers of neuroprotection
Kong *et al.* (2020)	Multiple sclerosis	Pharmacological intervention	Clinical trial	Investigated neuroprotective effects using a randomized controlled design
Chataway *et al.* (2020)	Multiple sclerosis	Pharmacological intervention	Multicentre, double-blind, placebo-controlled trial	Utilized neurological assessments, imaging techniques and biomarker analysis
Schoeps *et al.* (2022)	Dementia risk	Cognitive intervention	Longitudinal study	Investigated changes in brain activity, connectivity and metabolic patterns
Dos *et al.* (2022)	Parkinson's disease	Natural compound	Experimental study	Utilized behavioural tests (rotarod test, pole test) to evaluate motor function and coordination
Pena *et al.* (2022)	Various neurological conditions	Intervention not specified	Clinical study	Conducted comprehensive neurological assessments
Tang *et al.* (2020)	Intervention safety	Intervention not specified	Clinical trial	Monitored participants for adverse events using standardized protocols
Chen *et al.* (2022b)	Mild cognitive impairment	Herbal supplement	Clinical trial	Rigorous safety monitoring including vital signs, laboratory tests and patient-reported adverse events

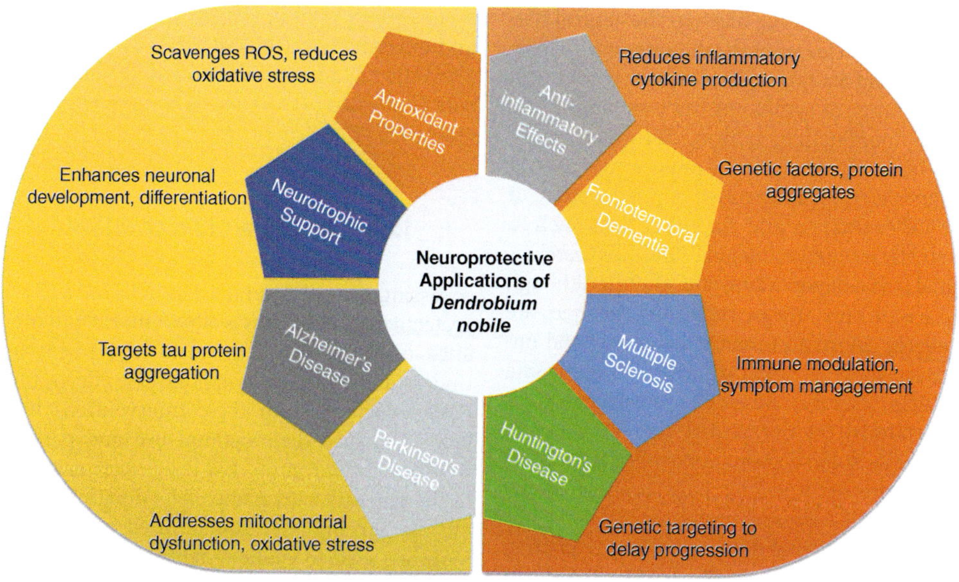

Fig. 12.2. Neuroprotective applications of *D. nobile* (authors' own figure).

Parkinson's disease is characterized by the loss of dopamine-producing neurons in the brain, resulting in debilitating motor symptoms, such as tremors, rigidity and slowness of movement. Additionally, non-motor symptoms such as cognitive impairment and mood disorders often accompany the disease, significantly diminishing a patient's quality of life. These findings underscore the importance of targeting these pathways for neuroprotection and for developing disease-modifying interventions (Ramesh *et al.*, 2023). By understanding and addressing mitochondrial dysfunction and oxidative stress, novel treatments may alleviate symptoms and slow disease progression in individuals with Parkinson's disease. This highlights the importance of ongoing research aimed at uncovering the underlying mechanisms of the disease and identifying effective therapeutic strategies (Prasuhn *et al.*, 2021).

Huntington's disease, which is genetically inherited, causes progressive nerve cell breakdown in the brain, leading to various cognitive, emotional and motor symptoms, particularly in the basal ganglia. This deterioration causes involuntary movements, cognitive decline and psychiatric symptoms, ultimately resulting in complete functional impairments. Research emphasizes the potential of transgenic models to delay disease progression and enhance outcomes in individuals affected by Huntington's disease (Pandey and Rajamma, 2018). By specifically targeting the underlying genetic mechanisms responsible for the condition, these transgenic models hold promise for mitigating symptoms and improving the quality of life of patients affected by this devastating disorder (Kim *et al.*, 2021). Continued exploration and refinement of transgenic models treatment modalities are crucial for advancing the field and providing effective interventions for individuals with Huntington's disease (Caron *et al.*, 2018).

ALS, also known as Lou Gehrig's disease, affects motor neurons in the brain and spinal cord, causing progressive muscle weakness and eventually paralysis, leading to progressive muscle weakness, paralysis and respiratory failure (Ray *et al.*, 2024). Despite the devastating prognosis, advances in supportive care have improved symptom management and extended survival in some cases (Tan *et al.*, 2020). The genetic and environmental factors contributing to ALS are discussed along with potential therapeutic strategies. Keon *et al.* (2021) emphasized the need for a comprehensive understanding of ALS pathogenesis to develop effective neuroprotective interventions.

Multiple sclerosis is an autoimmune disease in which the immune system attacks the myelin sheath covering nerve fibres in the central nervous system. This leads to inflammation, demyelination and damage to nerve fibres, resulting in a wide range of symptoms, such as fatigue, impaired coordination, sensory disturbances and cognitive impairment (Tahernia *et al.*, 2022). Diagnosis typically involves a combination of clinical evaluation and diagnostic tests such as magnetic resonance imaging (MRI) and cerebrospinal fluid analysis (Xiang *et al.*, 2022). Although there is no cure, disease-modifying therapies are available. Efforts to manage symptoms and impede disease progression are central to addressing multiple sclerosis. Research endeavours have concentrated on unravelling the fundamental mechanisms and devising innovative treatments to enhance outcomes for individuals grappling with multiple sclerosis (Uyyala, 2023).

Frontotemporal dementia (FTD) refers to a group of disorders characterized by progressive degeneration of the frontal and temporal lobes of the brain. This degeneration leads to changes in behaviour, personality and language skills, often presenting challenges in social interactions and daily functioning (Bhalla and Motos, 2023). Genetic factors significantly contribute to FTD, with mutations in genes such as *C9orf72*, *MAPT* and *GRN* linked to familial iterations of the disorder. Neuropathologically, FTD is characterized by neuronal loss, gliosis and the presence of abnormal protein aggregates such as tau inclusions or TDP-43 inclusions, depending on the subtype (Lee *et al.*, 2017).

Prion diseases, also referred to as transmissible spongiform encephalopathies (TSEs), are uncommon and lethal neurological conditions resulting from abnormal contagious proteins known as prions (Will and Ironside, 2017). Prion diseases can occur sporadically, genetically or through exposure to contaminated tissues or products. Instances encompass Creutzfeldt–Jakob disease (CJD), variant Creutzfeldt–Jakob disease (vCJD), Gerstmann–Sträussler–Scheinker syndrome (GSS) and fatal familial insomnia (FFI) (Goldman and Vallabh, 2022). These disorders are characterized by the build up of aberrantly folded prion proteins in the brain, resulting in neuronal depletion, spongiform alterations and neurological impairment. These diseases have a uniformly poor prognosis with rapid progression to severe disability and death (Tyler and Tyler, 2023). Despite ongoing research efforts, effective treatments for prion diseases remain elusive, and current management strategies focus on alleviating symptoms and providing supportive care to affected individuals (Bairadi *et al.*, 2023) (Table 12.3).

12.5 Safety and Tolerability

Assessing the safety and tolerability of neuroprotective interventions is of paramount importance in both clinical practice and research. Understanding the potential adverse effects and risks associated with these interventions will help to ensure their optimal use and patient well-being. Neuroprotective drugs were assessed in the patients with Alzheimer's disease. Rigorous monitoring and standardized assessment tools are used to evaluate adverse events, vital signs, laboratory measures and cognitive function (Cummings and Kinney, 2022). Multiple randomized controlled trials have analysed adverse events such as bleeding, infections and cardiovascular complications associated with the use of neuroprotective interventions (Goenka *et al.*, 2019). Moreover, neuroprotective therapies have been investigated in patients with traumatic brain injuries. This involved careful monitoring of adverse events, neurological examinations and laboratory assessments to evaluate the safety profile of the intervention (Hiskens, 2022). Assessing the safety and tolerability of neuroprotective interventions is essential for guiding clinical decision-making and ensuring patient safety. By systematically evaluating adverse events and monitoring patient well-being, researchers and healthcare professionals can make informed decisions regarding neuroprotective strategies.

12.5.1 Adverse reactions

Assessing adverse reactions associated with neuroprotective interventions is crucial for ensuring patient safety and optimizing treatment outcomes. Understanding potential side effects and adverse events helps healthcare professionals and researchers make informed decisions regarding the use of neuroprotective strategies. One review analysed data from multiple clinical

Table 12.3. Neurodegenerative diseases (authors' own table).

Disease	Description	Key characteristics	Current treatment strategies	Key references
Alzheimer's disease	Gradual decline in cognitive abilities and memory due to beta-amyloid plaques and tau protein tangles	Beta-amyloid plaques, tau protein tangles	Cholinesterase inhibitors, NMDA receptor antagonists, symptom management	Guha *et al.* (2020); Guo *et al.* (2022); Perumal and Pichaivel (2024)
Parkinson's disease	Loss of dopamine-producing neurons leading to motor symptoms like tremors, rigidity and slowness of movement	Dopamine depletion, Lewy bodies	Dopamine agonists, MAO-B inhibitors, deep brain stimulation	Prasuhn *et al.* (2021); Ramesh *et al.* (2023)
Huntington's disease	Progressive nerve cell breakdown causing involuntary movements, cognitive decline and psychiatric symptoms	CAG repeat expansion, degeneration of basal ganglia	Symptomatic management, antipsychotic drugs, genetic counselling	Caron *et al.* (2018); Pandey and Rajamma (2018); Kim *et al.* (2021)
Amyotrophic lateral sclerosis (ALS)	Affects motor neurons leading to progressive muscle weakness and eventual paralysis	Motor neuron degeneration, ubiquitin–proteasome system dysfunction	Riluzole, edaravone, respiratory support	Tan *et al.* (2020); Keon *et al.* (2021); Ray *et al.* (2024)
Multiple sclerosis	Autoimmune disease attacking the myelin sheath, leading to a range of symptoms	Inflammation, demyelination	Disease-modifying therapies, symptomatic management, rehabilitation	Tahernia *et al.* (2022); Xiang *et al.* (2022); Uyyala (2023)
Frontotemporal dementia (FTD)	Progressive degeneration of frontal and temporal lobes causing changes in behaviour, personality and language skills	Tau or TDP-43 protein pathology, frontal and temporal lobe atrophy	Symptomatic treatment, supportive care, behavioural therapy	Lee *et al.* (2017); Bhalla and Motos (2023)
Prion diseases	Rare neurological conditions caused by abnormal, contagious proteins called prions	Prion protein misfolding, formation of amyloid plaques	Symptomatic treatment, palliative care	Will and Ironside (2017); Goldman and Vallabh (2022); Bairadi *et al.* (2023); Tyler and Tyler (2023)

MAO-B, monoamine oxidase B.

trials and observational studies to identify and categorize the adverse events associated with neuroprotective interventions and to gain insights into the frequency, severity and types of adverse reactions observed across different interventions and patient populations (Schootemeijer *et al.*, 2020). Adverse reactions to neuroprotective agents in patients with stroke were assessed. A randomized controlled trial employed comprehensive monitoring and assessment methods to evaluate adverse events including cardiovascular events, bleeding complications and allergic reactions (Zavala-Tecuapetla *et al.*, 2020). Furthermore, adverse reactions to neuroprotective therapies have been investigated in patients with traumatic brain injury. A trial included systematic monitoring of adverse events, laboratory assessments and clinical examinations to determine the safety profile of the intervention. The frequency and severity of adverse reactions were observed during the study (Zoerle *et al.*, 2017). To ensure the safety and tolerability of neuroprotective interventions, it is important to assess and monitor the adverse reactions. By systematically evaluating and reporting adverse events, healthcare professionals and researchers can better understand the potential risks associated with these treatments and make informed decisions regarding their use.

12.5.2 Drug interactions

Evaluating potential drug interactions is crucial when considering the use of neuroprotective interventions, as the concurrent administration of multiple drugs can affect treatment efficacy and patient safety (Lang and Espay, 2018). Understanding potential interactions helps healthcare professionals make informed decisions and ensure optimal therapeutic outcomes. There are potential interactions between neuroprotective medications and commonly prescribed drugs in patients with neurodegenerative diseases (Koshatwar *et al.*, 2023). A comprehensive drug interaction database was used to analyse the pharmacokinetic and pharmacodynamic properties of neuroprotective agents to identify potential interactions (Supuran, 2020). The likelihood and clinical significance of drug interactions can guide clinicians in prescribing neuroprotective therapies. In another study, potential drug interactions of neuroprotective drugs in patients with stroke were assessed. The study involved a randomized controlled trial and included systematic monitoring and evaluation of medication profiles and adverse events (Frank *et al.*, 2022). The occurrence and clinical implications of drug interactions highlight the importance of considering potential interactions when prescribing neuroprotective agents for stroke patients (Solaimanian *et al.*,2023). The interactions of neuroprotective medications used in traumatic brain injuries were examined in a review that synthesized evidence from preclinical studies, clinical trials and case reports to identify potential interactions with commonly used medications in acute care settings (Hall *et al.*, 2019). The mechanisms of drug interactions provide recommendations for managing and avoiding the potential risks associated with co-administration (Goodlet *et al.*, 2019). By considering the pharmacokinetic and pharmacodynamic properties of neuroprotective agents and evaluating their potential interactions with other medications, healthcare professionals can optimize treatment plans and minimize the risk of adverse events.

12.5.3 Long-term effects

Understanding the long-term effects of neuroprotective interventions is essential to assess pivotal in comprehensively evaluating their efficacy and safety, particularly in the context of chronic neurological conditions (Paul and Jalil, 2021). Neuroprotective strategies encompass a diverse array of therapeutic approaches aimed at preserving the integrity and functionality of neurons in the central nervous system (Lee *et al.*, 2020). These interventions hold significant promise in mitigating the progression of neurodegenerative diseases, such as Alzheimer's disease, Parkinson's disease, Huntington's disease and ALS, which are characterized by gradual deterioration of neurons and subsequent decline in cognitive and motor functions (Pathak *et al.*, 2021). Over time, neuroprotective interventions may exert sustained beneficial effects on various neurological health aspects. One of the primary long-term benefits is preservation of cognitive function (Caruso *et al.*, 2022). By targeting the underlying mechanisms of neurodegeneration, such as oxidative stress, neuroinflammation and protein misfolding, these interventions can help maintain cognitive abilities such as memory, attention and executive function for an extended period (Villa and Perry, 2023). This preservation of cognitive function not only enhances the quality of life of individuals with neurodegenerative diseases but also delays the onset of severe cognitive impairment and dependency on caregivers (Silva *et al.*, 2020).

Furthermore, neuroprotective interventions have the potential to slow the progression of neurodegenerative processes. This slowing of disease progression may translate into prolonged periods of functional independence and improved overall prognosis for the affected individuals (Sandi *et al.*, 2021). By stabilizing neuronal structure and function, reducing neuronal damage and promoting neuronal repair and regeneration, these interventions can attenuate the gradual loss of neurons that is characteristic of neurodegenerative diseases (Wareham *et al.*, 2022). In addition to cognitive benefits and disease-modifying effects, neuroprotective interventions can enhance the quality of life of patients with chronic neurological conditions (Vijiaratnam *et al.*, 2021). By alleviating symptoms, improving mobility and motor function, and reducing the frequency and severity of complications, neuroprotective interventions enable individuals to engage more fully in daily activities, maintain social connections and experience greater overall well-being (Ren and Xiao, 2023). Moreover, by reducing the burden of disease progression and its associated healthcare costs, neuroprotective interventions may alleviate financial strain on individuals, families and healthcare systems in the long term (Duplantier and Gardner, 2021). Moreover, the sustained efficacy of neuroprotective interventions may extend beyond symptom management to encompass broader improvements in neurological health and function (Saceleanu *et al.*, 2023). By enhancing neuronal plasticity, promoting neurogenesis and fostering synaptic connectivity, these interventions can create a more resilient neurological environment that can withstand the challenges of ageing and disease (Marzola *et al.*, 2023). This enhanced neurological resilience may translate into greater resistance to future neurodegenerative processes and a reduced risk of developing additional neurological disorders later in life (Buchman and Bennett, 2022).

Although the potential benefits of neuroprotective interventions are considerable, their long-term safety and tolerability must be considered. Some interventions may carry risks of adverse effects or interactions with other medications, particularly when administered over extended periods (Ferensztajn-Rochowiak and Rybakowski, 2023). Therefore, ongoing monitoring and assessment of both efficacy and safety are crucial to ensure an optimal balance between therapeutic benefits and potential risks (Souto *et al.*, 2024). The study followed patients over a period of 5 years and assessed various clinical outcomes, including motor function, cognitive performance and quality of life (Limousin and Foltynie, 2019). Neuroprotective interventions provide insights into their impact on disease progression and long-term patient well-being (Muscaritoli, 2021). Neuroprotection has also been evaluated in patients with Alzheimer's disease. A study included a follow-up period of 3 years and assessed cognitive function, functional status and caregiver burden.

Overall, we have discussed the long-term effects of neuroprotective interventions in patients with ischaemic stroke. This chapter has synthesized evidence from randomized controlled trials and observational studies with long-term follow-up periods. Long-term functional outcomes, quality of life and mortality rates associated with neuroprotective interventions provide insight into overall effectiveness and safety. Neuroprotective interventions are essential to optimize the management of chronic neurological conditions. By preserving cognitive function, slowing disease progression, enhancing quality of life and promoting neurological resilience, these interventions offer promise for improving the outcomes and prognosis of individuals with neurodegenerative diseases. However, continued research and monitoring are needed to fully elucidate their long-term benefits and risks, and to refine their use in clinical practice.

12.6 Future Directions and Challenges

D. nobile, a traditional Chinese medicinal plant, has shown promising neuroprotective effects in preclinical and clinical studies. However, several future directions and challenges need to be addressed to fully understand the neuroprotective potential of *D. nobile*. One crucial aspect is the identification and characterization of active compounds responsible for the neuroprotective effects of *D. nobile*. This plant contains a wide array of bioactive constituents, including alkaloids, polysaccharides and phenolic compounds, which may contribute to its therapeutic properties. Further studies, such as phytochemical analyses and bioassay-guided fractionation, are needed to isolate and identify the key neuroprotective compounds in *D. nobile*. In addition, understanding the underlying molecular mechanisms of *D. nobile*-mediated neuroprotection is essential for its clinical translation. Elucidating the signalling pathways, cellular targets and molecular interactions involved in the neuroprotective effects of *D. nobile* can provide insights into its mode of action and facilitate the development of targeted interventions.

Furthermore, rigorous preclinical and clinical studies are required to establish the efficacy and safety profile of *D. nobile* for different neurological disorders. Another challenge is the standardization of *D. nobile* preparations. Variations in cultivation, harvesting, processing and formulation methods can affect the composition and quality of *D. nobile* products, leading to inconsistent therapeutic outcomes. Developing standardized protocols for the production and quality control of *D. nobile* extracts or formulations is crucial to ensure consistency, reproducibility and reliability in both preclinical and clinical studies (Wu *et al.*, 2020).

Finally, the translation of *D. nobile*-based neuroprotective therapies into clinical practice requires collaboration between researchers, clinicians, regulatory agencies and industry partners. Multidisciplinary approaches, including pharmacological, pharmacokinetic and pharmacogenetic studies, are necessary to establish the efficacy, safety and optimal use of *D. nobile* in various patient populations.

12.6.1 Novel formulations and delivery systems

D. nobile has great potential to enhance its therapeutic efficacy, improve bioavailability and optimize patient compliance. Various innovative approaches have been explored to overcome the challenges associated with the delivery of *D. nobile* compounds. One approach is the development of nanoformulations of *D. nobile* extracts. Nanotechnology-based delivery systems, such as nanoemulsions, nanocapsules and nanoparticles, have shown promise in enhancing the solubility, stability and targeted delivery of bioactive compounds. For instance, Liu *et al.* (2020) developed *D. nobile*-extract-loaded lipid nanoparticles and demonstrated improved oral bioavailability and sustained release of the active constituents. In addition to nanotechnology, other delivery systems, such as microencapsulation and liposomes, have been investigated for the efficient delivery of *D. nobile* compounds. Encapsulation of *D. nobile* extracts in microspheres or microcapsules can protect bioactive compounds from degradation, enhance their stability and enable controlled release. Zhang *et al.* (2019) developed a microencapsulation system for *D. nobile* polysaccharides and observed prolonged release and enhanced neuroprotective effects.

Moreover, the use of novel routes of administration has been explored to improve the delivery of *D. nobile* compounds. Transdermal patches, buccal films and nasal sprays are among the alternative routes investigated to bypass the gastrointestinal tract and enhance absorption of bioactive compounds. Furthermore, the combination of *D. nobile* with other delivery-enhancing agents or technologies has been explored to optimize its therapeutic potential. Co-administration with penetration enhancers, natural surfactants or drug delivery systems can enhance the permeability and absorption of *D. nobile* compounds. In conclusion, the development of novel formulations and delivery systems for *D. nobile* offers exciting opportunities to enhance therapeutic efficacy and overcome delivery challenges. Nanoformulations, microencapsulation, alternative routes of administration and combination approaches can improve the solubility, stability, bioavailability and targeted delivery of *D. nobile* compounds.

12.6.2 Targeted therapies

Targeted therapies in the field of neuroprotection aim to selectively modulate the specific molecular targets or pathways involved in neurodegeneration. These therapies offer the potential for enhanced efficacy and reduced off-target effects compared with conventional treatments. Targeted therapies in the realm of neuroprotection represent a significant advancement in the treatment of neurological disorders, offering precise interventions tailored to the specific molecular and cellular mechanisms underlying neurodegeneration. Unlike conventional approaches, which may have broader effects on neuronal function, targeted therapies aim to modulate specific pathways involved in neuronal damage and loss, thereby minimizing off-target effects and maximizing therapeutic efficacy. One area of targeted therapy involves the inhibition of pathological protein aggregation, a hallmark feature of many neurodegenerative diseases, such as Alzheimer's and Parkinson's diseases. By targeting enzymes involved in protein misfolding and aggregation, or by enhancing the clearance of aggregated proteins through mechanisms such as autophagy, the aim with targeted therapies is to prevent the accumulation of toxic protein aggregates and mitigate neuronal toxicity. Another promising approach is the modulation of the neuroinflammatory pathways implicated in the progression of neurodegenerative diseases. Chronic neuroinflammation is characterized by dysregulation of microglial activation. Targeted therapies designed to suppress aberrant neuroinflammatory responses while preserving the beneficial aspects of the immune system are promising for slowing disease progression and preserving neuronal function.

Moreover, targeted therapies may leverage advances in gene therapy and RNA-based technologies to correct the genetic mutations underlying hereditary forms of neurodegenerative diseases. These approaches aim to restore normal cellular functions and halt disease progression at the genetic level by delivering therapeutic genes or RNA molecules to specific brain regions. An example of targeted therapy is the use of monoclonal antibodies to target specific proteins implicated in neurodegenerative diseases. For instance, aducanumab, an anti-amyloid antibody, has been investigated as a targeted therapy for Alzheimer's disease. Aducanumab treatment led to a significant reduction in amyloid plaques in the brains of patients with prodromal or mild Alzheimer's disease, suggesting a potential neuroprotective effect that may slow disease progression and improve cognitive outcomes. Another targeted approach involves modulation of specific signalling pathways involved in neurodegeneration. For instance, the phosphoinositol 3-kinase (PI3K)/Akt pathway is pivotal for promoting cell survival and providing neuroprotection. Researchers have investigated potential therapeutic approaches for activating this pathway. The neuroprotective benefits of PI3K/Akt pathway activation were examined in a mouse model of Parkinson's disease using 1-methyl-4-phenyl-1,2,3,6-tetrahydropyridine (MPTP) and demonstrated improved neuronal survival and motor functions.

Furthermore, gene therapy offers targeted interventions for neuroprotection. Gene therapy can modulate specific cellular processes and promote neuronal survival by delivering therapeutic genes directly to affected brain regions. Targeted therapies may promote neuronal survival and regeneration through the activation of neurotrophic signalling pathways. By enhancing the production or activity of these factors, targeted therapies can support neuronal health and function and potentially reverse or delay neurodegenerative processes. Targeted therapies can also involve the use of small molecules that selectively target specific pathways. In the context of neuroprotection, these small molecules may act as agonists or antagonists for specific receptors or enzymes. For example, The neuroprotective effects of a specific P2X7 receptor antagonist were investigated in a cellular model of ischemic stroke, demonstrating reduced cell death and inflammation. Overall, targeted therapies hold great promise for the treatment of neurodegenerative diseases by offering precise interventions tailored to the underlying molecular and cellular mechanisms of the disease. While many of these therapies are still in the experimental stage, ongoing research efforts have the potential to translate advances in knowledge into effective treatments that can slow or even halt the progression of neurodegenerative diseases and improve outcomes for affected individuals. Targeted therapies hold promise for neuroprotection by selectively modulating the specific targets or pathways involved in neurodegeneration. Monoclonal antibodies, pathway modulation, gene therapy and small-molecule interventions are among the approaches explored for targeted neuroprotection.

12.6.3 Clinical trial design and standardization

Clinical trial design and standardization are essential to assess the efficacy and safety of neuroprotective interventions. Well-designed trials with standardized protocols ensure the reliability and comparability of results across different studies. Randomized controlled trials have enabled researchers to evaluate the effectiveness of neuroprotective interventions in controlled settings. For example, a randomized controlled trial conducted by Gravesteijn *et al.* (2020) assessed the neuroprotective effects of a novel drug in patients with multiple sclerosis. The trial employed rigorous randomization procedures and blinding techniques to minimize bias and ensure robust results. Standardized outcome measures and assessment tools were used to ensure comparability and consistency across trials. For example, in neurodegenerative disease trials, clinical rating scales are commonly used to assess cognitive function and motor symptoms. Standardized outcome measures enhance the interpretability and generalizability of findings.

In addition to outcome measures, trial protocols should adhere to standardized transparency and rigour in reporting. By following CONSORT guidelines, researchers can enhance the quality and reproducibility of their studies (Wu, 2024). Overall, randomized controlled trials with standardized protocols and outcome measures enhance the validity and comparability of study findings, leading to more robust evidence in the field of neuroprotection.

12.7 Conclusion

In conclusion, the neuroprotective effects of *D. nobile* have been extensively investigated and show promising potential in the field of neuroprotection. The diverse bioactive compounds present in *D. nobile*, including alkaloids, polysaccharides and other phytochemicals, contribute to its neuroprotective effects via multiple mechanisms.

Experimental studies and *in vitro* investigations have demonstrated that *D. nobile* exerts neuroprotective effects by reducing oxidative stress, inhibiting apoptosis, promoting neurotrophic factors and modulating inflammatory responses. These actions contribute to the preservation and enhancement of ischaemic stroke and have provided evidence of the neuroprotective effects of *D. nobile*, showing improvements in behavioural and cognitive outcomes. Preclinical studies have shown promising results, leading to understanding of the underlying molecular mechanisms and revealing the therapeutic potential of *D. nobile*. Clinical trials involving human subjects are required to validate their efficacy and safety in real-world settings. Standardized clinical trial designs and rigorous methodologies should be used to ensure reliable and reproducible results.

Challenges and limitations remain in the study of the neuroprotective effects of *D. nobile*. These include the need for standardized extraction protocols, identification and characterization of active compounds, determination of optimal dosing regimens and consideration of potential drug interactions and long-term effects. Additionally, the translation of findings from preclinical studies to clinical applications requires careful consideration of various factors including bioavailability, formulation development and patient heterogeneity. In summary, *D. nobile* shows substantial potential as a neuroprotective agent for neurodegenerative disorders. Its rich array of bioactive compounds and multifaceted mechanisms of action make it a compelling candidate for continued exploration. The neuroprotective properties exhibited by *D. nobile* inspire optimism for the emergence of innovative approaches to address neurodegenerative diseases and enhance neurological outcomes.

Acknowledgements

The authors express their gratitude for the financial support received through the Distinguished High-Level Talents Research Grant from the Guizhou Science and Technology Corporation Platform Talents Fund (grant nos. [2017]5733-001 and CK-1130-002), the National Natural Science Foundation of China (U1812403, 82373981 and 82060750), 2011 Collaborative Innovation Centre of Traditional Chinese Medicine in Guizhou Province (no. [2022]026) and the support provided by the Zunyi Medical University, China. Special appreciation is extended to all laboratory colleagues and research staff members for their valuable insights, constructive guidance and assistance with the contents of this chapter.

References

Baiardi, S., Mammana, A., Capellari, S. and Parchi, P. (2023) Human prion disease: Molecular pathogenesis, and possible therapeutic targets and strategies. *Expert Opinion on Therapeutic Targets* 27, 1271–1284.

Bhalla, A. and Motos, N. (2023) Psychiatric manifestations of neurodegenerative disorders: Focus on frontotemporal dementia. *Archives of Clinical Psychiatry* 50, 1–8.

Buchman, A.S. and Bennett, D.A. (2022) Mixed neuropathologies, neural motor resilience and target discovery for therapies of late-life motor impairment. *Frontiers in Human Neuroscience* 16, 853330.

Cabeza, R., Albert, M., Belleville, S., Craik, F.I.M., Duarte, A. *et al.* (2018) Maintenance, reserve and compensation: the cognitive neuroscience of healthy ageing. *Nature Reviews Neuroscience* 19, 701–710.

Cammisuli, D.M., Franzoni, F., Scarfò, G., Fusi, J., Gesi, M. *et al.* (2022) What does the brain have to keep working at its best? Resilience mechanisms such as antioxidants and brain/cognitive reserve for counteracting Alzheimer's disease degeneration. *Biology* 11, 650.

Caron, N.S., Dorsey, E.R. and Hayden, M.R. (2018) Therapeutic approaches to Huntington disease: from the bench to the clinic. *Nature Reviews Drug Discovery* 17, 729–750.

Carrera, I. and Cacabelos, R. (2019) Current drugs and potential future neuroprotective compounds for Parkinson's disease. *Current Neuropharmacology* 17, 295–306.

Caruso, G., Torrisi, S.A., Mogavero, M.P., Currenti, W., Castellano, S. *et al.* (2022) Polyphenols and neuroprotection: therapeutic implications for cognitive decline. *Pharmacology & Therapeutics* 232, 108013.

Carvajal, F.J., Mattison, H.A. and Cerpa, W. (2016) Role of NMDA receptor-mediated glutamatergic signaling in chronic and acute neuropathologies. *Neural Plasticity* 2016, 2701526.

Chataway, J., De Angelis, F., Connick, P., Parker, R.A., Plantone, D. *et al.* (2020) Efficacy of three neuroprotective drugs in secondary progressive multiple sclerosis (MS-SMART): a phase 2b, multiarm, double-blind, randomised placebo-controlled trial. *The Lancet Neurology* 19, 214–225.

Chen, B., Zhao, J., Zhang, R., Zhang, L., Zhang, Q. *et al.* (2022a) Neuroprotective effects of natural compounds on neurotoxin-induced oxidative stress and cell apoptosis. *Nutritional Neuroscience* 25, 1078–1099.

Chen, J., Shen, K., Fan, L., Hu, H., Li, T. *et al.* (2022b) Integrative medicine in treating post-stroke depression: study protocol for a multicenter, prospective, randomized, controlled trial. *Frontiers in Psychology* 13, 923506.

Cobianchi, S., Arbat-Plana, A., Lopez-Alvarez, V.M. and Navarro, X. (2017) Neuroprotective effects of exercise treatments after injury: The dual role of neurotrophic factors. *Current Neuropharmacology* 15, 495–518.

Colucci-D'Amato, L., Speranza, L. and Volpicelli, F. (2020) Neurotrophic factor BDNF, physiological functions and therapeutic potential in depression, neurodegeneration and brain cancer. *International Journal of Molecular Sciences* 21, 7777.

Crupi, R., Cordaro, M., Cuzzocrea, S. and Impellizzeri, D. (2020) Management of traumatic brain injury: from present to future. *Antioxidants* 9, 297.

Cummings, J. and Kinney, J. (2022) Biomarkers for Alzheimer's disease: context of use, qualification, and roadmap for clinical implementation. *Medicina (Kaunas, Lithuania)* 58, 952.

Davis, K.D., Aghaeepour, N., Ahn, A.H., Angst, M.S., Borsook, D. *et al.* (2020) Discovery and validation of biomarkers to aid the development of safe and effective pain therapeutics: Challenges and opportunities. *Nature Reviews Neurology* 16, 381–400.

Degan, D., Ornello, R., Tiseo, C., Carolei, A., Sacco, S. and Pistoia, F. (2018) The role of inflammation in neurological disorders. *Current Pharmaceutical Design* 24(14), 1485–1501.

Dos Santos, M.G., Schimith, L.E., André-Miral, C., Muccillo-Baisch, A.L., Arbo, B.D. *et al.* (2022) Neuroprotective effects of resveratrol in vivo and in vitro experimental models of Parkinson's disease: a systematic review. *Neurotoxicity Research* 40, 319–345.

Duplantier, S.C. and Gardner, C.D. (2021) A critical review of the study of neuroprotective diets to reduce cognitive decline. *Nutrients* 13, 2264.

Fan, C., Sun, X., Wang, X. and Yu, H. (2023) Therapeutic potential of the chemical composition of *Dendrobium nobile* Lindl. *Frontiers in Pharmacology* 14, 1163830.

Ferensztajn-Rochowiak, E. and Rybakowski, J.K. (2023) Long-term lithium therapy: side effects and interactions. *Pharmaceuticals* 16, 74.

Frank, D., Gruenbaum, B.F., Zlotnik, A., Semyonov, M., Frenkel, A. *et al.* (2022) Pathophysiology and current drug treatments for post-stroke depression: a review. *International Journal of Molecular Sciences* 23, 15114.

Ganapathy, A.A., Haripriya, V.M., Acharya, N., Somappa, S.B. and Kumaran, A. (2023) Ethnobotanical significance of medicinal plants: beta-amyloid and tau aggregation inhibitors against Alzheimer's disease. *Journal of Biochemical and Molecular Toxicology* 37, e23339.

Goenka, L., Uppugunduri Satyanarayana, C.R., S, S.K. and George, M. (2019) Neuroprotective agents in acute ischemic stroke—a reality check. *Biomedicine & Pharmacotherapy* 109, 2539–2547.

Goldman, J.S. and Vallabh, S.M. (2022) Genetic counseling for prion disease: updates and best practices. *Genetics in Medicine* 24, 1993–2003.

Goodlet, K.J., Zmarlicka, M.T. and Peckham, A.M. (2019) Drug-drug interactions and clinical considerations with co-administration of antiretrovirals and psychotropic drugs. *CNS Spectrums* 24, 287–312.

Gravesteijn, A.S., Beckerman, H., de Jong, B.A., Hulst, H.E. and de Groot, V. (2020) Neuroprotective effects of exercise in people with progressive multiple sclerosis (exercise PRO-MS): study protocol of a phase II trial. *BMC Neurology* 20, 177.

Guha, S., Johnson, G.V.W. and Nehrke, K. (2020) The crosstalk between pathological tau phosphorylation and mitochondrial dysfunction as a key to understanding and treating Alzheimer's disease. *Molecular Neurobiology* 57, 5103–5120.

Guo, Y., Li, S., Zeng, L.H. and Tan, J. (2022) Tau-targeting therapy in Alzheimer's disease: critical advances and future opportunities. *Ageing and Neurodegenerative Diseases* 2, 11.

Hall, E.D., Wang, J.A., Miller, D.M., Cebak, J.E. and Hill, R.L. (2019) Newer pharmacological approaches for antioxidant neuroprotection in traumatic brain injury. *Neuropharmacology* 145, 247–258.

Hashim, A.R., Bashir, D.W., Rashad, E., Galal, M.K. and Rashad, M.M. (2024) Neuroprotective assessment of betaine against copper oxide nanoparticle-induced neurotoxicity in the brains of albino rats: a histopathological, neurochemical, and molecular investigation. *ACS Chemical Neuroscience* 15, 1684–1701.

Hiskens, M.I. (2022) Targets of neuroprotection and review of pharmacological interventions in traumatic brain injury. *Journal of Pharmacology and Experimental Therapeutics* 382, 149–166.

Jain, A., Sarsaiya, S., Chen, J., Wu, Q., Lu, Y. *et al.* (2021) Changes in global Orchidaceae disease geographical research trends: recent incidences, distributions, treatment, and challenges. *Bioengineered* 12, 13–29.

Jarrahi, A., Braun, M., Ahluwalia, M., Gupta, R.V., Wilson, M. *et al.* (2020) Revisiting traumatic brain injury: from molecular mechanisms to therapeutic interventions. *Biomedicines* 8, 389.

Jurcau, A. (2021) Insights into the pathogenesis of neurodegenerative diseases: Focus on mitochondrial dysfunction and oxidative stress. *International Journal of Molecular Sciences* 22, 11847.

Keon, M., Musrie, B., Dinger, M., Brennan, S.E., Santos, J. *et al.* (2021) Destination amyotrophic lateral sclerosis. *Frontiers in Neurology* 12, 596006.

Kim, A., Lalonde, K., Truesdell, A., Gomes Welter, P., Brocardo, P.S. *et al.* (2021) New avenues for the treatment of Huntington's disease. *International Journal of Molecular Sciences* 22, 8363.

Kim, C.K., Sachdev, P.S. and Braidy, N. (2022) Recent neurotherapeutic strategies to promote healthy brain aging: are we there yet? *Aging and Disease* 13, 175–214.

Kong, X.-J., Liu, J., Li, J., Kwong, K., Koh, M. *et al.* (2020) Probiotics and oxytocin nasal spray as neuro-social-behavioral interventions for patients with autism spectrum disorders: a pilot randomized controlled trial protocol. *Pilot and Feasibility Studies* 6, 20.

Koshatwar, M., Acharya, S., Prasad, R., Lohakare, T. and Wanjari, M. (2023) Exploring the potential of antidiabetic agents as therapeutic approaches for Alzheimer's and Parkinson's diseases: a comprehensive review. *Cureus* 15, e44763.

Lang, A.E. and Espay, A.J. (2018) Disease modification in Parkinson's disease: current approaches, challenges, and future considerations. *Movement Disorders* 33, 660–677.

Lee, E.B., Porta, S., Michael Baer, G., Xu, Y. and Suh, E. (2017) Expansion of the classification of FTLD-TDP: distinct pathology associated with rapidly progressive frontotemporal degeneration. *Acta Neuropathologica* 134, 65–78.

Lee, K.H., Cha, M. and Lee, B.H. (2020) Neuroprotective effect of antioxidants in the brain. *International Journal of Molecular Sciences* 21, 7152.

Li, D.D., Zheng, C.Q., Zhang, F. and Shi, J.S. (2022a) Potential neuroprotection by *Dendrobium nobile* Lindl alkaloid in Alzheimer's disease models. *Neural Regeneration Research* 17, 972–977.

Li, R.L., Wang, L.Y., Duan, H.X., Zhang, Q., Guo, X. *et al.* (2022b) Regulation of mitochondrial dysfunction induced cell apoptosis is a potential therapeutic strategy for herbal medicine to treat neurodegenerative diseases. *Frontiers in Pharmacology* 13, 937289.

Limousin, P. and Foltynie, T. (2019) Long-term outcomes of deep brain stimulation in Parkinson disease. *Nature Reviews Neurology* 15, 234–242.

Liu, J., Liu, B., He, X., Qin, W. and Shi, J. (2020) *Dendorbium nobile* Lindl. alkaloids suppress NF-kb and NLRP3 signaling pathways to attenuate lipopolysaccharide-induced neuroinflammation. *Journal of Cellular Signaling* 1, 102–114.

Liu, J., van Beusekom, H., Bu, X.-L., Chen, G., Henrique Rosado de Castro, P. *et al.* (2023) Preserving cognitive function in patients with Alzheimer's disease: the Alzheimer's disease neuroprotection research initiative (ADNRI). *Neuroprotection* 1, 84–98.

Maharjan, R., Diaz Bustamante, L., Ghattas, K.N., Ilyas, S., Al-Refai, R. *et al.* (2020) Role of lifestyle in neuroplasticity and neurogenesis in an aging brain. *Cureus* 12, e10639.

Marcelli, S. (2018) A novel selective cell-permeable peptide disrupts presynaptic JNK2-stx1a interaction, providing neuroprotection and preventing glutamate-mediated excitotoxicity. *Scientific Reports* 9, 1–12.

Marchetti, B., Tirolo, C., L'Episcopo, F., Caniglia, S., Testa, N. *et al.* (2020) Parkinson's disease, aging and adult neurogenesis: Wnt/β-catenin signalling as the key to unlock the mystery of endogenous brain repair. *Aging Cell* 19, e13101.

Marzola, P., Melzer, T., Pavesi, E., Gil-Mohapel, J. and Brocardo, P.S. (2023) Exploring the role of neuroplasticity in development, aging, and neurodegeneration. *Brain Sciences* 13, 1610.

Muscaritoli, M. (2021) The impact of nutrients on mental health and well-being: insights from the literature. *Frontiers in Nutrition* 8, 656290.

Pandey, M. and Rajamma, U. (2018) Huntington's disease: the coming of age. *Journal of Genetics* 97, 649–664.

Pathak, N., Vimal, S.K., Tandon, I., Agrawal, L., Hongyi, C. *et al.* (2022) Neurodegenerative disorders of alzheimer, parkinsonism, amyotrophic lateral sclerosis and multiple sclerosis: an early diagnostic approach for precision treatment. *Metabolic Brain Disease* 37, 67–104.

Paul, S. and Jalil, E.C. (2021) Emerging neuroprotective strategies for the treatment of ischemic stroke: an overview of clinical and preclinical studies. *Experimental Neurology* 335, 113518.

Pena, M.T., Lahiri, R., Ebenezer, G.J., Wheat, S.W., Figarola, J. *et al.* (2022) The armadillo as a model for leprosy nerve function impairment: preventative and therapeutic interventions. *Frontiers in Medicine* 9, 879097.

Perumal, P. and Manimekalai, P. (2024) The different molecular targets and its significant role in Alzheimer's disease: a review. *Asian Journal of Medicine and Biomedicine* 8, 111–143.

Perus, L., Busto, G.U., Mangin, J.F., Le Bars, E. and Gabelle, A. (2022) Effects of preventive interventions on neuroimaging biomarkers in subjects at-risk to develop Alzheimer's disease: a systematic review. *Frontiers in Aging Neuroscience* 14, 1014559.

Plascencia-Villa, G. and Perry, G. (2023) Roles of oxidative stress in synaptic dysfunction and neuronal cell death in Alzheimer's disease. *Antioxidants* 12, 1628.

Prasuhn, J., Davis, R.L. and Kumar, K.R. (2021) Targeting mitochondrial impairment in Parkinson's disease: challenges and opportunities. *Frontiers in Cell and Developmental Biology* 8, 615461.

Ramesh, S. and Arachchige, A.S.P.M. (2023) Depletion of dopamine in Parkinson's disease and relevant therapeutic options: a review of the literature. *AIMS Neuroscience* 10, 200–231.

Ray, R.K., Yadav, P. and Chandra, A. (2024) Use of chat generative pretrained transformer in counseling patient of amyotrophic lateral sclerosis. *Journal of Primary Care Specialties* 5, 130–134.

Ren, J. and Xiao, H. (2023) Exercise for mental well-being: exploring neurobiological advances and intervention effects in depression. *Life* 13, 1505.

Saceleanu, V.M., Toader, C., Ples, H., Covache-Busuioc, R.-A., Costin, H.P. *et al.* (2023) Integrative approaches in acute ischemic stroke: from symptom recognition to future innovations. *Biomedicines* 11, 2617.

Sandi, D., Fricska-Nagy, Z., Bencsik, K. and Vécsei, L. (2021) Neurodegeneration in multiple sclerosis: symptoms of silent progression, biomarkers and neuroprotective therapy-Kynurenines are important players. *Molecules* 26, 3423.

Saragovi, H.U., Galan, A. and Levin, L.A. (2019) Neuroprotection: pro-survival and anti-neurotoxic mechanisms as therapeutic strategies in neurodegeneration. *Frontiers in Cellular Neuroscience* 13, 231.

Schoeps, V.A., Graves, J.S., Stern, W.A., Zhang, L., Nourbakhsh, B. *et al.* (2022) N-acetyl cysteine as a neuroprotective agent in progressive multiple sclerosis (NACPMS) trial: study protocol for a randomized, double-blind, placebo-controlled add-on phase 2 trial. *Contemporary Clinical Trials* 122, 106941.

Schootemeijer, S., van der Kolk, N.M., Bloem, B.R. and de Vries, N.M. (2020) Current perspectives on aerobic exercise in people with Parkinson's disease. *Neurotherapeutics* 17, 1418–1433.

Shohag, S., Akhter, S., Islam, S., Sarker, T., Sifat, M.K. *et al.* (2022) Perspectives on the molecular mediators of oxidative stress and antioxidant strategies in the context of neuroprotection and neurolongevity: an extensive review. *Oxidative Medicine and Cellular Longevity* 2022, 7743705.

Silva, R., Bobrowicz-Campos, E., Cardoso, D., Costa, P., Couto, F. *et al.* (2020) Effects of caregiver-provided individual cognitive interventions on cognition, social functioning and quality of life in older adults with major neurocognitive disorders: a systematic review. *JBI Evidence Synthesis* 18, 743–806.

Sindhu, R.K., Kaur, P., Kaur, P., Singh, H., Batiha, G.E.-S. *et al.* (2022) Exploring multifunctional antioxidants as potential agents for management of neurological disorders. *Environmental Science and Pollution Research International* 29, 24458–24477.

Singh, A., Kukreti, R., Saso, L. and Kukreti, S. (2019) Oxidative stress: a key modulator in neurodegenerative diseases. *Molecules* 24, 1583.

Singh, E. and Devasahayam, G. (2020) Neurodegeneration by oxidative stress: a review on prospective use of small molecules for neuroprotection. *Molecular Biology Reports* 47, 3133–3140.

Solaimanian, S., Rahmani, B. and Shiri-Shahsavar, M.R. (2023) The impact of hemp seed oil on drug-drug interactions: a comprehensive review of potential mechanisms and clinical implications. In *Review*.

Souto, E.B., Blanco-Llamero, C., Krambeck, K., Kiran, N.S., Yashaswini, C. *et al.* (2024) Regulatory insights into nanomedicine and gene vaccine innovation: safety assessment, challenges, and regulatory perspectives. *Acta Biomaterialia* 180, 1–17.

Supuran, C.T. (2020) An update on drug interaction considerations in the therapeutic use of carbonic anhydrase inhibitors. *Expert Opinion on Drug Metabolism & Toxicology* 16, 297–307.

Tahernia, H., Esnaasharieh, F., Amani, H., Milanifard, M. and Mirakhori, F. (2022) Diagnosis and treatment of MS in patients suffering from various degrees of the disease with a clinical approach: the orginal article. *Journal of Pharmaceutical Negative Results* 13, 1908–1921.

Tan, A.C., Ashley, D.M., López, G.Y., Malinzak, M., Friedman, H.S. *et al.* (2020) Management of glioblastoma: State of the art and future directions. *CA: A Cancer Journal for Clinicians* 70, 299–312.

Tang, W., Cao, Z., Han, M., Wang, Z., Chen, J. *et al.* (2020) Hydroxychloroquine in patients with mainly mild to moderate coronavirus disease 2019: open label, randomised controlled trial. *BMJ (Clinical Research Ed.)* 369, m1849.

Teleanu, D.M., Niculescu, A.-G., Lungu, I.I., Radu, C.I., Vladâcenco, O. *et al.* (2022) An overview of oxidative stress, neuroinflammation, and neurodegenerative diseases. *International Journal of Molecular Sciences* 23, 5938.

Tyler, S.E. and Tyler, L.D. (2023) Pathways to healing: plants with therapeutic potential for neurodegenerative diseases. *IBRO Neuroscience Reports* 14, 210–234.

Uyyala, S. (2023) The development of new treatments for neurological disorders: insights, innovations, and ethical foundations. *Indus Journal of Medical and Health Sciences* 1, 45–55.

Vijiaratnam, N., Simuni, T., Bandmann, O., Morris, H.R. and Foltynie, T. (2021) Progress towards therapies for disease modification in Parkinson's disease. *The Lancet. Neurology* 20, 559–572.

Villoslada, P. and Steinman, L. (2020) New targets and therapeutics for neuroprotection, remyelination and repair in multiple sclerosis. *Expert Opinion on Investigational Drugs* 29, 443–459.

Voet, S., Srinivasan, S., Lamkanfi, M. and van Loo, G. (2019) Inflammasomes in neuroinflammatory and neurodegenerative diseases. *EMBO Molecular Medicine* 11, e10248.

Wang, F., Wan, J., Liao, Y., Liu, S., Wei, Y. *et al.* (2023) *Dendrobium* species regulate energy homeostasis in neurodegenerative diseases: a review. *Food Science and Human Wellness* 12, 2151–2174.

Wang, G., Wang, J., Deng, Y., Qin, L., He, Y. *et al.* (2022) Chemical constituents and nutritional health functions of *Dendrobium nobile*: a review. *Food Science and Technology* 42, e84522.

Wankhede, N.L., Kale, M.B., Upaganlawar, A.B., Taksande, B.G., Umekar, M.J. *et al.* (2022) Involvement of molecular chaperone in protein-misfolding brain diseases. *Biomedicine & Pharmacotherapy* 147, 112647.

Wareham, L.K., Liddelow, S.A., Temple, S., Benowitz, L.I., Di Polo, A. *et al.* (2022) Solving neurodegeneration: common mechanisms and strategies for new treatments. *Molecular Neurodegeneration* 17, 23.

Will, R.G. and Ironside, J.W. (2017) Sporadic and infectious human prion diseases. *Cold Spring Harbor Perspectives in Medicine* 7, a024364.

Wu, S. (2024) Integrative medicine for the treatment of stroke. *The Lancet* 404, 2135–2137.

Xiang, Y., Zeng, C., Liu, B., Tan, W., Wu, J. *et al.* (2022) Deep learning-enabled identification of autoimmune encephalitis on 3D multi-sequence MRI. *Journal of Magnetic Resonance Imaging* 55(4), 1082–1092.

Zarneshan, S.N., Fakhri, S. and Khan, H. (2022) Targeting Akt/CREB/BDNF signaling pathway by ginsenosides in neurodegenerative diseases: a mechanistic approach. *Pharmacological Research* 177, 106099.

Zavala-Tecuapetla, C., Cuellar-Herrera, M. and Luna-Munguia, H. (2020) Insights into potential targets for therapeutic intervention in epilepsy. *International Journal of Molecular Sciences* 21, 8573.

Zhang, W., Zhang, M., Wu, Q. and Shi, J. (2022) *Dendrobium nobile* Lindl. Alkaloids ameliorate Aβ25-35-induced synaptic deficits by targeting Wnt/β-catenin pathway in Alzheimer's disease models. *Journal of Alzheimer's Disease* 86, 297–313.

Zhao, Y., Zhang, X., Chen, X. and Wei, Y. (2022) Neuronal injuries in cerebral infarction and ischemic stroke: from mechanisms to treatment (review). *International Journal of Molecular Medicine* 49, 1–9.

Zhu, T., Wang, L., Wang, L. and Wan, Q. (2022) Therapeutic targets of neuroprotection and neurorestoration in ischemic stroke: applications for natural compounds from medicinal herbs. *Biomedicine & Pharmacotherapy* 148, 112719.

Zoerle, T., Carbonara, M., Zanier, E.R., Ortolano, F., Bertani, G. *et al.* (2017) Rethinking neuroprotection in severe traumatic brain injury: toward bedside neuroprotection. *Frontiers in Neurology* 8, 354.

13 *Dendrobium nobile* and Skin Health: From Traditional Use to Modern Applications

Surendra Sarsaiya, Archana Jain, Qihai Gong, Qin Wu and Jingshan Shi*
Key Laboratory of Basic Pharmacology and Joint International Research Laboratory of Ethnomedicine of Ministry of Education, Zunyi Medical University, Zunyi, Guizhou, China

Abstract
This chapter explores the multifaceted applications of *Dendrobium nobile* in skincare, tracing its journey from traditional Chinese medicine to modern dermatological science. Historically revered for its ability to nourish the skin, *D. nobile* contains bioactive compounds like polysaccharides, which exhibit antioxidant, anti-inflammatory and immunomodulatory properties. Contemporary research has substantiated these benefits, demonstrating the plant's efficacy in enhancing skin hydration and elasticity, and reducing signs of ageing. The chapter also addresses safety considerations and the potential for adverse reactions, ensuring comprehensive coverage of its therapeutic profile. By integrating traditional knowledge with modern scientific inquiry, the chapter illuminates the promising role of *D. nobile* in promoting skin health and guiding future research directions, including advanced formulation strategies and synergistic use with other botanical ingredients.

13.1 Introduction

Dendrobium nobile Lindl. is an orchid species that has been traditionally used in Chinese medicine for its various health benefits. In recent years, there has been a growing interest in the potential applications of *D. nobile* in skincare and its effects on skin health (Wang, 2021). This has led to extensive research and exploration of bioactive compounds and their therapeutic effects on skin. One of the key components of *D. nobile* polysaccharides possesses antioxidant, anti-inflammatory and immunomodulatory properties (Zhang *et al.*, 2021). These properties make *D. nobile* an intriguing ingredient for skincare formulations, as oxidative stress, inflammation and immune dysregulation are key factors in the development of various skin conditions. In the realm of traditional medicine, the convergence of nature's bounty and human ingenuity has led to remarkable discoveries with profound implications for health and well-being. Among the myriad treasures found in the rich tapestry of traditional practices, *D. nobile*, a species native to East Asia, is a symbol of botanical elegance and therapeutic efficacy (Tiwari *et al.*, 2023).

D. nobile, commonly known as the noble dendrobium, 'shi hu' or 'shi hu gen' in traditional Chinese medicine, has a rich history of use spanning centuries in East Asian cultures

*Corresponding author: shijs@zmu.edu.cn

© Jingshan Shi, Archana Jain, Surendra Sarsaiya, Qin Wu and Qihai Gong 2025. Dendrobium nobile: *Research and Therapeutic Applications* (J. Shi, A. Jain, S. Sarsaiya, Q. Wu and Q. Gong)
DOI: 10.1079/9781800627987.0013

(Zhang *et al.*, 2021). *D. nobile* has been revered in traditional Chinese medicine because of its multifaceted health benefits, particularly in promoting skin health. Its usage traces back to ancient texts where it was celebrated for its ability to nourish the skin, enhance complexion and mitigate various dermatological ailments (Cakova *et al.*, 2017). The intricate interplay of bioactive compounds in this orchid has fascinated healers and herbalists, prompting further exploration of its medicinal properties (Wang *et al.*, 2023). In recent years, the surge of scientific inquiry and technological advancements has provided fertile grounds for unlocking the secrets of *D. nobile*'s potential. Modern research endeavours have delved into its phytochemical composition, pharmacological actions and clinical applications, shedding new light on its traditional uses and revealing novel avenues for therapeutic innovation (Huang *et al.*, 2023). This exploration delves into the rich history and traditional uses of *D. nobile* in maintaining skin health, tracing its journey from a cultural staple to scientific curiosity. Furthermore, it examines recent advancements in phytochemistry and pharmacology that have shed light on the mechanisms underlying its purported benefits (Wu *et al.*, 2023). By bridging the realms of tradition and modernity, investigations aim to elucidate the potential of *D. nobile* as a valuable asset in the pursuit of radiant and healthy skin (Singh *et al.*, 2021). By bridging the wisdom of traditional medicine with cutting-edge scientific inquiry, studies aim to illuminate the promising role of *D. nobile* in promoting skin health and inspire further investigation into its diverse therapeutic potential (Hu *et al.*, 2021; Ghai *et al.*, 2022).

13.1.1 Importance of *D. nobile* in skin health

D. nobile, an orchid species, has gained significant attention owing to its potential benefits in promoting skin health. Traditional medicinal practices have long recognized the therapeutic properties of *D. nobile* in the treatment of various skin conditions. In recent years, scientific research has shed light on the mechanisms underlying its effects, further validating its importance to skin health (Li *et al.*, 2023). In the pursuit of radiant and healthy skin, the importance of *D. nobile* emerges as a beacon of promise in both traditional practices and modern skincare formulations. This orchid, celebrated for centuries in traditional Chinese medicine, encapsulates a wealth of therapeutic potential deeply intertwined with skin health (Ghai *et al.*, 2022). Across generations, *D. nobile* has held a revered status in East Asian cultures, revered not only for its ornamental beauty but also for its profound impact on skin wellness. From ancient texts to contemporary herbal compendiums, its multifaceted properties in promoting skin health have persisted as a testament to its enduring efficacy (Li *et al.*, 2022). The intricate blend of bioactive compounds in *D. nobile* has captivated healers and skincare enthusiasts alike, inspiring a wealth of research aimed at unlocking their secrets.

As modern science continues to illuminate the intricate mechanisms underlying its skin-nourishing effects, the significance of *D. nobile* in skincare has become increasingly apparent (Wang, 2021). Moreover, the integration of *D. nobile* into contemporary skincare formulations represents a harmonious fusion of tradition and innovation. Although the specific mechanisms related to the regulation of pigmentation remain under-researched, their overall contribution to skin tone uniformity and luminosity is promising. Studies suggest that incorporating *D. nobile* into skincare products can provide natural solutions to age-related skin concerns and boost skin resilience against environmental damage. In conclusion, *D. nobile* holds great importance in promoting skin health owing to its wound-healing, anti-ageing and anti-inflammatory properties.

13.2 Traditional Use of *D. nobile* in Skincare

D. nobile has been traditionally used in skincare because of its various beneficial properties. It has long been revered for its time-honoured contributions to skincare practices. Passed down through generations, its traditional use encompasses a spectrum of beneficial properties tailored to nurture the skin. Central to its traditional skincare application is its role as a moisturizing agent. Extracts from *D. nobile* have been shown to reduce melanin formation, which is crucial for the management of skin darkening and ageing. Additionally, the presence of stilbenoids in *D. nobile* contributes to its effectiveness in skin protection, as these compounds help mitigate the generation of reactive oxygen species, which can lead to skin damage (Meng *et al.*, 2020). Moreover, *D. nobile* has earned acclaim for its anti-ageing properties, cherished for its capacity to diminish the appearance of fine lines, wrinkles and age-related imperfections. Through its skin rejuvenating properties, it fosters the regeneration of skin cells, unveiling a revitalized complexion characterized by enhanced luminosity and smoothness (Li *et al.*, 2022; Jain *et al.*, 2024). Furthermore, *D. nobile* can be a relief for sensitive or irritated skin, offering soothing and calming properties that alleviate redness, inflammation and discomfort (Long *et al.*, 2023a). In the quest for a brighter and more even complexion, *D. nobile* emerges as a cherished ally. Its complexion-enhancing properties work synergistically to diminish the appearance of dark spots and hyperpigmentation, bestowing on the skin a luminous radiance (He *et al.*, 2023). Finally, *D. nobile* can help guard against the effects of environmental stress, fortified with natural antioxidants that shield the skin from the detrimental effects of free radicals (Fig. 13.1) (Wang *et al.*, 2022).

13.2.1 Traditional formulations and practices of *D. nobile*

D. nobile boasts a rich heritage of traditional usage spanning diverse cultures, where its medicinal virtues and skincare benefits have been cherished and cultivated over generations. Traditional formulations and practices revolving around *D. nobile* have evolved and endured, offering a glimpse into the profound wisdom of ancient healing traditions. In the annals of traditional Chinese medicine *D. nobile* is prized for its nourishing and tonifying properties. Traditional Chinese medicine finds its place in herbal formulations and decoctions aimed at fostering holistic health and vitality, with skincare benefits being a notable focus (Duan *et al.*, 2024). *D. nobile* stems and extracts serve as key ingredients in the preparation of herbal teas and tonics, cherished for their potential to rejuvenate the body and support skin health. These traditional remedies are consumed with reverence, and their purported effects extend beyond mere sustenance to holistic wellness (Wu *et al.*, 2023). The topical application of *D. nobile* extracts or infusions represents a time-honoured practice in skincare rituals. From creams and ointments to facial masks, these formulations moisturize, calm irritation and enhance overall complexion (Chen, 2020). In the intricate tapestry of traditional formulations, *D. nobile* is often combined with other herbs, each contributing synergistically to amplify the therapeutic effects. These harmonious blends are meticulously crafted to address specific skincare concerns, such as smoothing wrinkles, restoring skin elasticity and alleviating conditions such as acne or eczema (Ramesh *et al.*, 2019). It is important to note that while traditional formulations and practices have been widely used, scientific research is still ongoing to validate

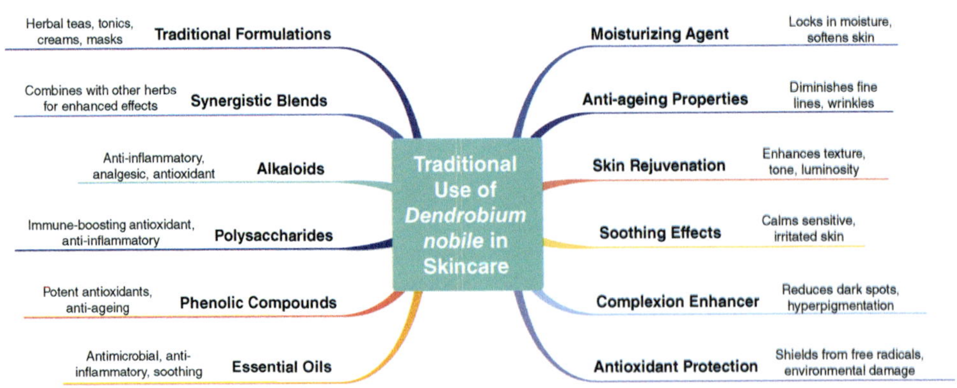

Fig. 13.1. Traditional use of *D. nobile* in skincare (authors' own figure).

their effectiveness and understand their mechanisms of action. As with any herbal or traditional remedy, it is recommended patients consult a qualified healthcare professional before using *D. nobile* or any related formulations.

13.2.2 Key components and benefits for skincare

D. nobile contains bioactive components that form the foundation of its medicinal properties and potential benefits for skincare. These key constituents unveil a tapestry of therapeutic effects, each contributing to their unique role in enhancing skin health and vitality. Among the myriad constituents, *D. nobile* alkaloids, including dendrobine and dendroxine, have a multifaceted arsenal of benefits including anti-inflammatory, analgesic and antioxidant effects. By soothing inflammation, relieving pain and combating oxidative stress, they offer a holistic approach to skincare that promotes healing and rejuvenation (Duan *et al.*, 2022). Derived from *D. nobile*, polysaccharides have emerged as potent immunomodulators with the capacity to enhance immune function. Additionally, they exhibit antioxidant and anti-inflammatory properties, and safeguard the skin against oxidative damage and inflammation. By fortifying the skin's natural defences and quelling inflammatory responses, polysaccharides bestow upon the skin a shield against resilience and vitality (Fu *et al.*, 2023). Phenolic compounds found in *D. nobile*, such as flavonoids and phenanthrenes, offer a formidable line of defence against oxidative stress. Through their protective embrace, phenolic compounds nurture the skin, preserving youthful radiance for years (Zhu *et al.*, 2021). *D. nobile* essential oils infuse skincare formulations with aromatic essence and therapeutic potential. Rich in volatile compounds, these oils exhibit antimicrobial and anti-inflammatory properties, thus offering a gentle yet effective solution for addressing skin infections and inflammation (Natta *et al.*, 2022). Through the intricate interplay of these bioactive components, *D. nobile* has emerged as a botanical powerhouse, offering a synergistic blend of healing and rejuvenating properties for the skin. The combination of these bioactive components in *D. nobile* contributes to its overall health-promoting effects, including the potential benefits for skincare.

13.3 Scientific Research into *D. nobile*'s Effects on Skin Health

Scientific inquiry into the effects of *D. nobile* on skin health has unveiled a wealth of compelling evidence, shedding light on its multifaceted therapeutic potential and validating its traditional usage. Through rigorous investigation and meticulous experimentation, researchers have unraveled the intricate mechanisms underlying its beneficial effects, offering insights that pave the way for innovative skincare solutions. Studies have demonstrated that *D. nobile* extracts possess significant moisturizing and hydrating properties, effectively replenishing the skin's moisture barrier and preventing trans-epidermal water loss. By enhancing skin hydration levels, *D. nobile* promotes a smoother, more refined skin texture, reducing the appearance of dryness and roughness (Meng *et al.*, 2020). Research suggests that *D. nobile* extracts exhibit potent anti-ageing effects, including reduction of fine lines, wrinkles and age-related pigmentation. These effects are attributed to the antioxidant properties of *D. nobile*, which help neutralize free radicals and protect the skin from oxidative stress-induced ageing (Cruz *et al.*, 2023). Scientific studies have revealed that *D. nobile* extracts stimulate the proliferation of skin cells and promote collagen synthesis, leading to improved skin texture, elasticity and firmness. Additionally, *D. nobile* accelerates wound healing and tissue repair, promoting skin regeneration and overall recovery (Li *et al.*, 2022). Research indicates that *D. nobile* possesses potent anti-inflammatory properties, which help alleviate skin inflammation, redness and irritation. These effects are mediated by the inhibition of pro-inflammatory cytokines and modulation of inflammatory pathways, making it suitable for soothing sensitive and irritated skin (Long *et al.*, 2023a). Scientific investigations have highlighted the antimicrobial properties of *D. nobile* extracts, which inhibit the growth of harmful bacteria and fungi on the skin. Additionally, *D. nobile* offers protective effects against environmental agents, such as ultraviolet (UV) radiation and pollution, safeguarding the skin from damage and premature ageing (Fan *et al.*, 2023).

Emerging research suggests that *D. nobile* extracts facilitate wound healing by promoting the migration and proliferation of skin cells, stimulating collagen synthesis and reducing inflammation at the wound site. These effects accelerate healing, leading to improved wound closure and tissue regeneration (Cakova *et al.*, 2017). Overall, the findings from scientific research underscore the significant potential of *D. nobile* as a valuable ingredient in skincare formulations. By harnessing its moisturizing, anti-ageing, rejuvenating, soothing, antimicrobial, protective and wound-healing properties, *D. nobile* offers a holistic approach to skin health and wellness, paving the way for innovative and effective skincare solutions.

13.4 Dermatological Applications of *D. nobile*

D. nobile, a traditional medicinal plant with a rich history of use, has gained attention in dermatology because of its potential therapeutic applications in various skin conditions. Plants possess a wide range of bioactive compounds that exert beneficial effects on the skin. *D. nobile* has garnered significant attention for its diverse dermatological applications owing to its rich array of bioactive compounds and multifaceted therapeutic properties. From traditional remedies to modern skincare formulations, *D. nobile* has a wide range of dermatological applications, offering solutions for various skin concerns. *D. nobile* extracts are widely used in moisturizing and hydrating skincare products such as creams, lotions and serums (Sharma and Pathak, 2020). *D. nobile*, which is rich in polysaccharides and other hydrating compounds, helps to attract and retain moisture in the skin, preventing dehydration and restoring optimal hydration levels. These moisturizing treatments provide immediate relief for dry, rough or dehydrated skin, leaving it soft, smooth and supple (Tiwari *et al.*, 2022). The antioxidant properties of *D. nobile* make it a valuable ingredient in anti-ageing skincare formulations (Zhang *et al.*, 2023a). By neutralizing free radicals and reducing oxidative stress, *D. nobile* protects the skin from environmental damage and signs of ageing, such as fine lines, wrinkles and age spots (Ye *et al.*, 2023). Anti-ageing products enriched with *D. nobile* extracts promote skin renewal, firmness and elasticity, restoring a youthful and radiant complexion (Nie *et al.*, 2020).

D. nobile possesses potent anti-inflammatory properties, making it an effective ingredient in soothing and calming skincare remedies (Chen, 2020). These formulations help alleviate skin inflammation, redness and irritation caused by environmental stressors, allergic reactions or skin conditions such as eczema and dermatitis. By calming sensitive and irritated skin, *D. nobile*-based treatments provide relief and comfort and restore skin health and comfort (Ramesh *et al.*, 2019). *D. nobile* is often incorporated into skin-brightening and complexion-enhancing skincare products because of its ability to reduce the appearance of dark spots, hyperpigmentation and uneven skin tone (Duan *et al.*, 2024). By inhibiting melanin production and promoting skin cell turnover, *D. nobile* helps to achieve a brighter, more luminous complexion. These products effectively address dullness and discoloration, revealing a more even and radiant skin tone (He *et al.*, 2023).

The wound-healing properties of *D. nobile* make it a valuable ingredient in skincare formulations designed to promote tissue repair and regeneration (Cruz *et al.*, 2023). *D. nobile* extract accelerates wound closure, stimulates collagen synthesis and reduces inflammation, facilitating the healing of cuts, abrasions and other skin injuries (Hong *et al.*, 2022). Wound-healing products containing *D. nobile* extracts support the natural healing process, minimize scarring and promote healthy skin recovery (Firoozbahr *et al.*, 2023). *D. nobile* exhibits potential in dermatology owing to its ability to regulate melanogenesis and the process of melanin production in the skin (Swainson *et al.*, 2023). It has garnered attention for its role in modulating skin pigmentation, offering a promising avenue for addressing conditions such as hyperpigmentation and skin discoloration. Studies have highlighted its inhibitory effects on key enzymes involved in melanin synthesis, presenting a novel approach for managing melanogenesis and achieving a more balanced and even skin tone (Meng *et al.*, 2020). Overall, the dermatological applications of *D. nobile* include a wide range of skincare solutions, from moisturizing and anti-ageing treatments to soothing, skin brightening and wound healing. By harnessing its therapeutic properties, *D. nobile* offers effective solutions for achieving and maintaining healthy radiant skin. The dermatological applications of *D. nobile* highlight its potential as a natural ingredient in skincare formulations (Fig. 13.2).

13.5 Clinical Studies and Evaluation of Efficacy

Clinical studies are pivotal in determining the efficacy and safety of products such as *D. nobile* for enhancing skin health. These studies involved carefully designed experiments conducted with human participants to evaluate specific outcomes related to skin health. Clinical studies can employ objective measurements to assess changes in skin parameters such as hydration levels, elasticity, firmness and wrinkle depth. These measurements provide quantitative data for evaluating the effectiveness of *D. nobile* in improving skin health (He *et al.*, 2023). Participants in clinical studies can provide subjective feedback on their experiences with *D. nobile*-containing products. This feedback may include perceptions of skin texture, smoothness and overall satisfaction with the product, thus providing valuable insights into its efficacy (Gogoi and Sen, 2023). Long-term clinical studies can assess the sustained effects of *D. nobile* on skin health over time. By monitoring participants for an extended period, researchers can determine whether the benefits of *D. nobile* are maintained or enhanced with continued use (Liu *et al.*, 2024). Clinical studies have also evaluated the safety profile of *D. nobile*-containing products. Participants have been monitored for adverse reactions or side effects to ensure that the product was safe for use on the skin (Wu *et al.*, 2023). Comparative clinical studies can compare the efficacy of *D. nobile* with other skincare ingredients or formulations. These studies provide valuable information on the relative effectiveness of *D. nobile* compared with existing skincare products (Sharma and Pathak, 2020).

Ethnobotanical research involves documenting the traditional uses of plants, including *D. nobile*, in indigenous cultures. Such studies can provide valuable insights into the historical use of *D. nobile* for skin health and other purposes (Ghai *et al.*, 2022). Preclinical studies involve laboratory-based experiments using cell cultures and animal models to explore the potential mechanisms of action and the effects of *D. nobile* on skin health. These studies provide preliminary evidence of its efficacy and safety before proceeding to human trials (Nugraha *et al.*, 2020). To evaluate the efficacy, safety and tolerability of

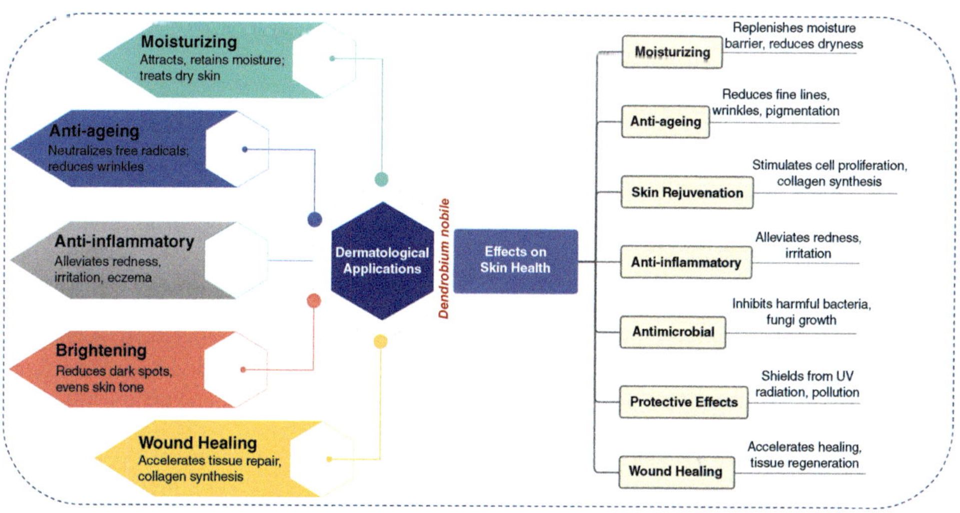

Fig. 13.2. Dermatological applications of *D. nobile* and effect on skin health (authors' own figure).

D. nobile in human clinical trials studies may involve participants with specific skin conditions, such as dryness, ageing or inflammation, and assess outcomes such as hydration, elasticity, wrinkle reduction and improvement in skin texture (Zhang *et al.*, 2023a). Based on the findings of preclinical and clinical studies, skincare formulations containing *D. nobile* extracts or derivatives may be developed. These formulations are optimized for stability, efficacy and sensory attributes to maximize consumer benefits.

In some studies skincare products containing *D. nobile* have undergone rigorous testing to ensure safety and efficacy. This includes stability testing, irritation testing and consumer perception studies to validate performance and acceptability (Peng *et al.*, 2021). Before commercialization, skincare products containing *D. nobile* must meet regulatory requirements for safety and efficacy. This may involve obtaining approval from regulatory agencies such as the Food and Drug Administration (FDA) in the USA or the European Medicines Agency (EMA) in Europe (Tiwari *et al.*, 2024). After the product is launched, post-market surveillance involves monitoring the safety and efficacy of *D. nobile* in real-world use. This helps identify any potential adverse effects or issues and ensures ongoing compliance with regulatory requirements (Shahrajabian *et al.*, 2020).

Randomized controlled trials (RCTs) are the gold standard for evaluating intervention efficacy. In the context of *D. nobile* and skin health, RCTs could involve comparing a *D. nobile*-containing skincare product to a placebo or another active comparator to assess its effects on hydration, elasticity, wrinkle reduction and skin texture (Su *et al.*, 2020). Longitudinal studies followed participants over an extended period to assess the long-term effects of the interventions. For *D. nobile*, longitudinal studies could investigate whether the benefits observed in short-term trials are sustained or enhanced with continued use over weeks or months (Peng *et al.*, 2021).

Comparative studies have compared the efficacy of different interventions. In the context of skincare, comparative studies have compared the effects of *D. nobile*-containing products with those of other botanical extracts or synthetic ingredients commonly used in skincare formulations (Wang, 2021). Dose–response studies have been used to assess the relationship between the dose of an intervention and its effects. For *D. nobile*, dose–response studies could investigate whether higher concentrations or frequencies of application result in greater improvements in skin health outcomes (Ketsa and Warrington, 2023). Population-specific studies have focused on specific demographic groups, such as individuals with particular skin conditions or age groups. For example, clinical trials could target individuals with dry, ageing, sensitive or inflammatory skin conditions to assess the effects of *D. nobile* in these populations (He *et al.*, 2023). Safety and tolerability studies have evaluated the adverse effects and tolerability of these interventions. In the case of *D. nobile*, safety studies could assess its potential for irritation, allergic reactions or other adverse effects when topically applied to the skin (Cruz *et al.*, 2023). Overall, clinical studies provide robust scientific evidence to support the efficacy and safety of *D. nobile* in improving skin health. By conducting well-designed trials with rigorous methodologies, researchers can establish the role of *D. nobile* in skincare and provide consumers with confidence in its benefits. Through these stages, the journey of *D. nobile* from traditional use to modern applications for skin health involves a systematic approach of scientific inquiry, validation and translation into safe and effective skincare products for consumers (Table 13.1).

13.6 Formulations and Products of Skincare

Skincare formulations and products are pivotal in nurturing and maintaining healthy skin and offer a spectrum of solutions tailored to diverse needs and concerns. Topical skincare formulations, such as creams, lotions, serums and gels, offer specialized solutions for various skin concerns. Creams are rich and blend water with oil-based ingredients, providing hydration and nourishment, while protecting the skin. Lotions, lighter in consistency, are absorbed quickly and are suitable for daily use, and deliver hydration along with the targeted active ingredients. Serums, known for their high concentration of active compounds, penetrate deeply to address specific concerns, such as hydration, skin brightening

Table 13.1. Clinical studies and efficacy evaluation of *D. nobile* in skincare (authors' own table).

Stage/type of study	Description	Key findings	Outcome measures	Participants	Reference
Ethnobotanical studies	Ethnobotanical research involves documenting the traditional uses of *D. nobile* in indigenous cultures. These studies provide valuable insights into the historical use of *D. nobile* for skin health and other purposes.	Provides historical context and traditional uses of *D. nobile* for skin health	–	Indigenous cultures	Ghai *et al.* (2022)
Preclinical research	Preclinical studies involve laboratory-based experiments using cell cultures and animal models to explore the potential mechanisms of action and the effects of *D. nobile* on skin health. These studies provide preliminary evidence of its efficacy and safety before moving on to human trials.	Demonstrates potential mechanisms of action and initial safety profile of *D. nobile* for skin health	*In vitro* cell culture assays, animal models	Cell cultures, animal models	Nugraha *et al.* (2020)
Clinical trials	Clinical trials are conducted to evaluate the efficacy, safety and tolerability of *D. nobile* in humans. These studies may involve participants with specific skin conditions, such as dryness, ageing skin or inflammation, and assess outcomes such as hydration, elasticity, wrinkle reduction and improvement in skin texture.	Assesses efficacy, safety and tolerability of *D. nobile* in human subjects with various skin conditions	Hydration levels, elasticity, wrinkle reduction, improvement in skin texture	Human participants with skin conditions	Zhang *et al.* (2023a)
Formulation development	Based on the findings of preclinical and clinical studies, skincare formulations containing *D. nobile* extracts or derivatives may be developed. These formulations are optimized for stability, efficacy and sensory attributes to maximize the benefits for consumers.	Develops skincare formulations containing *D. nobile* extracts with optimized stability and efficacy	Stability testing, sensory attributes, efficacy testing	Formulation scientists	Zhang *et al.* (2023a)
Product testing and validation	Skincare products containing *D. nobile* undergo rigorous testing to ensure safety and efficacy. This includes stability testing, irritation testing and consumer perception studies to validate their performance and acceptability.	Validates safety, efficacy and consumer acceptability of skincare products containing *D. nobile*	Stability testing, irritation testing, consumer perception studies	Skincare product testers	Peng *et al.* (2021)
Market access and regulatory approval	Before commercialization, skincare products containing *D. nobile* must meet regulatory requirements for safety and efficacy. This may involve obtaining approvals from regulatory agencies such as the US FDA or the EMA.	Navigates regulatory requirements and obtains approvals for commercialization of skincare products containing *D. nobile*	Regulatory approvals, compliance with safety and efficacy standards	Regulatory agencies, product developers	Tiwari *et al.* (2024)

Continued

Table 13.1. Continued.

Stage/type of study	Description	Key findings	Outcome measures	Participants	Reference
Post-market surveillance	After the product is launched, post-market surveillance involves monitoring the safety and efficacy of *D. nobile* in real-world use. This helps to identify any potential adverse effects or issues and ensures ongoing compliance with regulatory requirements.	Monitors safety and efficacy of *D. nobile*-containing products in real-world use post-commercialization	Surveillance of adverse effects, compliance with regulatory requirements	Consumers, regulatory agencies	Shahrajabian *et al.* (2020)
RCTs	RCTs are the gold standard for evaluating intervention efficacy. In the context of *D. nobile* and skin health, RCTs could involve comparing a *D. nobile*-containing skincare product to a placebo or another active comparator to assess its effects on hydration, elasticity, wrinkle reduction and skin texture.	Provides high-quality evidence of efficacy of *D. nobile*-containing skincare products compared to control interventions	Hydration levels, elasticity, wrinkle reduction, improvement in skin texture	Human participants with skin conditions	Su *et al.* (2020)
Longitudinal studies	Longitudinal studies follow participants over an extended period to assess the long-term effects of the interventions. In the case of *D. nobile*, longitudinal studies could investigate whether the benefits observed in short-term trials are sustained or enhanced with continued use over weeks or months.	Explores the sustainability and enhancement of *D. nobile* benefits over time in real-world use	Long-term changes in hydration levels, elasticity, wrinkle reduction, skin texture enhancement	Human participants with skin conditions	Peng *et al.* (2021)
Comparative studies	Comparative studies compare the efficacy of different interventions. In the context of skincare, comparative studies could compare the effects of *D. nobile*-containing products to other botanical extracts or synthetic ingredients commonly used in skincare formulations.	Compares efficacy of *D. nobile*-containing products to other skincare interventions	Comparative analysis of efficacy, safety and tolerability	Human participants with skin conditions	Wang (2021)
Dose–response studies	Dose–response studies assess the relationship between the dose of an intervention and its effects. For *D. nobile*, dose–response studies could investigate whether higher concentrations or frequencies of application result in greater improvements in skin health outcomes.	Determines optimal dosage of *D. nobile* for maximizing improvements in skin health outcomes	Relationship between dosage and changes in skin parameters	Human participants with varying dosages	Ketsa and Warrington (2023)
Population-specific studies	Population-specific studies focus on specific demographic groups, such as individuals with particular skin conditions or age groups. For example, clinical trials could target individuals with dry skin, ageing skin, sensitive skin or inflammatory skin conditions to assess the effects of *D. nobile* in these populations.	Investigates efficacy and safety of *D. nobile* in specific population groups with varying skin conditions	Changes in skin parameters specific to targeted population groups	Human participants with specific skin conditions	He *et al.* (2023)

Safety and tolerability studies	Safety and tolerability studies have evaluated the adverse effects and tolerability of these interventions. In the case of *D. nobile*, safety studies could assess its potential for irritation, allergic reactions or other adverse effects when applied topically to the skin.	Evaluates safety profile of *D. nobile*-containing products, including potential adverse effects and tolerability	Incidence of adverse effects, tolerance levels	Human participants with skin conditions	Cruz *et al.* (2023)
Objective measurements	Clinical studies can employ objective measurements to assess changes in skin parameters such as hydration levels, elasticity, firmness and wrinkle depth These measurements provide quantitative data to evaluate the effectiveness of *D. nobile* in improving skin health.	Provides quantitative data on changes in skin parameters	Hydration levels, elasticity, firmness, wrinkle depth	Human participants with skin conditions	He *et al.* (2023)
Subjective assessments	Participants in clinical studies can provide subjective feedback on their experiences with *D. nobile*-containing products. This feedback may include perceptions of skin texture, smoothness and overall satisfaction with the product, providing valuable insights into its efficacy.	Provides qualitative feedback on user experience with *D. nobile*-containing products	Perception of skin texture, smoothness, overall satisfaction with the product	Human participants with skin conditions	Gogoi and Sen (2023)
Long-term effects	Long-term clinical studies can assess the sustained effects of *D. nobile* on skin health over time. By monitoring participants for an extended period, researchers can determine whether the benefits of *D. nobile* are maintained or enhanced with continued use.	Evaluates sustainability and enhancement of *D. nobile* benefits over time	Long-term changes in skin parameters	Human participants with skin conditions	Liu *et al.* (2024)
Safety profile	Clinical studies have also evaluated the safety profile of *D. nobile*-containing products. Participants are monitored for adverse reactions or side effects, helping to ensure that the product is safe for use on the skin.	Assesses safety profile of *D. nobile*-containing products	Incidence of adverse reactions or side effects	Human participants with skin conditions	Wu *et al.* (2023)
Comparative studies	Comparative clinical studies can compare the efficacy of *D. nobile* with other skincare ingredients or formulations. These studies provide valuable information on the relative effectiveness of *D. nobile* compared to existing skincare products.	Compares efficacy of *D. nobile* with other skincare ingredients or formulations	Comparative analysis of efficacy, safety and tolerability	Human participants with skin conditions	Sharma and Pathak (2020)

and anti-ageing effects. Gels formulated with water-based ingredients such as hyaluronic acid or salicylic acid offer a cooling sensation ideal for oily or acne-prone skin, helping to maintain clarity and balance. Each formulation caters to different skin needs, from intensive moisture to specific treatments, ensuring a tailored approach to skincare routines. These topical formulations are available in various skincare products, catering to a spectrum of skincare needs, from moisturization to anti-ageing and acne treatments. Opting for products compatible with individual skin types and concerns is essential for achieving optimal results.

Facial and sheet masks are integral to skincare routines, each offering distinct benefits and enhancing the skin in unique ways. Facial masks, which are available in clay, gel or cream formulations, provide deep cleansing, exfoliation and hydration when applied and left on for a specified time (Cruz *et al.*, 2023). On the other hand, sheet masks are infused with serum or essences that adhere closely to the skin, delivering active compounds directly for maximum absorption. Renowned for their convenience and ability to lock in moisture, sheet masks are ideal for promoting hydration and revitalizing skin appearance. Both types of mask are enriched with ingredients such as hyaluronic acid, vitamins and botanical extracts; they cater to diverse skincare needs to enhance hydration and skin brightening to treat acne and combat signs of ageing. Incorporating facial and sheet masks into skincare regimens provides an additional layer of nourishment, contributing to overall skin health and a rejuvenated complexion (Li *et al.*, 2023; Makgobole *et al.*, 2024).

Cosmetic products are designed to effectively address a wide range of skin concerns and offer tailored solutions for various needs. For acne-prone skin, formulations containing ingredients such as salicylic acid, benzoyl peroxide or tea tree oil unclog pores, reduce inflammation and regulate oil production, promoting clearer skin (Zhang *et al.*, 2023b). Dry or dehydrated skin benefits from moisturizers and hydrating serum enriched with hyaluronic acid, ceramides and glycerine, which restore moisture levels and enhance skin suppleness and comfort (Ganapathy *et al.*, 2023). Hyperpigmentation can be tackled with skin-brightening products that incorporate hydroquinone, kojic acid or vitamin C to fade dark spots and achieve a more even skin tone (Meng *et al.*, 2020). Sensitive skin types benefit from gentle, fragrance-free products that minimize potential irritants, preserve the skin barrier and reduce reactions (Li *et al.*, 2023). Ageing skin concerns, such as fine lines, wrinkles and loss of elasticity, are targeted by anti-ageing formulations containing retinoids, peptides, antioxidants and hyaluronic acid, helping to rejuvenate and revitalize the complexion (Cruz *et al.*, 2023). Careful consideration and patch testing of products tailored to individual skin needs to ensure optimal effectiveness and compatibility, thereby enhancing overall skin health and appearance.

Harnessing the potential of *D. nobile* extract in skincare offers a range of promising benefits rooted in its traditional use and therapeutic properties. Known for its anti-ageing effects, the extract stimulates collagen synthesis, improves skin elasticity and reduces wrinkles and fine lines (Long *et al.*, 2023b). Its moisturizing properties contribute to balancing skin moisture levels, combating dryness and promoting a smooth, supple complex (Meng *et al.*, 2020). *D. nobile* extract also acts as an antioxidant that protects the skin from oxidative stress and free radical damage, thereby supporting a youthful and radiant appearance (Wang *et al.*, 2022). Furthermore, its soothing properties make it suitable for sensitive skin and for alleviating irritation, redness and inflammation (Ganapathy *et al.*, 2023). This versatile extract can be integrated into a variety of skincare formulations, including creams, lotions, serums and masks, either alone or in combination with other ingredients, to enhance product efficacy and deliver comprehensive benefits (Zhang *et al.*, 2023b). Incorporating *D. nobile* extract into skincare routines offers a holistic approach to achieving healthier and more resilient skin. Careful consideration of *D. nobile* extract concentration, product formulation and individual skin sensitivities is imperative when selecting skincare products. Patch testing is recommended to ascertain compatibility and minimize adverse reactions, ensuring a harmonious integration into skincare routines (Table 13.2).

Table 13.2. Formulations and products for skincare (authors' own table).

Category	Product/formulation	Description	Reference
Topical creams, lotions and serums	Creams	Rich formulations providing intensive moisture and nourishment, blending water and oil-based ingredients for hydration and protection	Sharma and Pathak (2020)
	Lotions	Lighter, fluid alternatives to creams, absorbing quickly and delivering hydration and active ingredients effectively	Sharma and Pathak (2020)
	Serums	High-concentration formulations penetrating deep into the skin, offering targeted benefits such as hydration, skin brightening and anti-ageing effects	Sharma and Pathak (2020)
	Gels	Water-based, refreshing formulations ideal for oily or acne-prone skin types, maintaining balance and clarity	Cruz *et al.* (2023)
Facial masks and sheet masks	Facial masks	Treatments available in various forms offering deep cleansing, exfoliation and hydration, nourishing and rejuvenating the skin	Makgobole *et al.* (2024)
	Sheet masks	Adhering to the skin and delivering active ingredients directly, promoting hydration and revitalization	Li *et al.* (2023)
Cosmetic products for specific skin concerns	Acne-prone skin	Products containing salicylic acid, benzoyl peroxide or tea tree oil, unclogging pores, reducing inflammation and controlling oil production	Zhang *et al.* (2023b)
	Dry or dehydrated skin	Moisturizers and hydrating serums infused with hyaluronic acid, ceramides and glycerine, replenishing moisture levels and alleviating dryness	Ganapathy *et al.* (2023)
	Hyperpigmentation	Skin-brightening products containing hydroquinone, kojic acid or vitamin C, diminishing hyperpigmentation and evening out skin tone	Meng *et al.* (2020)
	Sensitive skin	Gentle, fragrance-free products suitable for sensitive skin types, minimizing reactions and preserving skin barrier integrity	Li *et al.* (2023)
	Ageing skin	Anti-ageing formulations enriched with retinoids, peptides, antioxidants and hyaluronic acid, targeting fine lines, wrinkles and loss of elasticity	Cruz *et al.* (2023)
Integration of *D. nobile* extract in skincare routines	Anti-ageing properties	Stimulates collagen synthesis, enhances skin elasticity and reduces the appearance of wrinkles and fine lines	Long *et al.* (2023b)

Continued

Table 13.2. Continued.

Category	Product/formulation	Description	Reference
	Moisturizing and hydrating effects	Improves skin's moisture balance, combating dryness and promoting a smooth, supple complexion	Meng *et al.* (2020)
	Antioxidant activity	Shields the skin from oxidative stress and free radical damage, contributing to a youthful and radiant complexion	Wang *et al.* (2022)
	Soothing and calming benefits	Alleviates skin irritation, redness and inflammation, suitable for sensitive skin types	Ganapathy *et al.* (2023)
	Integration into skincare products	Seamlessly incorporated into various skincare products such as creams, lotions, serums and masks, enhancing efficacy and delivering holistic skincare benefits	Zhang *et al.* (2023b)

13.7 Safety Considerations and Adverse Reactions

Ensuring safety and being mindful of potential adverse reactions are crucial aspects when integrating skincare products into a routine, particularly those infused with *D. nobile* extract. Safety and adverse reactions with skincare products containing *D. nobile* extract are important to consider. Conducting a patch test is a crucial step before using any new skincare product, particularly those containing *D. nobile* extract. This involves applying a small amount of the product to a small area of the skin, such as the inner forearm, and monitoring for any signs of allergic reactions or sensitivities, such as redness, itching or swelling. Patch testing helps to identify potential adverse reactions early on and allows the determination of whether the product is suitable for the skin (Lin *et al.*, 2020). Individuals with known allergies should pay close attention to ingredient lists when selecting skincare products. Ingredients such as *D. nobile* extract may trigger allergic reactions in some people. Consulting healthcare professionals or dermatologists before use can help assess the suitability of a product for specific needs and minimize the risk of adverse reactions (Fan *et al.*, 2021).

Fragrances from *D. nobile* extracts are common causes of skin irritation and allergic reactions. If a user has sensitive skin or is prone to fragrance allergies, choosing fragrance-free skincare products is advisable. Fragrance-free formulations reduce the likelihood of irritation and provide a gentler option for sensitive individuals (Ketsa and Warrington, 2023). Excessive use of skincare products, even those with natural extracts such as *D. nobile*, can disrupt the skin's natural balance and lead to irritation. Therefore, it is essential to follow the recommended usage guidelines provided by the manufacturer to avoid overuse and potential adverse effects. The use of directed products helps maintain skin health and minimizes the risk of irritation or other complications (Ganapathy *et al.*, 2023).

Some skincare ingredients, such as retinoids or exfoliants, may interact with *D. nobile* extract or other active ingredients in skincare products. It is important to be cautious when combining different products to avoid adverse effects and diminished efficacy. Consulting skincare professionals can provide guidance on suitable product combinations and minimize the risk of potential interactions (Long *et al.*, 2023b). The high pH level of skincare products, even those with natural extracts such as *D. nobile*, can impact their effectiveness and compatibility with other products. Ensuring pH compatibility between different skincare products helps to maintain their efficacy and prevents potential irritation or adverse reactions. Checking the pH levels of skincare products and using products with complementary pH levels can optimize their performance and enhance the overall skin health (Jędrejko *et al.*, 2021). Before using natural extracts such as *D. nobile*, seeking advice from dermatologists or healthcare providers is advisable, especially if a consumer has specific skin concerns or medical conditions. They can offer personalized recommendations tailored to the skin type and address any concerns or questions about skincare product selection and usage. Consulting professionals ensures that informed decisions are made about skincare routines and prioritizes skin safety and health (Ramesh *et al.*, 2019).

Overall, prioritizing safety and being mindful of potential adverse reactions is essential when incorporating skincare products containing *D. nobile* extract into a routine. According to the outlined guidelines, including patch testing, considering allergies and sensitivities, opting for fragrance-free options when necessary, adhering to recommended usage guidelines, being cautious of interactions with other skincare ingredients, ensuring pH compatibility and seeking professional advice when needed, a skincare regimen that promotes healthy, radiant skin while minimizing the risk of adverse reactions can be created. Every individual's skin is unique, so adjusting a routine accordingly is key to achieving optimal skincare results.

13.8 Future Perspectives and Research Directions

D. nobile, a plant with a long history of traditional use in skincare, has gained significant attention in modern research because of its potential benefits for skin health. Although existing studies have provided valuable insights, further research is needed to explore its full potential and uncover new applications. Future research on *D. nobile* in the context of skin health holds significant promise for advancing our understanding of its potential benefits. Delving deeper into the phytochemical composition of *D. nobile* presents an opportunity to uncover specific bioactive compounds with the potential to enhance skin health. Advanced analytical techniques can be used to identify novel molecules within plant extracts that may exert therapeutic effects. Investigating the antioxidant properties of *D. nobile* extracts and compounds is crucial, as antioxidants play a key role in protecting skin cells from oxidative stress and maintaining skin health. Understanding how these extracts combat oxidative damage can inform their application in skincare formulations aimed at promoting young-looking and radiant skin. Additionally, studies exploring the anti-inflammatory potential of *D. nobile* in skin conditions have indicated that inflammation plays a significant role in various dermatological issues. Identifying natural remedies with anti-inflammatory properties from *D. nobile* extract could lead to innovative therapeutic strategies for managing skin inflammation and improving overall skin conditions. Integrating these research insights into skincare development may offer effective solutions to enhance skin health and address common skin concerns.

Research on *D. nobile* extracts focuses on several key areas with potential implications for skincare and dermatology. Assessing its wound-healing properties involves studying its ability to promote tissue regeneration and collagen synthesis, which could lead to the development of effective treatments for wounds and skin injuries. Investigations of its capabilities to protect against UV radiation aim to determine if *D. nobile* extracts offer photoprotective benefits against UV-induced skin damage, potentially serving as a natural sunscreen or complementary therapy for skin cancer prevention. Evaluating the moisturizing effects of *D. nobile* on the skin seeks to understand its ability to enhance the skin's barrier function and prevent moisture loss, which is critical for maintaining overall skin health. To validate these benefits, rigorous clinical trials are essential to establish the efficacy and safety of *D. nobile*-based formulations for various skin conditions, ensuring that promising preclinical findings translate into effective clinical treatments. Integrating findings from these research areas could pave the way for innovative skincare solutions and therapeutic strategies, leveraging the natural properties of *D. nobile* to enhance skin wellness and resilience.

Exploring innovative formulations incorporating *D. nobile* extracts, such as creams, lotions or serums, optimized for effective skin delivery and therapeutic efficacy is crucial for advancing skincare product development. Research into formulation development aims to harness the therapeutic benefits of *D. nobile*, ensuring its practical application and efficacy in addressing various skin concerns. Concurrently, investigating the mechanistic insights of *D. nobile* on skin health at cellular and molecular levels provides an essential understanding of its interactions with skin cells and modulation of key signalling pathways. This mechanistic understanding is pivotal for elucidating the therapeutic potential of botanicals and informing targeted skincare approaches. Additionally, conducting bioavailability studies to examine how active compounds from *D. nobile* are absorbed and distributed within the body will help optimize dosage regimens and formulation strategies. Enhancing bioavailability is critical to maximize the therapeutic effects of botanical extracts in skincare formulations. By pursuing these comprehensive research directions, scientists can significantly advance our understanding of the potential benefits of *D. nobile* for skin health and contribute to the development of innovative skincare interventions that meet diverse consumer needs (Table 13.3).

13.8.1 Development of innovative formulations and delivery systems

The development of innovative formulations and delivery systems plays a significant role in maximizing the efficacy and potential of *D. nobile* in skincare. These advancements aim to enhance the stability, bioavailability and targeted delivery of active compounds from *D. nobile* extract to the skin. Nanoemulsions are colloidal systems with small droplets that improve the solubility and permeability of active compounds. Studies have explored the development of *D. nobile* nanoemulsions to enhance skin penetration and improve efficacy. Encapsulation techniques involve enclosing *D. nobile* extract in microscopic carriers, such as liposomes or microspheres, to protect the active compounds and enhance their stability and controlled release. Encapsulation can improve the bioavailability and sustained release of *D. nobile* in topical formulations, and provides a controlled and sustained delivery of active compounds through the skin. Researchers have explored the development of *D. nobile*-infused transdermal patches to deliver beneficial components directly into the skin layers for targeted effects. Hydrogels are three-dimensional networks of hydrophilic polymers that can encapsulate *D. nobile* extract and deliver it to the skin in a gel-like form. Hydrogels provide a cooling and moisturizing effect and can improve the stability and prolonged release of *D. nobile* extract. Microneedle systems create microchannels in the skin, allowing for enhanced delivery of active compounds. Microneedle patches loaded with *D. nobile* extract have been investigated for their potential in delivering *D. nobile* components into the deeper layers of the skin. These innovative formulations and delivery systems hold promise for optimizing the delivery and efficacy of *D. nobile* in skincare applications, allowing for the targeted and controlled release of its beneficial components.

13.8.2 Exploration of combination approaches with other botanicals or active ingredients

In recent years, there has been growing interest in exploring the synergistic effects of combining *D. nobile* extract with other botanicals or active ingredients in skincare formulations. These combination approaches aim to enhance overall efficacy and provide comprehensive benefits for skin health. *D. nobile* and *Centella asiatica*, both traditional botanicals with skincare benefits, have been combined into formulations to target multiple aspects of skin health. This combination has shown potential for promoting wound healing, reducing inflammation and improving collagen synthesis. Niacinamide, a form of vitamin B3, has been combined with *D. nobile* extract in skincare products. This combination has demonstrated promising effects in improving skin barrier function, reducing hyperpigmentation and enhancing overall skin appearance and texture (Turcov *et al.*, 2023). Hyaluronic acid, a hydrating ingredient known for its moisture-retaining properties, has been combined with *D. nobile* extract. This combination offers enhanced moisturization, plumping effects and improved skin hydration, resulting in smoother and more youthful complexion (Meng *et al.*, 2020). Combining *D. nobile* extract with peptides, such as palmitoyl pentapeptide-4, has shown

Table 13.3. Future research directions for *D. nobile* in skin health (authors' own table).

Research avenue	Description	Potential benefits	Methodologies
Phytochemical profiling	Delve deeper into the phytochemical composition of *D. nobile* to identify specific compounds relevant to skin health	Identification of bioactive molecules for skin benefits	Advanced analytical techniques
Antioxidant activity	Investigate the antioxidant properties of *D. nobile* extracts and compounds	Protection against oxidative stress	*In vitro* assays, antioxidant assays
Anti-inflammatory potential	Explore the anti-inflammatory effects of *D. nobile* on skin conditions	Reduction of inflammation in skin	*In vivo* models, cytokine assays
Wound-healing properties	Assess the wound-healing capabilities of *D. nobile* extracts, focusing on tissue regeneration and collagen synthesis	Enhanced tissue repair and regeneration	Animal models, histological analysis
UV protection	Investigate whether *D. nobile* extracts offer photoprotective benefits against UV-induced skin damage	Prevention of UV-induced skin damage	UV irradiation studies, sun protection factor (SPF) assays
Moisturizing effects	Evaluate the moisturizing and hydrating effects of *D. nobile* on the skin	Improved skin barrier function and hydration	Transepidermal water loss (TEWL) measurement, hydration assays
Clinical trials	Validate the efficacy and safety of *D. nobile*-based formulations for various skin conditions	Clinical validation of efficacy and safety	RCTs, patient assessment
Formulation development	Explore innovative formulations incorporating *D. nobile* extracts optimized for skin delivery and efficacy	Development of practical skincare products	Formulation optimization, skin penetration studies
Mechanistic insights	Investigate the underlying mechanisms of action of *D. nobile* on skin health at cellular and molecular levels	Understanding molecular pathways for skin benefits	Cell culture studies, gene expression analysis
Bioavailability studies	Examine the bioavailability and pharmacokinetics of active compounds from *D. nobile* to optimize dosage regimens and formulations	Optimization of dosage and formulation strategies	Pharmacokinetic studies, bioavailability assays

promising results in promoting collagen synthesis, improving skin firmness and reducing the appearance of wrinkles and fine lines. The combination of *D. nobile* extract with antioxidants, such as vitamin C and green tea extract, has been explored for its synergistic antioxidant effects. This combination provides enhanced protection against oxidative stress and helps combat the signs of ageing caused by free radicals. These combined approaches offer exciting possibilities for formulating skincare products that harness the synergistic effects of multiple botanicals or active ingredients. Further research is needed to fully understand the mechanisms and optimize combinations for maximum efficacy and safety.

13.9 Conclusion

In conclusion, the exploration of *D. nobile* in the context of skin health, from its traditional use to its modern applications, has revealed its potential as a valuable ingredient in skincare formulations. The phytochemical composition of plants, including polysaccharides, phenolic compounds and alkaloids, contributes to their diverse bioactive properties and offers a range of benefits for the skin. Research has demonstrated the antioxidant properties of *D. nobile*, which help protect the skin from oxidative stress and ageing. Its anti-inflammatory effects contribute to soothing and calming of the skin, making it suitable for managing conditions such as eczema and dermatitis. The moisturizing and hydrating effects of *D. nobile* help maintain the moisture balance of the skin, promoting plumpness and suppleness.

Furthermore, *D. nobile* has shown promise in stimulating collagen synthesis, enhancing skin elasticity and reducing wrinkles and fine lines. It also regulates melanogenesis and contributes to the management of hyperpigmentation and uneven skin tone. While there is still a need for further research to fully understand the underlying mechanisms of action and optimize the extraction and formulation processes, the integration of *D. nobile* extract in skincare routines holds significant potential. Its incorporation into innovative formulations, combination approaches with other botanicals or active ingredients, and the development of advanced delivery systems can further enhance its efficacy and usability for skincare products. It is important to note that safety considerations, such as patch testing and monitoring for potential adverse reactions, should be considered when using *D. nobile* in skincare products. Clinical studies and efficacy evaluations are necessary to establish the long-term safety and effectiveness of various skincare applications. Overall, the utilization of *D. nobile* in skincare represents a bridge between traditional knowledge and modern scientific advancements. With continued research and development, *D. nobile* has the potential to revolutionize the skincare industry, offering natural and effective solutions for promoting and maintaining skin health.

Acknowledgements

The authors express their gratitude for the financial support received through the Distinguished High-Level Talents Research Grant from the Guizhou Science and Technology Corporation Platform Talents Fund (grant nos. [2017]5733-001 and CK-1130-002), the National Natural Science Foundation of China (U1812403, 82373981 and 82060750), 2011 Collaborative Innovation Centre of Traditional Chinese Medicine in Guizhou Province (no. [2022]026) and the support provided by the Zunyi Medical University, China. Special appreciation is extended to all laboratory colleagues and research staff members for their valuable insights, constructive guidance and assistance with the contents of this chapter.

References

Cakova, V., Bonte, F. and Lobstein, A. (2017) *Dendrobium:* sources of active ingredients to treat age-related pathologies. *Aging and Disease* 8(6), 827–849.

Chen, J.T. (2020) Orchid biochemistry. *International Journal of Molecular Sciences* 21(7), 2338.

Cruz, A.M., Gonçalves, M.C., Marques, M.S., Veiga, F., Paiva-Santos, A.C. *et al.* (2023) In vitro models for anti-aging efficacy assessment: a critical update in dermocosmetic research. *Cosmetics* 10(2), 66.

Duan, H., Er-Bu, A., Dongzhi, Z., Xie, H., Ye, B. *et al.* (2022) Alkaloids from dendrobium and their biosynthetic pathway, biological activity and total synthesis. *Phytomedicine* 102, 154132.

Duan, Z.Y., Sun, Y.P., Wang, Z.B. and Kuang, H.X. (2024) Moslae herba: botany, traditional uses, phytochemistry, and pharmacology. *Molecules* 29(8), 1716.

Fan, C., Sun, X., Wang, X. and Yu, H. (2023) Therapeutic potential of the chemical composition of *dendrobium nobile* Lindl. *Frontiers in Pharmacology* 14, 1163830.

Fan, Y., Jiang, T., Chun, Z., Wang, G., Yang, K. *et al.* (2021) Zinc affects the physiology and medicinal components of *Dendrobium nobile* Lindl. *Plant Physiology and Biochemistry* 162, 656–666.

Firoozbahr, M., Kingshott, P., Palombo, E.A. and Zaferanloo, B. (2023) Recent advances in using natural antibacterial additives in bioactive wound dressings. *Pharmaceutics* 15(2), 644.

Fu, X., Chen, S., Xian, S., Wu, Q., Shi, J. *et al.* (2023) Dendrobium and its active ingredients: emerging role in liver protection. *Biomedicine & Pharmacotherapy* 157, 114043.

Ganapathy, A.A., Haripriya, V.M., Acharya, N., Somappa, S.B. and Kumaran, A. (2023) Ethnobotanical significance of medicinal plants: beta-amyloid and tau aggregation inhibitors against Alzheimer's disease. *Journal of Biochemical and Molecular Toxicology* 37(6), e23339.

Ghai, D., Kaur, A., Kahlon, P.S., Pawar, S.V. and Sembi, J.K. (2022) A walk through the maze of secondary metabolism in orchids: a transcriptomic approach. *Frontiers in Plant Science* 13, 837563.

Gogoi, P. and Sen, S. (2023) Orchids of Dibru-Saikhowa: a systematic review on their traditional use, pharmacological activity and phytochemistry. *Current Topics in Medicinal Chemistry* 23(24), 2277–2299.

He, X., Gao, X. and Xie, W. (2023) Research progress in skin aging, metabolism, and related products. *International Journal of Molecular Sciences* 24(21), 15930.

Hong, S., Kim, E.-Y., Lim, S.-E., Kim, J.-H., Sohn, Y. *et al.* (2022) *Dendrobium nobile* Lindley administration attenuates atopic dermatitis-like lesions by modulating immune cells. *International Journal of Molecular Sciences* 23(8), 4470.

Hu, Y., Ye, Z., Wu, M., She, Y., Li, L. *et al.* (2021) The communication between intestinal microbiota and ulcerative colitis: an exploration of pathogenesis, animal models, and potential therapeutic strategies. *Frontiers in Medicine* 8, 766126.

Huang, W., Wang, J., Kuang, M., Xiao, Z., Fan, B. *et al.* (2023) Exploring global research status and trends in anti-obesity effects of traditional Chinese medicine through intestinal microbiota: a bibliometric study. *Frontiers in Cellular and Infection Microbiology* 13, 1271473.

Jain, A., Sarsaiya, S., Gong, Q., Wu, Q. and Shi, J. (2024) Pharmacological and Therapeutic Biofunction of *Dendrobium nobile*: A Critical Review. *Natural Product Communications* 19(11), doi:10.1177/1934578X241300010.

Jędrejko, K., Lazur, J. and Muszyńska, B. (2021) Risk associated with the use of selected ingredients in food supplements. *Chemistry & Biodiversity* 18(2), e2000686.

Ketsa, S. and Warrington, I.J. (2023) The dendrobium orchid: botany, horticulture, and utilization. *Crop Science* 63(4), 1829–1888.

Li, K., Wu, F., Chen, M., Xiao, Z., Xu, Y. *et al.* (2023) Identification, biological function profiling and biosynthesis of secondary metabolites in medicinal orchids. *Metabolites* 13(7), 1–25.

Li, R., Liu, K., Huang, X., Li, D., Ding, J., Liu, B., Chen, X. (2022) Bioactive Materials Promote Wound Healing through Modulation of Cell Behaviors. *Advanced Science* 9(10), e2105152.

Lin, W., Wang, J., Xu, X., Wu, Y., Qiu, D. *et al.* (2020) rapid propagation in vitro and accumulation of active substances of endangered *Dendrobium cariniferum* Rchb. f. *Bioengineered* 11(1), 386–396.

Liu, X.M., Shi, H. and Li, W. (2024) Review on the potential roles of traditional Chinese medicines in the prevention, treatment, and postoperative recovery of age-related cataract. *Journal of Ethnopharmacology* 324, 117786.

Long, Y., Wang, W., Zhang, Y., Du, F., Zhang, S. *et al.* (2023a) Photoprotective effects of *dendrobium nobile* Lindl. polysaccharides against UVB-induced oxidative stress and apoptosis in hacat cells. *International Journal of Molecular Sciences* 24(7), 6120.

Long, Y., Wang, W., Zhang, Y., Zhang, S. and Li, Z. (2023b) *Dendrobium nobile* Lindl polysaccharides attenuate UVB-induced photodamage by regulating oxidative stress, inflammation and mmps expression in mice model. *Photochemistry and Photobiology* 99(5), 1269–1281.

Makgobole, M.U., Onwubu, S., Khathi, A., Mpofana, N. and Mkhwanazi, B. (2024) Cosmeceuticals from marine: the prospect of marine products in skin rejuvenation and care. *Egyptian Journal of Basic and Applied Sciences* 11(1), 297–317.

Meng, H., Li, J., Dong, Y., He, Y., Ren, H. *et al.* (2020) Poly traditional Chinese medicine formulation prepared with skin moisturizing properties. *Dermatologic Therapy* 33(6), e14105.
Natta, S., Mondol, M.S.A., Pal, K., Mandal, S., Sahana, N. *et al.* (2022) Chemical composition, antioxidant activity and bioactive constituents of six native endangered medicinal orchid species from north-eastern Himalayan region of India. *South African Journal of Botany* 150, 248–259.
Nie, X., Chen, Y., Li, W. and Lu, Y. (2020) Anti-aging properties of *Dendrobium nobile* Lindl.: from molecular mechanisms to potential treatments. *Journal of Ethnopharmacology* 257, 112839.
Nugraha, A.S., Triatmoko, B., Wangchuk, P. and Keller, P.A. (2020) Vascular epiphytic medicinal plants as sources of therapeutic agents: their ethnopharmacological uses, chemical composition, and biological activities. *Biomolecules* 10(2), 181.
Peng, Y., Tao, H., Wang, S., Xiao, J., Wang, Y. *et al.* (2021) Dietary intervention with edible medicinal plants and derived products for prevention of Alzheimer's disease: a compendium of time-tested strategy. *Journal of Functional Foods* 81, 104463.
Ramesh, T., Koperuncholan, M., Praveena, R., Ganeshkumari, K., Vanithamani, J. *et al.* (2019) Medicinal properties of some dendrobium orchids – a review. *Journal of Applied and Advanced Research* 4(4), 119–128.
Shahrajabian, M.H., Sun, W. and Cheng, Q. (2020) Traditional herbal medicine for the prevention and treatment of cold and flu in the autumn of 2020, overlapped with COVID-19. *Natural Product Communications* 15(8), doi:10.1177/1934578X20951431.
Sharma, A. and Pathak, P. (2020) The budding potential of orchids in the cosmeceutical sector: role of orchids in skincare and health. *The Journal of The Orchid Society of India* 34, 79–85.
Singh, P.A., Bajwa, N. and Baldi, A. (2021) Possible role of traditional systems of medicine to manage COVID-19: a review. *Israel Journal of Plant Sciences* 68(1–2), 3–28.
Su, X.-L., Wang, J.-W., Che, H., Wang, C.-F., Jiang, H. *et al.* (2020) Clinical application and mechanism of traditional Chinese medicine in treatment of lung cancer. *Chinese Medical Journal* 133(24), 2987–2997.
Swainson, N.M., Pengoan, T., Khonsap, R., Meksangsee, P., Hagn, G. *et al.* (2023) *In vitro* inhibitory effects on free radicals, pigmentation, and skin cancer cell proliferation from *Dendrobium* hybrid extract: a new plant source of active compounds. *Heliyon* 9(9), 1–13.
Tiwari, P., Bose, S.K., Gautam, A. and Chen, J.T. (2023) Emerging trends and insights into the cultivation strategies, ethnomedicinal uses, and socio-economic attributes of orchids. *The Journal of Horticultural Science and Biotechnology* 98(3), 273–298.
Tiwari, P., Sharma, A., Bose, S.K. and Gautam, A. (2022) Biotechnological interventions in orchids: recent updates, translational success, and commercial outcomes. *Research Square* 1, 1–38.
Tiwari, P., Sharma, A., Bose, S.K. and Park, K.I. (2024) Advances in orchid biology: biotechnological achievements translational success, and commercial outcomes. *Horticulturae* 10(2), 152.
Turcov, D., Zbranca-Toporas, A. and Suteu, D. (2023) Bioactive compounds for combating oxidative stress in dermatology. *International Journal of Molecular Sciences* 24(24), 17517.
Wang, F., Wan, J., Liao, Y., Liu, S., Wei, Y. *et al.* (2023) Dendrobium species regulate energy homeostasis in neurodegenerative diseases: a review. *Food Science and Human Wellness* 12(6), 2151–2174.
Wang, G., Wang, J., Deng, Y., Qin, L., He, Y. *et al.* (2022) Chemical constituents and nutritional health functions of *Dendrobium nobile*: a review. *Food Science and Technology* 42, e84522.
Wang, Y.H. (2021) Traditional uses and pharmacologically active constituents of dendrobium plants for dermatological disorders: a review. *Natural Products and Bioprospecting* 11(5), 465–487.
Wu, W., Lin, Y., Farag, M.A., Li, Z. and Shao, P. (2023) Dendrobium as a new natural source of bioactive for the prevention and treatment of digestive tract diseases: a comprehensive review with future perspectives. *Phytomedicine* 114, 154784.
Ye, G., Zhang, J., Xu, X., Zeng, C., Ye, Q. *et al.* (2023) Comparative analysis of water-soluble polysaccharides from *Dendrobium* second love "Tokimeki" and *Dendrobium nobile* in structure, antioxidant, and anti-tumor activity in vitro. *International Journal of Molecular Sciences* 24(12), 10361.
Zhang, J., Xu, H.X., Zhao, Z.L., Xian, Y.F. and Lin, Z.X. (2021) *Dendrobium nobile* Lindl: a review on its chemical constituents and pharmacological effects. *Chinese Medicine and Culture* 4(4), 235–242.
Zhang, Q., Fu, X., Liu, H., Chen, Y., Chen, S. *et al.* (2023a) A systematical review on ethnobotanical, phytochemical and pharmacological aspects of *Dendrobium nobile* Lindl. *Phytochemistry Reviews* 22(3), 743–780.
Zhang, Y., Lu, J., Li, H. and Song, H. (2023b) Advances in dietary polysaccharides as hypoglycemic agents: mechanisms, structural characteristics, and innovative applications. *Critical Reviews in Food Science and Nutrition* 5, 1–21.
Zhu, A.-L., Hao, J.-W., Liu, L., Wang, Q., Chen, N.-D. *et al.* (2021) Simultaneous quantification of 11 phenolic compounds and consistency evaluation in four *Dendrobium* species used as ingredients of the traditional Chinese medicine shihu. *Frontiers in Nutrition* 8, 771078.

14 *Dendrobium nobile*: A Potential Therapeutic Approach for Dry Eye Disease

Surendra Sarsaiya, Archana Jain, Qihai Gong, Qin Wu and Jingshan Shi*
Key Laboratory of Basic Pharmacology and Joint International Research Laboratory of Ethnomedicine of Ministry of Education, Zunyi Medical University, Zunyi, Guizhou, China

Abstract

Dry eye disease (DED) is a multifactorial condition characterized by tear film instability and ocular surface inflammation leading to discomfort and visual disturbances. Despite the availability of conventional treatments such as artificial tears and anti-inflammatory drugs, many patients experience inadequate relief. This chapter explores the potential of *Dendrobium nobile*, a traditional Chinese medicinal orchid, as an alternative therapeutic approach for DED. *D. nobile* contains bioactive compounds with anti-inflammatory, antioxidant and mucin-stimulating properties that may address multiple aspects of the pathophysiology of DED. Preclinical and clinical studies have suggested that *D. nobile* is effective in enhancing tear film stability, reducing ocular surface inflammation and alleviating dry eye symptoms. Although promising, further research is needed to elucidate the mechanisms of action and long-term safety. This chapter provides a comprehensive overview of the potential therapeutic benefits of *D. nobile*, highlighting its potential as a natural remedy for DED management and underscoring the need for further investigation.

14.1 Introduction

Dry eye disease (DED) is a multifactorial disease of the tears and ocular surface that results in symptoms of discomfort, visual disturbance and tear film instability with potential damage to the ocular surface. It is accompanied by increased osmolarity of the tear film and inflammation of the ocular surface (Shimazaki, 2018). The tear film, consisting of water, lipid and mucin components, is vital for upholding the integrity of the eye surface and furnishing lubrication and shielding to the eyes (Labetoulle *et al.*, 2022). Disruptions in tear film composition or function can lead to tear film instability, ocular surface inflammation and epithelial damage, contributing to the development of DED (Kumar *et al.*, 2023a). Common symptoms of DED include dryness, burning or stinging sensations, redness and foreign body sensation. The burden of DED on a patient's quality of life can be significant, impacting activities of daily living, productivity and overall well-being (Ma *et al.*, 2022). Despite its prevalence and impact, DED remains a challenging condition to manage, with many individuals experiencing inadequate relief from their symptoms despite common conventional treatments like artificial tears, anti-inflammatory drugs and punctal plugs (Kim *et al.*, 2021). However, there is increasing awareness of the necessity for

*Corresponding author: shijs@zmu.edu.cn

© Jingshan Shi, Archana Jain, Surendra Sarsaiya, Qin Wu and Qihai Gong 2025. Dendrobium nobile: *Research and Therapeutic Applications* (J. Shi, A. Jain, S. Sarsaiya, Q. Wu and Q. Gong)
DOI: 10.1079/9781800627987.0014

alternative therapies that target the root causes of DED and offer better and longer-lasting relief to patients. Despite the range of treatment choices, many people still find their symptoms persisting, underscoring the demand for alternative approaches.

Lately there has been a surge in interest in investigating natural remedies from plants for managing DED (Messmer *et al.*, 2019). DED arises from alterations in the tear film, leading to uncomfortable symptoms and epithelial irregularities. Tear insufficiency and excessive evaporation are the main factors that cause DED (Rodriguez-Garcia *et al.*, 2022). DED is categorized into two main types: tear-deficient (or aqueous-deficient) dry eye and evaporative dry eye, further branching into various intrinsic and extrinsic factors (Harrell *et al.*, 2023). The preference for the term "disorder" over "disease" in the context of dry eye conditions reflects a nuanced understanding of the condition's multifactorial and chronic nature, which may not align with the traditional implications of the term "disease." This distinction has influenced research and clinical guidelines, such as those by the American Academy of Ophthalmology in 2013. However, the persistence of "disease" in the nomenclature—Dry Eye Disease (DED)—is indeed common in the literature, likely due to historical usage and the broad acceptance of the term within the medical and scientific communities. (Sheppard *et al.*, 2023).

One botanical that has garnered attention is *Dendrobium nobile* Lindl., a species of orchid traditionally used in Chinese medicine for its purported medicinal properties. This chapter describes the potential of *D. nobile* as a therapeutic agent for DED (Ling *et al.*, 2022). It begins with an overview of the pathophysiology of DED and the limitations of the current treatments. Next, *D. nobile* is introduced, discussing its taxonomy, distribution and traditional uses in Chinese medicine. The chemical composition of *D. nobile* is explored, focusing on its bioactive compounds and potential therapeutic properties (Yang *et al.*, 2022). Subsequently, the mechanisms by which these bioactive compounds may exert therapeutic effects in DED are investigated by exploring their potential interactions with relevant pathways (Han *et al.*, 2023). Preclinical and clinical studies evaluating the safety and efficacy of *D. nobile* in DED are reviewed, and key findings are summarized. Additionally, safety considerations associated with the use of *D. nobile*, including potential side effects and interactions with medications, are discussed. Finally, areas for future research are identified and the implications of the findings are discussed. This chapter aims to provide a comprehensive overview of *D. nobile* as a potential therapeutic approach for DED, highlighting its promise and the need for further investigation.

14.2 Pathophysiology of DED

DED presents as a multifaceted condition characterized by alterations in tear volume or composition, leading to discomfort, visual disturbances and potential harm to the eye's surface. The mechanisms underlying DED encompass tear film instability, inflammation and damage to the ocular surface (Giannaccare *et al.*, 2021). The tear film comprises three layers: a lipid layer, an aqueous layer and a mucin layer. Each layer is essential for preserving the health of the eye's surface and ensuring clear vision. DED can result from decreased tear production, known as aqueous-deficient dry eye, or from increased tear evaporation, referred to as evaporative dry eye. Dysfunction in the lacrimal glands, meibomian glands or accessory glands can lead to inadequate tear production or alterations in tear composition, contributing to DED (Gorimanipalli *et al.*, 2023). Irregularities in tear film composition or distribution can cause tear film instability, characterized by quick tear breakup and inadequate coverage of the eye's surface. This instability often results in symptoms such as burning, stinging and foreign body sensation (Kojima *et al.*, 2020).

Persistent inflammation is crucial in the development of DED, arising in response to tear film instability and surface damage, with the release of pro-inflammatory cytokines, chemokines and other inflammatory mediators, leading to further tissue damage and perpetuating the inflammatory cascade (Rao *et al.*, 2022). Additionally, T cell activation and the infiltration of inflammatory cells into the conjunctiva and lacrimal glands exacerbate the inflammatory response in DED, with T cell-mediated immune responses playing a significant role in enhancing and prolonging inflammation on the eye's surface

(Yu *et al.*, 2021). Cytokine dysregulation presents a complex scenario where the finely tuned balance of immune signalling molecules is disrupted, leading to a cascade of inflammatory events with profound implications for health. Whether through overproduction or underproduction, aberrant cytokine activity contributes to a spectrum of disorders, from autoimmune diseases to severe infections (Camacho *et al.*, 2021). Understanding the intricate mechanisms driving cytokine dysregulation is pivotal for developing targeted therapies. Precision medicine approaches, coupled with advancements in biomarker identification, offer promise in tailoring interventions to individual cytokine profiles, thereby optimizing treatment outcomes while minimizing adverse effects (Lamb *et al.*, 2022). Continued research into cytokine dysregulation holds the key to unravelling the complexities of immune-mediated disorders and advancing therapeutic strategies to alleviate their burden on human health. Prolonged exposure to desiccating stress and inflammation in DED can cause significant damage to the corneal and conjunctival epithelium, leading to squamous metaplasia and a reduction in goblet cells, which weakens the integrity of the eye's surface and exacerbates dry eye symptoms (Bu *et al.*, 2024). Additionally, neurogenic inflammation mediated by sensory nerves plays a role in the pathophysiology of DED; activation of sensory nerve fibres in the cornea and conjunctiva by noxious stimuli triggers the release of neuropeptides, further intensifying inflammation and discomfort (Asiedu, 2022).

Environmental factors such as low humidity, wind, air pollution and prolonged screen use exacerbate dry eye symptoms by increasing tear evaporation and ocular surface desiccation (Calonge *et al.*, 2017). Systemic conditions and medications also impact tear production and ocular surface health; diseases like Sjögren's syndrome, autoimmune disorders, hormonal imbalances and vitamin deficiencies are associated with a higher risk of DED (Gorimanipalli *et al.*, 2023). Understanding the complex interplay of tear film dysfunction, inflammation, ocular surface damage, and environmental and systemic factors is crucial for effectively diagnosing and managing DED (Rao *et al.*, 2022). Targeted interventions that restore tear film stability, suppress inflammation and promote ocular surface healing are key components of DED management strategies (Fig. 14.1) (Suárez-Cortés *et al.*, 2022).

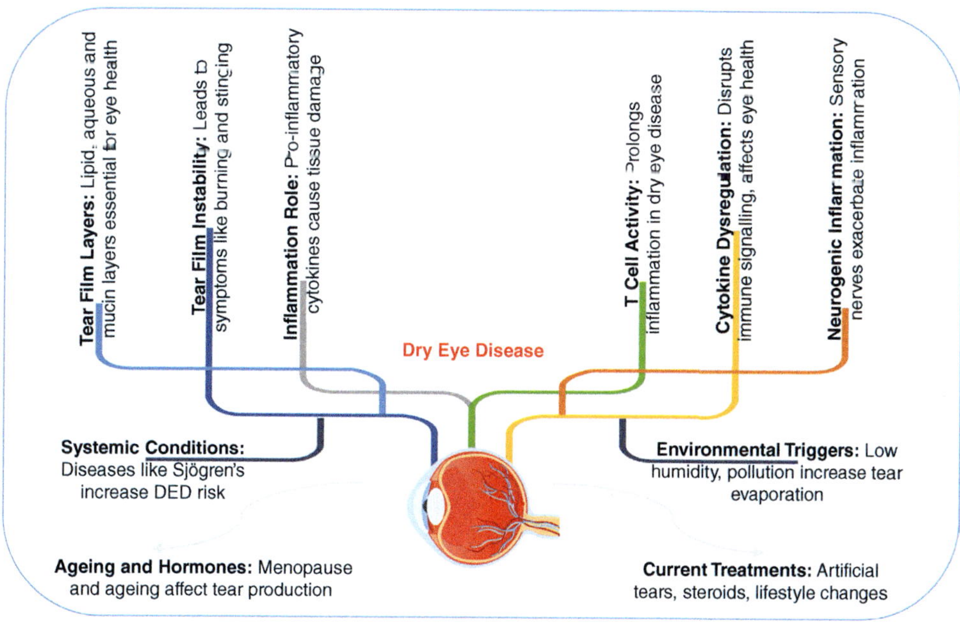

Fig. 14.1. A description of DED (authors' own figure).

14.2.1 Tear film homeostasis

Tear film homeostasis is a delicate balance maintained by the dynamic interplay of tear production, distribution and clearance mechanisms essential for ocular health (Pflugfelder and Stern, 2020). Comprising aqueous, lipid and mucin layers, the tear film acts as a protective barrier, nourishing and lubricating the ocular surface while defending against pathogens and foreign particles (Silva *et al.*, 2021). Disruptions in tear film homeostasis, whether due to decreased tear production, altered composition or impaired clearance, can result in ocular surface disorders such as DED (Koh *et al.*, 2022). Understanding the multifactorial nature of tear film dynamics is crucial for developing targeted therapies aimed at restoring homeostasis and alleviating the symptoms associated with ocular surface diseases. The interplay among these layers is crucial for tear film stability and function. Imbalances in tear film equilibrium can result in DED, marked by symptoms like dryness, irritation, burning and blurred vision (Kawahara, 2023). Factors such as tear film instability, increased tear evaporation and decreased tear production can all contribute to the development of DED (Sheppard and Nichols, 2023). Understanding the mechanisms underlying tear film homeostasis is crucial for diagnosing and managing DED effectively (Rolando and Merayo-Lloves, 2022).

14.2.2 Factors contributing to DED

Several factors contribute to the development of DED, ranging from environmental triggers to underlying systemic conditions (Liu *et al.*, 2021). Environmental elements like dry or windy weather conditions, air conditioning and extended screen use can heighten tear evaporation and worsen dryness on the eye's surface. Additionally, indoor heating, air pollution and smoke exposure can further aggravate symptoms of DED by inducing inflammation and oxidative stress on the ocular surface (Alves *et al.*, 2023). Hormonal changes also play a significant role in DED, particularly in women experiencing the menopause. Fluctuations in oestrogen levels can lead to alterations in tear film composition and stability, predisposing individuals to dry eye symptoms (Deo *et al.*, 2024). Autoimmune conditions like Sjögren's syndrome, rheumatoid arthritis and lupus are commonly associated with DED due to immune-mediated destruction of lacrimal gland tissue and decreased tear production (Bustamante-Arias *et al.*, 2022).

Chronic inflammation is another key contributor to DED pathogenesis, with pro-inflammatory cytokines and immune cells infiltrating the ocular surface and exacerbating tissue damage (Rao *et al.*, 2022). Other factors, including ageing, contact lens wear, certain medications (e.g. antihistamines, antidepressants) and systemic diseases (e.g. diabetes, thyroid disorders), can also increase the risk of developing DED.

14.2.3 Current treatments and limitations

DED is a prevalent ailment marked by inadequate tear production or diminished tear quality, resulting in discomfort, inflammation and possible harm to the eye's surface. Although there is ongoing research into various treatments for DED, including pharmaceuticals and lifestyle interventions, the use of *D. nobile* specifically for this condition is not well established.

Current treatments for DED include artificial tears, which are lubricating eye drops that supplement natural tear production and alleviate mild to moderate symptoms (Gomes and Santo, 2019). For moderate to severe cases, prescription eye drops containing corticosteroids or immunomodulators are used to reduce inflammation and improve tear film stability (Prinz *et al.*, 2022). Lifestyle modifications, such as maintaining proper hydration, avoiding dry environments, taking breaks during prolonged screen use and using humidifiers, can also help manage symptoms (Donthineni *et al.*, 2021). Nutritional supplements, particularly omega-3 fatty acids and vitamins A and D, have been studied for their potential benefits in managing DED (Downie *et al.*, 2019). Additionally, in certain instances, small plugs can be placed into the tear ducts to obstruct drainage and retain natural tears on the eye's surface (Best *et al.*, 2019).

D. nobile, traditionally used for various health purposes including eye health, faces several limitations regarding its use for DED. There is limited scientific evidence specifically demonstrating its efficacy for this condition, as no well-designed clinical trials have been conducted (Hsu *et al.*, 2022). Additionally, the variability in formulations, including extracts, teas and supplements, complicates the assessment of their consistency and therapeutic potential (Ramesh *et al.*, 2019). Safety concerns also arise due to the lack of adequate studies on its ophthalmic use, posing risks of irritation, allergic reactions or other adverse effects, especially with improperly formulated or contaminated products (Fu *et al.*, 2023). Regulatory concerns further complicate the use of herbal supplements like *D. nobile*, as they are not subjected to the stringent testing and oversight standards applied to pharmaceutical drugs, leading to potential variations in product quality and safety (Song *et al.*, 2017). Given the availability of established treatments for DED with proven efficacy and safety profiles, such as artificial tears and prescription medications, there is little incentive for patients and healthcare providers to explore alternative therapies like *D. nobile* without sufficient evidence of benefit (Liu *et al.*, 2024).

14.3 Chemical Composition of *D. nobile*

D. nobile, a species of orchid traditionally used in Chinese medicine, contains a diverse array of chemical compounds that contribute to its medicinal properties. Among the bioactive compounds identified in *D. nobile* are polysaccharides, flavonoids, bibenzyls, phenanthrenes, alkaloids and amino acids. These compounds have been isolated and characterized through various analytical techniques, including chromatography and spectroscopy.

14.3.1 Bioactive compounds usable in DED

Various bioactive substances present in *D. nobile* exhibit promising potential for treating DED. Polysaccharides derived from *D. nobile* have displayed anti-inflammatory and antioxidant characteristics, which may help alleviate ocular surface inflammation and oxidative stress associated with DED (Guo *et al.*, 2024). Flavonoids present in *D. nobile* exhibit antioxidant activity and have been implicated in modulating inflammatory pathways, making them promising candidates for DED therapy (Yang *et al.*, 2022). Additionally, bibenzyls and phenanthrenes isolated from *D. nobile* have shown anti-inflammatory and immunomodulatory effects, which could contribute to their therapeutic efficacy in DED (Ling *et al.*, 2022). Furthermore, alkaloids and amino acids found in *D. nobile* may play a role in promoting ocular surface health and supporting tear film stability (Yang *et al.*, 2022). The chemical composition of *D. nobile* is complex and diverse, with over a hundred identified compounds, including alkaloids, flavonoids, polysaccharides, phenols and amino acids (Wang *et al.*, 2022). Many of these substances have demonstrated diverse pharmacological properties, encompassing anti-inflammatory, antioxidant, immunomodulatory and neuroprotective effects.

Among the bioactive compounds identified in *D. nobile*, alkaloids such as dendrobine and nobiline have received particular attention for their potential therapeutic properties. Alkaloids found in *D. nobile* exhibit anti-inflammatory effects by inhibiting the production of pro-inflammatory cytokines and modulating immune cell activity (Zhang *et al.*, 2023). Additionally, flavonoids like luteolin and quercetin possess antioxidant and anti-inflammatory properties, potentially contributing to their efficacy in managing DED (Chen *et al.*, 2022). Extracted polysaccharides from *D. nobile* have been observed to stimulate mucin production, an essential component of the tear film, indicating a potential role in enhancing tear film stability and lubricating the ocular surface (Firoozbahr *et al.*, 2023). The rich assortment of bioactive compounds in *D. nobile* offers promise for innovative therapeutic strategies in addressing DED (Kumar *et al.*, 2023b).

14.3.2 Potential therapeutic properties for eye disease

The bioactive compounds present in *D. nobile* exhibit a range of potential therapeutic properties that may be beneficial for treating various eye diseases, including DED. *D. nobile* contains compounds such as flavonoids, bibenzyls and phenanthrenes that exhibit anti-inflammatory properties, which can help mitigate ocular surface inflammation associated with DED (Yang *et al.*, 2022). Additionally, its polysaccharides, flavonoids and other constituents demonstrate antioxidant activity, scavenging free radicals and protecting ocular tissues from oxidative damage (Wang *et al.*, 2022). The immunomodulatory effects of polysaccharides and alkaloids found in *D. nobile* help regulate immune responses and modulate inflammatory pathways involved in eye diseases (Ling *et al.*, 2022). Furthermore, polysaccharides and other mucilaginous compounds contribute to its hydrating and lubricating properties, aiding in maintaining tear film stability and alleviating dryness and discomfort (Ramesh *et al.*, 2019). Some bioactive compounds in *D. nobile* are also implicated in promoting tissue regeneration and wound healing, which could be beneficial for repairing damage to the ocular surface in DED (Yang *et al.*, 2022). Overall, the diverse chemical composition of *D. nobile* offers a range of potential therapeutic properties that make it a promising candidate for the treatment of DED and other eye conditions.

14.4 Mechanisms of Action in DED

DED is a multifactorial condition characterized by disruptions in tear film homeostasis and ocular surface inflammation. Understanding the underlying mechanisms involved in DED pathogenesis is essential for developing effective therapeutic strategies (Periman *et al.*, 2020). *D. nobile* may offer several therapeutic benefits for DED, including anti-inflammatory properties that help reduce ocular surface inflammation (Yang *et al.*, 2022). This may include antioxidant activity that neutralizes harmful free radicals, protecting ocular tissues from oxidative damage. Compounds in *D. nobile* may stimulate tear production, addressing tear deficiency, a major contributor to DED symptoms (Ling *et al.*, 2022). Additionally, *D. nobile* might enhance tear film stability, reducing tear evaporation and maintaining proper ocular surface lubrication (Ti *et al.*, 2021). Neuroprotective effects on corneal nerves could help preserve their integrity and function, alleviating ocular discomfort associated with DED (Liu *et al.*, 2020). Furthermore, *D. nobile* compounds could modulate pro-inflammatory cytokine levels on the ocular surface, aiding in the resolution of inflammation and improvement of DED symptoms (Hsu *et al.*, 2024).

D. nobile offers multiple potential therapeutic benefits for DED, including facilitating the repair and regeneration of the corneal epithelium, which promotes healing of ocular surface damage (Philip, 2023). Its compounds may enhance mucin secretion, crucial for maintaining tear film stability and adherence to the ocular surface (Han *et al.*, 2023), and regulate meibomian gland function, improving the lipid layer quality of the tear film to address evaporative dry eye (Kawahara, 2023).

Additionally, *D. nobile* might modulate corneal nerve sensitivity, reducing discomfort and irritation (Yang *et al.*, 2022), and strengthen the ocular surface barrier, preventing further epithelial damage (Yu *et al.*, 2015). It could also help normalize tear film pH, creating a healthier environment for ocular surface health and comfort (Sun *et al.*, 2021). Furthermore, its antioxidant properties may decrease oxidative stress by scavenging reactive oxygen species, thereby preventing cellular damage (Chao *et al.*, 2018). Certain ingredients of *D. nobile* might increase tear film volume, improving lubrication and reducing symptoms of dryness and irritation (Yang *et al.*, 2022). These mechanisms include modulation of inflammatory pathways, promotion of tear film stability, enhancement of ocular surface lubrication and protection against oxidative stress. Dendrobine and nobiline, alkaloids present in *D. nobile*, have demonstrated the ability to suppress the production of pro-inflammatory cytokines like interleukin-6 (IL-6) and tumour necrosis factor alpha (TNF-α). This action leads to a decrease in ocular surface inflammation and helps relieve symptoms associated with dry eye. Flavonoids present in *D. nobile*, such as luteolin and quercetin, exhibit antioxidant properties and may protect against oxidative damage to ocular surface tissues. Polysaccharides extracted from *D. nobile* have been found to stimulate the production of mucin, a key component of the tear film that helps maintain ocular surface lubrication and barrier function. By addressing multiple aspects of the pathophysiology of DED, *D. nobile* may offer a comprehensive and effective therapeutic approach for the management of dry eye symptoms (Table 14.1 and Fig. 14.2).

14.4.1 Interactions with relevant pathways

Bioactive compounds present in *D. nobile* have the potential to interact with multiple pathways

Table 14.1. Mechanisms of action of *D. nobile* in DED (authors' own table).

Mechanism of action	Description	Reference
Anti-inflammatory properties	Reduces ocular surface inflammation, common in DED	Yang *et al.* (2022)
Antioxidant activity	Protects ocular tissues from oxidative damage	Periman *et al.* (2020)
Promotion of tear production	Stimulates tear production, addressing deficiency	Ling *et al.* (2022)
Enhancement of tear film stability	Improves tear film stability, reducing evaporation	Ti *et al.* (2021)
Neuroprotective effects	Protects corneal nerves, alleviating discomfort	Liu *et al.* (2020)
Modulation of cytokine levels	Regulates pro-inflammatory cytokines, resolving inflammation	Hsu *et al.* (2024)
Support of epithelial repair	Facilitates corneal epithelium repair, promoting healing	Philip (2023)
Mucin secretion enhancement	Enhances mucin secretion, maintaining tear film stability	Han *et al.* (2023)
Regulation of meibomian gland function	Regulates meibomian gland function, improving lipid layer quality	Kawahara (2023)
Modulation of nerve sensitivity	Modulates corneal nerve sensitivity, reducing discomfort	Yang *et al.* (2022)
Improvement of ocular surface barrier function	Strengthens ocular surface barrier, preventing damage	Yu *et al.* (2015)
Normalization of tear film pH	Normalizes tear film pH, improving ocular surface health	Sun *et al.* (2021)
Reduction of oxidative stress	Decreases oxidative stress, protecting cells	Chao *et al.* (2018)
Enhancement of tear film volume	Increases tear film volume, reducing dryness	Yang *et al.* (2022)

involved in the development of DED. Specifically, the flavonoids present in *D. nobile* can influence inflammatory signalling pathways such as nuclear factor kappa B (NF-κB) and mitogen-activated protein kinase (MAPK) pathways. By inhibiting the activation of these pathways, flavonoids may attenuate the production of pro-inflammatory cytokines and reduce ocular surface inflammation in DED. Additionally, polysaccharides present in *D. nobile* may exert antioxidant effects by scavenging free radicals and inhibiting oxidative stress pathways, thereby protecting ocular tissues from damage. Furthermore, alkaloids and phenanthrenes found in *D. nobile* may interact with neurotransmitter receptors or ion channels involved in tear secretion and tear film stability, contributing to their therapeutic effects in DED.

14.4.2 Putative therapeutic mechanisms

The putative therapeutic mechanisms of *D. nobile* in DED may involve multiple pathways and biological processes. Compounds in *D. nobile* may offer several therapeutic benefits for DED by suppressing ocular surface inflammation through the inhibition of pro-inflammatory cytokine production and modulation of immune responses (Zhang *et al.*, 2023). Its bioactive compounds possess antioxidant properties that scavenge free radicals, reduce oxidative stress and protect ocular tissues from damage, thereby promoting ocular surface health (Zhang *et al.*, 2021). Additionally, *D. nobile*'s immunomodulatory effects help regulate immune responses and modulate inflammatory pathways, reducing inflammation and improving ocular surface integrity (Ling *et al.*, 2022). Mucilaginous compounds in the plant enhance tear film stability and promote hydration, alleviating dryness and discomfort associated with DED. Furthermore, certain bioactive compounds exert neuroprotective effects on ocular surface nerves, preserving their function and reducing sensitivity in DED. Overall, the therapeutic effects of *D. nobile* in DED may be mediated through its interactions with relevant pathways involved in inflammation, oxidative stress, tear film stability and neuroprotection. Further research is needed to elucidate

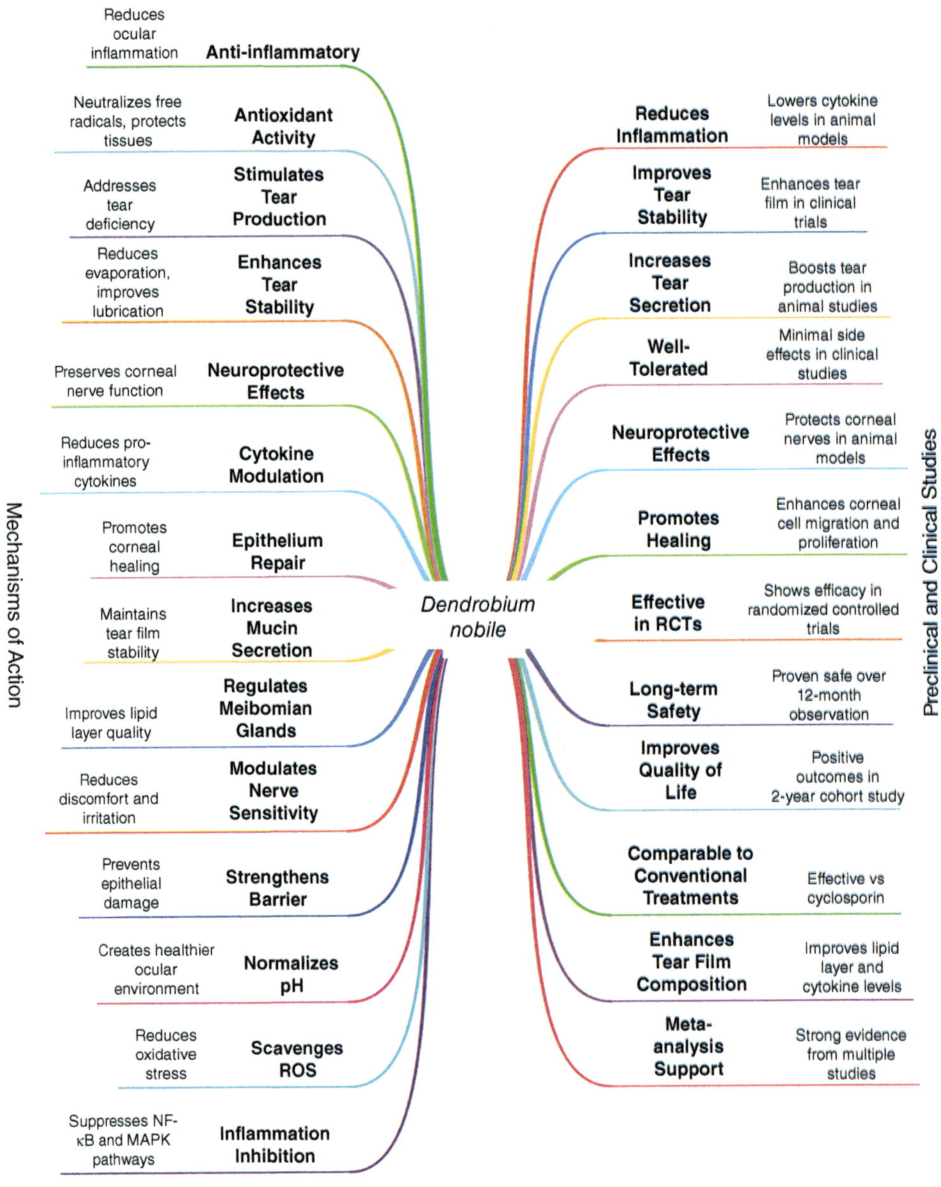

Fig. 14.2. Mechanisms of action of *D. nobile* and preclinical and clinical studies (authors' own figure).

the specific mechanisms of action underlying the therapeutic effects of *D. nobile* in DED and to validate its efficacy in clinical settings.

14.5 Preclinical and Clinical Studies

Preclinical and clinical studies evaluating the safety and efficacy of *D. nobile* in DED have yielded promising results. In animal studies, *D. nobile* extracts have been found to reduce ocular surface inflammation, improve tear film stability and increase tear secretion. These effects were associated with a reduction in inflammatory cytokine levels, modulation of immune cell activity and enhancement of ocular surface barrier function. Common symptoms of dry eye include sensations of dryness, burning and a feeling of foreign objects in the eye. Additionally, objective assessments of tear film stability, tear secretion and the integrity of the ocular surface have shown improvement after treatment with *D. nobile*. Furthermore, *D. nobile* has been well-tolerated with minimal side effects reported in clinical studies. These findings suggest that *D. nobile* may hold promise as a safe and effective therapeutic option for the management of dry eye symptoms.

The therapeutic potential of *D. nobile* for DED has been explored through various studies. Ling *et al.* (2022) investigated the anti-inflammatory properties of *D. nobile* extracts using an *in vitro* model simulating ocular surface inflammation. Their study focused on how the extracts influenced the expression of inflammatory cytokines and chemokines in cultured corneal epithelial cells exposed to pro-inflammatory mediators. In another study by Liu *et al.* (2024), the antioxidant properties of *D. nobile* polysaccharides were evaluated in a murine model of dry eye. Researchers assessed their ability to reduce oxidative stress markers in ocular tissues, highlighting their potential protective effects against oxidative damage in dry eye conditions. Additionally, Cakova *et al.* (2017) conducted a comparative study in animal models to assess the efficacy of *D. nobile* extracts compared to conventional treatments like artificial tears and cyclosporin eye drops. They evaluated parameters such as tear film stability, corneal epithelial integrity and inflammatory markers, providing insights into the potential of *D. nobile* as a novel therapeutic option for managing DED.

In recent studies exploring the therapeutic potential of *D. nobile* for DED, researchers have investigated several aspects of its biological effects. Yang *et al.* (2022) conducted research using a murine model of dry eye induced by desiccating stress to assess the anti-inflammatory properties of *D. nobile* extracts. Their study focused on how these extracts influenced inflammatory markers such as TNF-α, IL-6 and cyclooxygenase-2 (COX-2) in mice, providing insights into its potential to mitigate inflammation on the ocular surface. He *et al.* (2023) explored the neuroprotective properties of *D. nobile* compounds in an animal model of DED, examining their ability to protect corneal nerves from degeneration and improve sensory function, highlighting its potential benefits for alleviating ocular discomfort. Additionally, Chao *et al.* (2018) investigated the wound-healing potential of *D. nobile* extracts on corneal epithelial cells, focusing on their capacity to enhance cell migration and proliferation, which are crucial processes in corneal wound healing that are particularly relevant in the context of DED. These studies collectively contribute to understanding the multifaceted therapeutic effects of *D. nobile* in addressing various aspects of DED pathology.

In a randomized controlled trial, researchers investigated the potential of *D. nobile* eye drops as a treatment for moderate to severe DED (Huang *et al.*, 2023). Conducted across multiple centres, the study aimed to compare the effectiveness of *D. nobile* eye drops against a placebo in improving both objective and subjective dry eye symptoms, including measures like tear film breakup time, corneal staining and symptom severity scores (Philip, 2023). In a prospective observational study, the long-term safety and efficacy of *D. nobile* eye drops were evaluated in individuals with chronic dry eye. Over a 12-month period, participants underwent follow-up assessments to track the sustained impact of the treatment (Yang *et al.*, 2018).

Several recent studies have explored the potential of *D. nobile* as a therapeutic option for DED through different research approaches. Han *et al.* (2023) conducted a prospective cohort study focusing on patients with chronic dry eye treated with *D. nobile* eye drops over a 2-year period, aiming to evaluate long-term outcomes in clinical parameters and quality of life measures. Philip (2023) undertook a randomized controlled trial across multiple centres to compare the effectiveness of *D. nobile* eye drops against conventional treatments like artificial tears or cyclosporin in patients diagnosed with severe DED. This trial utilized objective measures of tear film stability and subjective symptom scores to assess treatment efficacy. Additionally, Li *et al.* (2022a,b) conducted an exploratory study investigating the impact of *D. nobile* supplementation on tear film composition in patients with DED. This pilot study examined changes in tear film parameters such as lipid layer thickness and cytokine levels before and after supplementation with *D. nobile* extracts, providing insights into its potential mechanisms of action. Together, these studies contribute valuable data to the growing body of research on *D. nobile* as a potential therapeutic agent for managing DED, highlighting both its clinical outcomes and underlying physiological effects. Furthermore, a meta-analysis was conducted to consolidate findings from various clinical studies on *D. nobile* for DED. This systematic review and meta-analysis aimed to provide comprehensive insights into the overall efficacy and safety of therapies based on *D. nobile*, offering evidence-based recommendations for clinical practice. Including these studies would further enrich the discussion on the preclinical and clinical evidence supporting the potential therapeutic role of *D. nobile* in DED management (Table 14.2).

14.6 Evaluation of Safety and Efficacy

Extensive investigations have been conducted to assess the safety and efficacy of *D. nobile* as a potential therapeutic approach for DED. These evaluations span preclinical and clinical studies, offering comprehensive insights into the impact of *D. nobile* on ocular health and overall well-being. Preclinical studies have been pivotal in elucidating the safety profile and potential therapeutic mechanisms of *D. nobile*. These studies, often conducted *in vitro* and in animal models, have consistently demonstrated the anti-inflammatory, antioxidant and neuroprotective properties of *D. nobile* extracts. By inhibiting inflammatory cytokines, scavenging free radicals and protecting corneal nerves, *D. nobile* exhibits promising therapeutic effects in DED pathogenesis. Clinical trials have further corroborated the efficacy of *D. nobile*-based therapies in managing dry eye symptoms and improving objective measures of ocular health. These trials, ranging from pilot studies to randomized controlled trials, have reported significant improvements in tear film stability, tear production and subjective symptoms of dryness and discomfort following treatment with *D. nobile* eye drops or supplements. Importantly, these improvements have been observed across diverse patient populations, highlighting the broad applicability of *D. nobile* in DED management.

14.6.1 Key findings and insights

The key findings from preclinical and clinical studies underscore the therapeutic potential of *D. nobile* in alleviating dry eye symptoms and restoring ocular surface health. Preclinical investigations have provided valuable insights into the underlying mechanisms of action of *D. nobile*, elucidating its anti-inflammatory, antioxidant and neuroprotective effects. Clinical trials have further validated these findings, demonstrating the safety and efficacy of *D. nobile*-based therapies in improving both subjective and objective measures of dry eye. Insights gleaned from these studies have highlighted the multifaceted nature of *D. nobile*'s therapeutic effects in DED. Beyond its anti-inflammatory and antioxidant properties, *D. nobile* may also exert hydrating, lubricating and immunomodulatory effects on the ocular surface, contributing to its overall efficacy in dry eye management. Moreover, the favourable safety profile of *D. nobile*, coupled with its natural origin, makes it an attractive candidate for long-term use in chronic conditions such as DED.

14.6.2 Potential side effects and safety considerations

Although *D. nobile* has shown promise as a therapeutic agent for DED, it is important to consider potential side effects and safety considerations associated with its use. Although *D. nobile* is generally regarded as safe when consumed in moderate amounts as a dietary supplement or functional food ingredient, some individuals may experience adverse reactions such as gastrointestinal upset, allergic reactions or drug interactions. Additionally, the safety of long-term or high-dose *D. nobile* supplementation has not been extensively studied, and further research is needed to establish its safety profile in various populations, including pregnant or breastfeeding women, children, and individuals with underlying health conditions. Furthermore, quality control measures and standardized extraction methods should be employed to ensure the purity, potency and safety of *D. nobile* products.

14.6.3 Interaction with medications

Limited information is available regarding potential interactions between *D. nobile* and medications commonly used to treat DED or

Table 14.2. Preclinical and clinical studies (authors' own table).

Publication year	Study topic	Study type	Objective	Main findings
Preclinical studies				
2017	Comparison of the efficacy of *D. nobile* extracts with conventional dry eye treatments	Animal model	Compare therapeutic effects of *D. nobile* extracts with artificial tears and cyclosporin eye drops	*D. nobile* extracts showed comparable efficacy to conventional treatments in an experimental animal model of dry eye.
2018	Investigating the wound-healing potential of *D. nobile* extracts on corneal epithelial cells	*In vitro*	Assess the capacity of *D. nobile* extracts to enhance migration and proliferation of corneal epithelial cells	*D. nobile* extracts demonstrated potential to promote corneal epithelial cell migration and proliferation, essential for corneal wound healing.
2022	The anti-inflammatory potential of *D. nobile* extracts	*In vitro*	Investigate the effect of *D. nobile* extracts on inflammatory cytokines and chemokines in cultured corneal epithelial cells	*D. nobile* extracts showed potential to modulate inflammatory markers in an *in vitro* model simulating ocular surface inflammation.
2022	Assessing the anti-inflammatory properties of *D. nobile* extracts	Animal model	Examine the influence of *D. nobile* extracts on inflammatory markers in a murine model of DED	*D. nobile* extracts demonstrated anti-inflammatory effects by modulating TNF-α, IL-6 and COX-2 in mice with dry eye induced by desiccating stress.
2023	Assessment of the neuroprotective properties of *D. nobile* compounds on corneal nerves	Animal model	Investigate the potential of *D. nobile* compounds to protect corneal nerves from degeneration	*D. nobile* compounds showed promise in preserving corneal nerves and enhancing sensory function in an animal model of dry eye.
2024	Antioxidant properties of *D. nobile* polysaccharides	Animal model	Evaluate the antioxidant activity of *D. nobile* polysaccharides in a murine model of dry eye	*D. nobile* polysaccharides exhibited antioxidant effects and reduced oxidative stress markers in ocular tissues in an animal model of dry eye.
Clinical studies				
2017	Long-term outcomes of *D. nobile* eye drops in patients with chronic dry eye	Prospective cohort	Follow patients with chronic dry eye treated with *D. nobile* eye drops over 2 years	*D. nobile* eye drops showed long-term benefits in improving clinical parameters and quality-of-life outcomes in patients with chronic dry eye.

Continued

Table 14.2. Continued.

Publication year	Study topic	Study type	Objective	Main findings
2018	Long-term safety and efficacy of *D. nobile* eye drops in chronic dry eye	Prospective observational study	Assess long-term safety and efficacy of *D. nobile* eye drops in patients with chronic dry eye	*D. nobile* eye drops demonstrated sustained effectiveness and safety over a 12-month period in individuals with chronic dry eye.
2022	Impact of *D. nobile* supplementation on tear film composition	Exploratory	Investigate changes in tear film composition in dry eye patients before and after *D. nobile* supplementation	*D. nobile* supplementation showed potential to impact tear film composition, including lipid layer thickness and cytokine levels, in dry eye patients.
2023	Efficacy of *D. nobile* eye drops in moderate to severe DED	Randomized controlled trial	Evaluate effectiveness of *D. nobile* eye drops compared to placebo in improving dry eye symptoms	*D. nobile* eye drops showed potential in improving both objective and subjective measures of dry eye symptoms in patients with moderate to severe DED.
2023	Effectiveness of *D. nobile* eye drops compared to conventional treatments in severe DED	Randomized controlled trial	Compare effectiveness of *D. nobile* eye drops with standard treatments in severe dry eye patients	*D. nobile* eye drops exhibited comparable efficacy to conventional treatments in patients diagnosed with severe DED.

other ocular conditions. However, caution should be exercised when using *D. nobile* supplements or eye drops concurrently with other medications, particularly those with known ocular or systemic side effects. Patients should inform their healthcare providers about any medications, supplements or herbal products they are currently taking to minimize the risk of potential drug interactions. Healthcare providers should carefully evaluate the safety and appropriateness of *D. nobile*-based therapies in patients with complex medication regimens or underlying medical conditions, ensuring optimal patient care and treatment outcomes. Overall, safety and efficacy evaluations of *D. nobile* underscore its potential as a promising therapeutic approach for DED. Further research is warranted to elucidate the long-term safety profile and potential interactions of *D. nobile*-based therapies and to optimize their use in clinical practice. By addressing safety considerations and potential side effects, healthcare providers can ensure the safe and effective integration of *D. nobile* into comprehensive dry eye management strategies, ultimately improving patient outcomes and quality of life.

14.7 Future Directions and Implications

Despite the promising findings from preclinical and clinical studies, there are several areas for future research to further elucidate the therapeutic potential of *D. nobile* for DED. Research into *D. nobile* for DED is advancing across several fronts. Studies are elucidating the specific mechanisms of action underlying its therapeutic effects, particularly its interactions with pathways involved in tear film stability, inflammation and ocular surface protection. Efforts are also focused on optimizing formulations and dosage regimens of *D. nobile* products to enhance efficacy, improve patient adherence and minimize side effects. Explorations into potential synergistic effects with other botanicals or conventional treatments like artificial tears and anti-inflammatory medications are underway to maximize treatment outcomes for DED. Additionally, there is a push towards conducting large-scale, multicentre clinical trials to further assess the safety, efficacy and long-term outcomes of *D. nobile* across diverse populations with DED. Researchers are also investigating its potential as a preventive or adjunctive therapy for related ocular conditions such as meibomian gland dysfunction and other ocular surface disorders, aiming to expand its therapeutic applications in ophthalmology. Overall, continued research in these areas will help to advance our understanding of the therapeutic potential of *D. nobile* and its role in the management of DED.

14.7.1 Areas for further research

As research on *D. nobile* as a potential therapeutic approach for DED continues to progress, several future directions and implications emerge, offering valuable insights into its clinical applications and research priorities. These aspects are crucial for advancing our understanding of *D. nobile*'s potential efficacy and safety profile in DED management, as well as guiding future research endeavours and clinical practice. Further investigation into the underlying mechanisms of action of *D. nobile* in the pathogenesis of DED is crucial for advancing its therapeutic potential. Detailed studies are needed to elucidate how *D. nobile* interacts with specific cellular receptors, signalling pathways and inflammatory mediators that contribute to ocular surface inflammation and tear film instability. Understanding these mechanisms will provide valuable insights into the precise therapeutic effects of *D. nobile* and guide the development of targeted interventions for DED. Additionally, comprehensive research on the bioavailability and pharmacokinetics of *D. nobile* compounds is essential. Studies should focus on determining the absorption, distribution, metabolism and excretion of its constituents to optimize dosage regimens and formulation strategies. Improved knowledge of *D. nobile*'s pharmacokinetic profile will facilitate the development of effective delivery systems, ensuring enhanced bioavailability and therapeutic efficacy in the management of DED. To ascertain the efficacy and safety of therapies based on *D. nobile* for individuals with DED, it is imperative to conduct rigorous clinical trials. Large-scale randomized controlled trials, spanning diverse patient cohorts and employing standardized outcome measures, are indispensable for establishing the clinical efficacy of *D. nobile* and comparing its effectiveness with current treatments. Moreover, long-term follow-up investigations are critical for evaluating the sustained efficacy and safety of *D. nobile*-based interventions over extended periods.

Continued monitoring of the long-term safety profile of *D. nobile*-based therapies is essential. Longitudinal studies should evaluate the incidence and severity of adverse effects associated with prolonged use of *D. nobile* supplements or topical formulations. Additionally, post-marketing surveillance and pharmacovigilance programmes can help identify and mitigate any potential safety concerns that may arise in real-world clinical practice. Exploration of combination therapies involving *D. nobile* and other therapeutic agents used in DED management is warranted. Investigating potential synergistic effects between *D. nobile* and existing treatments such as artificial tears, anti-inflammatory agents or tear film stabilizers may enhance overall treatment outcomes and improve patient satisfaction. Additionally, studies assessing the compatibility and safety of combining *D. nobile* with other botanical extracts or nutraceuticals could provide novel therapeutic strategies for DED management. In summary, future research on *D. nobile* in the context of DED should focus on mechanistic elucidation, bioavailability and pharmacokinetics, clinical validation, long-term safety profiling and exploration of combination therapies.

14.7.2 Potential clinical applications

D. nobile-based therapies may serve as primary treatment options for mild to moderate cases of DED, particularly in patients seeking natural or botanical alternatives to conventional medications. Its anti-inflammatory, antioxidant and neuroprotective properties make it a promising option for addressing underlying pathophysiological mechanisms. *D. nobile* can be integrated as adjunctive therapy in comprehensive DED management protocols. Combining *D. nobile* with existing treatments such as artificial tears, lubricating ointments or anti-inflammatory agents may provide synergistic benefits and improve overall treatment outcomes. Prophylactic use of *D. nobile* supplements or eye drops may be considered in individuals at risk of developing DED due to environmental factors, ageing or systemic conditions. Its potential to preserve tear film integrity, reduce oxidative stress and protect corneal nerves may mitigate the onset or progression of DED in susceptible populations. Tailoring *D. nobile*-based therapies to individual patient characteristics, including disease severity, aetiology and treatment preferences, can optimize therapeutic efficacy and patient satisfaction. Personalized treatment approaches guided by comprehensive clinical assessments and biomarker profiling may enhance treatment outcomes and minimize adverse effects. Overall, future research endeavours and clinical applications of *D. nobile* hold significant promise for advancing the management of DED. By addressing key research gaps, exploring novel therapeutic avenues and embracing personalized treatment strategies, *D. nobile* has the potential to revolutionize DED care and improve the quality of life for millions of affected individuals worldwide.

14.8 Conclusion and Recommendations

In conclusion, *D. nobile* holds promise as a potential therapeutic approach for DED, offering a natural and alternative option for patients who may not respond adequately to conventional treatments. The bioactive compounds present in *D. nobile* have been shown to possess anti-inflammatory, antioxidant and mucin-stimulating properties, which may address multiple aspects of the pathophysiology of DED and provide comprehensive relief from dry eye symptoms. Both preclinical and clinical investigations have shown that *D. nobile* is safe and effective in enhancing tear film stability, diminishing inflammation of the ocular surface and easing symptoms of dry eye. Nevertheless, additional research is warranted to comprehensively understand its mechanisms of action, enhance its therapeutic benefits and ascertain its safety over the long term. With careful consideration of potential side effects and safety considerations, *D. nobile* has the potential to emerge as a valuable addition to the armamentarium of treatments for DED, improving the quality of life for patients affected by this common and burdensome ocular condition.

The exploration of *D. nobile* as a potential therapeutic approach for DED holds significant promise, with emerging evidence suggesting its efficacy in alleviating ocular surface inflammation, improving tear film stability and enhancing overall ocular health. Through mechanistic elucidation, clinical validation and long-term safety profiling, *D. nobile* has demonstrated potential as a natural remedy for DED management, offering a promising alternative or adjunctive therapy to conventional treatments. Based on current research and clinical evidence, several recommendations emerge for the use of *D. nobile* in managing DED. Clinicians are advised to integrate *D. nobile*-based therapies into comprehensive DED management protocols, particularly for patients with mild to moderate symptoms or those preferring natural alternatives to conventional medications. Close monitoring of treatment response and potential adverse effects is critical to ensure optimal patient outcomes. Personalizing *D. nobile*-based therapies based on individual patient characteristics such as disease severity, aetiology and treatment preferences can enhance therapeutic efficacy and patient satisfaction through shared decision-making. Continued surveillance of the long-term safety profile of *D. nobile* supplements or topical formulations is essential, with clinicians remaining vigilant for any adverse effects associated with prolonged use. Collaborative research efforts involving clinicians, scientists and industry partners are needed to deepen understanding of *D. nobile*'s therapeutic mechanisms, refine formulation strategies and validate clinical efficacy in larger patient cohorts. Increasing public health awareness through educational initiatives aimed at patients, healthcare professionals and policy makers is crucial to ensure informed decisions regarding the benefits, safety and appropriate use of *D. nobile* in DED management.

Acknowledgements

The authors express their gratitude for the financial support received through the Distinguished High-Level Talents Research Grant from the Guizhou Science and Technology Corporation Platform Talents Fund (grant nos. [2017]5733-001 and CK-1130-002), the National Natural Science Foundation of China (U1812403, 82373981 and 82060750), 2011 Collaborative Innovation Centre of Traditional Chinese Medicine in Guizhou Province (no. [2022]026) and the support provided by the Zunyi Medical University, China. Special appreciation is extended to all laboratory colleagues and research staff members for their valuable insights, constructive guidance and assistance with the contents of this chapter.

References

Alves, M., Asbell, P., Dogru, M., Giannaccare, G., Grau, A. *et al.* (2023) TFOS lifestyle report: impact of environmental conditions on the ocular surface. *The Ocular Surface* 29, 1–52.

Asiedu, K. (2022) Role of ocular surface neurobiology in neuronal-mediated inflammation in dry eye disease. *Neuropeptides* 95, 102266.

Best, A.-L., Labetoulle, M., Legrand, M., M'garrech, M., Barreau, E. *et al.* (2019) Punctal and canalicular plugs: indications, efficacy and safety. *Journal Francais d'Ophtalmologie* 42(3), e95–e104.

Bu, J., Liu, Y., Zhang, R., Lin, S. and Zhuang, J. (2024) Potential new target for dry eye disease-oxidative stress. *Antioxidants* 13(4), 1–22.

Bustamante-Arias, A., Ruiz Lozano, R.E. and Rodriguez-Garcia, A. (2022) Dry eye disease, a prominent manifestation of systemic autoimmune disorders. *European Journal of Ophthalmology* 32(6), 3142–3162.

Cakova, V., Bonte, F. and Lobstein, A. (2017) *Dendrobium:* sources of active ingredients to treat age-related pathologies. *Aging and Disease* 8(6), 827–849.

Calonge, M., Pinto-Fraga, J., González-García, M.J., Enríquez-de-Salamanca, A., López-de la Rosa, A. *et al.* (2017) Effects of the external environment on dry eye disease. *International Ophthalmology Clinics* 57(2), 23–40.

Camacho, V., Kuznetsova, V. and Welner, R.S. (2021) Inflammatory cytokines shape an altered immune response during myeloid malignancies. *Frontiers in Immunology* 12, 1–16.

Chao, W.-H., Lai, M.-Y., Pan, H.-T., Shiu, H.-W., Chen, M.-M. *et al.* (2018) *Dendrobium nobile* Lindley and its bibenzyl component moscatilin are able to protect retinal cells from ischemia/hypoxia by dowregulating placental growth factor and upregulating Norrie disease protein. *BMC Complementary and Alternative Medicine* 18(1), 16.

Chen, C., Yu, L.-T., Cheng, B.-R., Xu, J.-L., Cai, Y. *et al.* (2022) Promising therapeutic candidate for myocardial ischemia/reperfusion injury: what are the possible mechanisms and roles of phytochemicals? *Frontiers in Cardiovascular Medicine* 8, 1–31.

Deo, N. and Nagrale, P. (2024) Gnosis, and prevalence by age, sex, and race. *Cureus* 16(2), 1–8.

Donthineni, P.R., Shanbhag, S.S. and Basu, S. (2021) An evidence-based strategic approach to prevention and treatment of dry eye disease, a modern global epidemic. *Healthcare* 9(1), 89.

Downie, L.E., Ng, S.M., Lindsley, K.B. and Akpek, E.K. (2019) Omega-3 and omega-6 polyunsaturated fatty acids for dry eye disease. *The Cochrane Database of Systematic Reviews* 12, 1465–1858.

Firoozbahr, M., Kingshott, P., Palombo, E.A. and Zaferanloo, B. (2023) Recent advances in using natural antibacterial additives in bioactive wound dressings. *Pharmaceutics* 15(2), 644.

Fu, X., Chen, S., Xian, S., Wu, Q., Shi, J. *et al.* (2023) Dendrobium and its active ingredients: emerging role in liver protection. *Biomedicine & Pharmacotherapy* 157, 1–13.

Giannaccare, G., Ghelardini, C., Mancini, A., Scorcia, V. and Di Cesare Mannelli, L. (2021) New perspectives in the pathophysiology and treatment of pain in patients with dry eye disease. *Journal of Clinical Medicine* 11(1), 1–8.

Gomes, J.A.P. and Santo, R.M. (2019) The impact of dry eye disease treatment on patient satisfaction and quality of life: a review. *The Ocular Surface* 17(1), 9–19.

Gorimanipalli, B., Khamar, P., Sethu, S. and Shetty, R. (2023) Hormones and dry eye disease. *Indian Journal of Ophthalmology* 71(4), 1276–1284.

Guo, Y., Kong, Y., Sun, J., Jiao, Y., Hong, Y. *et al.* (2024) Alleviation of ultraviolet-B radiation-induced photoaging using *Saussurea medusa* maxim polysaccharide. *Photochemistry and Photobiology* 100(3), 622–632.

Han, Y., Guo, S., Li, Y., Li, J., Zhu, L. *et al.* (2023) Berberine ameliorate inflammation and apoptosis via modulating PI3K/AKT/NFκB and MAPK pathway on dry eye. *Phytomedicine* 121, 155081.

Harrell, C.R., Feulner, L., Djonov, V., Pavlovic, D. and Volarevic, V. (2023) The molecular mechanisms responsible for tear hyperosmolarity-induced pathological changes in the eyes of dry eye disease patients. *Cells* 12(23), 2755.

He, Q., Xiao, L., Shi, Y., Li, W. and Xin, X. (2023) Natural products: protective effects against ischemia-induced retinal injury. *Frontiers in Pharmacology* 14, 1149708.

Hsu, W.-H., Sangkhathat, C., Lu, M.-K., Lin, W.-Y., Liu, H.-P. *et al.* (2024) *Dendrobium nobile* polysaccharide attenuates blue light-induced injury in retinal cells and in vivo in *Drosophila*. *Antioxidants* 13(5), 603.

Huang, Q., Zhan, M. and Hu, Z. (2023) Auricular acupressure for dry eye disease: a systematic review and meta-analysis of randomized controlled trials. *Medicina (Kaunas, Lithuania)* 59(1), 177.

Kawahara, A. (2023) Treatment of dry eye disease (DED) in Asia: strategies for short tear film breakup time-type DED. *Pharmaceutics* 15(11), 1–17.

Kim, M., Lee, Y., Mehra, D., Sabater, A.L. and Galor, A. (2021) Dry eye: why artificial tears are not always the answer. *BMJ Open Ophthalmology* 6(1), e000697.

Koh, S., Rao, S.K., Srinivas, S.P., Tong, L. and Young, A.L. (2022) Evaluation of ocular surface and tear function - A review of current approaches for dry eye. *Indian Journal of Ophthalmology* 70(6), 1883–1891.

Kojima, T., Dogru, M., Kawashima, M., Nakamura, S. and Tsubota, K. (2020) Advances in the diagnosis and treatment of dry eye. *Progress in Retinal and Eye Research* 78, 100842.

Kumar, A., Nirmal, P., Kumar, M., Jose, A., Tomer, V. *et al.* (2023a) Major phytochemicals: recent advances in health benefits and extraction method. *Molecules* 28(2), 887.

Kumar, N.R., Praveen, M., Narasimhan, R., Khamar, P., D'Souza, S. *et al.* (2023b) Tear biomarkers in dry eye disease: progress in the last decade. *Indian Journal of Ophthalmology* 71(4), 1190–1202.

Labetoulle, M., Benitez-Del-Castillo, J.M., Barabino, S., Herrero Vanrell, R., Daull, P. *et al.* (2022) Artificial tears: biological role of their ingredients in the management of dry eye disease. *International Journal of Molecular Sciences* 23(5), 2434.

Lamb, C.A., Saifuddin, A., Powell, N. and Rieder, F. (2022) The future of precision medicine to predict outcomes and control tissue remodeling in inflammatory bowel disease. *Gastroenterology* 162(5), 1525–1542.

Li, W., Mu, X., Wu, X., He, W., Liu, Y. *et al.* (2022a) *Dendrobium nobile* protects retinal cells from UV-induced oxidative stress damage via nrf2/HO-1 and MAPK pathways. *Journal of Ethnopharmacology* 288, 1–10.

Li, W., Mu, X., Wu, X., He, W., Liu, Y. *et al.* (2022b) *Dendrobium nobile* Lindl. polysaccharides protect fibroblasts against UVA-induced photoaging via JNK/c-Jun/MMPs pathway. *Journal of Ethnopharmacology* 298, 115590.

Ling, J., Chan, C.-L., Ho, C.-Y., Gao, X., Tsang, S.-M. *et al.* (2022) The extracts of dendrobium alleviate dry eye disease in rat model by regulating aquaporin expression and MAPKs/NF-κB signalling. *International Journal of Molecular Sciences* 23(19), 11195.

Liu, D., Dong, Z., Xiang, F., Liu, H., Wang, Y. *et al.* (2020) Dendrobium alkaloids promote neural function after cerebral ischemia-reperfusion injury through inhibiting pyroptosis induced neuronal death in both in vivo and in vitro models. *Neurochemical Research* 45(2), 437–454.

Liu, M., Kumar, A., Bukkapatnam, S. and Kuttolamadom, M. (2021) A review of the anomalies in directed energy deposition (DED) processes & potential solutions - part quality & defects. *Procedia Manufacturing* 53, 507–518.

Liu, X., Shi, H. and Li, W. (2024) Review on the potential roles of traditional chinese medicines in the prevention, treatment, and postoperative recovery of age-related cataract. *Journal of Ethnopharmacology* 324, 117786.

Ma, J., Issa, M., Varma, D., Ahmed, I.I.K. (2022) Urgent Virtual Eye Assessments During the COVID-19 Pandemic. *Clinical Opthalmology* 16, 2069–2078.

Messmer, E., Chan, C., Asbell, P., Johnson, G., Sloesen, B. *et al.* (2019) Comparing the needs and preferences of patients with moderate and severe dry eye symptoms across four countries. *BMJ Open Ophthalmology* 4(1), 1–8.

Periman, L.M., Perez, V.L., Saban, D.R., Lin, M.C. and Neri, P. (2020) The immunological basis of dry eye disease and current topical treatment options. *Journal of Ocular Pharmacology and Therapeutics* 36(3), 137–146.

Pflugfelder, S.C. and Stern, M.E. (2020) Biological functions of tear film. *Experimental Eye Research* 197, 108115.

Philip, A.K. (2023) Nanotechnology-based therapeutic strategies for dry eye disease. *Journal of Exploratory Research in Pharmacology* 8(3), 253-262

Prinz, J., Maffulli, N., Fuest, M., Walter, P., Bell, A. *et al.* (2022) Efficacy of topical administration of corticosteroids for the management of dry eye disease: systematic review and meta-analysis. *Life* 12(11), 1932.

Ramesh, T., Koperuncholan, M., Praveena, R., Ganeshkumari, K., Vanithamani, J. *et al.* (2019) Medicinal properties of some dendrobium orchids – a review. *Journal of Applied and Advanced Research* 4(4), 119–128.

Rao, S.K., Mohan, R., Gokhale, N., Matalia, H. and Mehta, P. (2022) Inflammation and dry eye disease-where are we? *International Journal of Ophthalmology* 15(5), 820–827.

Rodriguez-Garcia, A., Babayan-Sosa, A., Ramirez-Miranda, A., Santa Cruz-Valdes, C., Hernandez-Quintela, E. *et al.* (2022) A practical approach to severity classification and treatment of dry eye disease: a proposal from the Mexican Dry Eye disease expert panel. *Clinical Ophthalmology* 16, 1331–1355.

Rolando, M. and Merayo-Lloves, J. (2022) Management strategies for evaporative dry eye disease and future perspective. *Current Eye Research* 47(6), 813–823.

Sheppard, J., Shen Lee, B. and Periman, L.M. (2023) Dry eye disease: identification and therapeutic strategies for primary care clinicians and clinical specialists. *Annals of Medicine* 55(1), 241–252.

Sheppard, J.D. and Nichols, K.K. (2023) Dry eye disease associated with meibomian gland dysfunction: focus on tear film characteristics and the therapeutic landscape. *Ophthalmology and Therapy* 12(3), 1397–1418.

Shimazaki, J. (2018) Definition and diagnostic criteria of dry eye disease: historical overview and future directions. *Investigative Opthalmology & Visual Science* 59(14), DES7.

Silva, B., São Braz, B., Delgado, E. and Gonçalves, L. (2021) Colloidal nanosystems with mucoadhesive properties designed for ocular topical delivery. *International Journal of Pharmaceutics* 606, 120873.

Song, C., Kang, J., Ji, X., Fu, C., Wen, X. *et al.* (2017) Efficient plant regeneration and genetic fidelity assessment of in vitro-derived plants of *Dendrobium nobile*-an endangered medicinal and ornamental herb. *Orchid Science and Biotechnology* 1(2), 51–55.

Suárez-Cortés, T., Merino-Inda, N. and Benitez-Del-Castillo, J.M. (2022) Tear and ocular surface disease biomarkers: a diagnostic and clinical perspective for ocular allergies and dry eye disease. *Experimental Eye Research* 221, 109121.

Sun, Y., Geng, J. and Wang, D. (2021) Cardioprotective effects of ginsenoside compound-mc1 and *Dendrobium nobile* Lindl against myocardial infarction in an aged rat model: involvement of TLR4/NF-κB signaling pathway. *European Journal of Inflammation* 19, 1–9.

Ti, H., Zhuang, Z., Yu, Q. and Wang, S. (2021) Progress of plant medicine derived extracts and alkaloids on modulating viral infections and inflammation. *Drug Design, Development and Therapy* 15, 1385–1408.

Wang, G., Wang, J., Deng, Y., Qin, L., He, Y. *et al.* (2022) Chemical constituents and nutritional health functions of *Dendrobium nobile*: a review. *Food Science and Technology* 42, e84522.

Yang, C.C., Chien, J.Y., Chou, Y.Y., Ciou, J.W. and Huang, S.P. (2022) The effects of *Lycium chinense*, *Cuscuta chinensis*, *Senna tora*, *Ophiopogon japonicus*, and *Dendrobium nobile* decoction on a dry eye mouse model. *Medicina* 58(8), 1134.

Yang, L.C., Liao, J.W., Wen, C.L. and Lin, W.C. (2018) Subchronic and genetic safety assessment of a new medicinal *Dendrobium* species: *Dendrobium* Taiseed Tosnobile in rats. *Evidence-Based Complementary and Alternative Medicine* 2018, 1–12.

Yu, L., Yu, C., Dong, H., Mu, Y., Zhang, R. *et al.* (2021) Recent developments about the pathogenesis of dry eye disease: based on immune inflammatory mechanisms. *Frontiers in Pharmacology* 12, 732887.

Yu, Z., Gong, C., Lu, B., Yang, L., Sheng, Y. *et al.* (2015) *Dendrobium chrysotoxum* Lindl. alleviates diabetic retinopathy by preventing retinal inflammation and tight junction protein decrease. *Journal of Diabetes Research* 2015, 518317.

Zhang, J., Xu, H.X., Zhao, Z.L., Xian, Y.F. and Lin, Z.X. (2021) *Dendrobium nobile* Lindl: a review on its chemical constituents and pharmacological effects. *Chinese Medicine and Culture* 4(4), 235–242.

Zhang, Q., Fu, X., Liu, H., Chen, Y., Chen, S. *et al.* (2023) A systematical review on ethnobotanical, phytochemical and pharmacological aspects of *Dendrobium nobile* Lindl. *Phytochemistry Reviews* 22(3), 743–780.

15 Future Research Directions for *Dendrobium nobile*

Surendra Sarsaiya, Archana Jain, Qihai Gong, Qin Wu and Jingshan Shi*
Key Laboratory of Basic Pharmacology and Joint International Research Laboratory of Ethnomedicine of Ministry of Education, Zunyi Medical University, Zunyi, Guizhou, China

Abstract

Dendrobium nobile, a medicinal orchid widely recognized for its bioactive compounds, holds significant potential in therapeutic and ecological research. This chapter explores key areas for future investigation, including the biosynthesis of its polysaccharides and alkaloids, pharmacological activities and environmental adaptability. Multi-omics approaches, integrating genomics, proteomics and metabolomics, are highlighted as tools to uncover the genetic and biochemical foundations of its medicinal properties, enabling targeted breeding and conservation strategies. Advances in clinical research are discussed, focusing on evaluating the efficacy, safety and dosage optimization of *D. nobile* in various health conditions. The chapter also emphasizes the importance of sustainable cultivation, ecological conservation and quality control to ensure the plant's availability for research and medicinal applications. Through interdisciplinary collaborations and advanced experimental methods, the chapter outlines a roadmap for enhancing the therapeutic applications of *D. nobile* while addressing challenges in biodiversity conservation and translational research.

15.1 Introduction

Dendrobium nobile Lindl., a prominent medicinal orchid, has garnered significant attention due to its bioactive compounds and traditional applications in Chinese medicine. Future research directions for this species can be categorized into several key areas, each contributing to a deeper understanding of its therapeutic potential and ecological sustainability. One promising avenue is the exploration of the biosynthesis of bioactive compounds. Understanding the metabolic pathways that lead to the production of polysaccharides and alkaloids in *D. nobile* can enhance the efficiency of extracting these valuable substances for medicinal use (Zhang *et al.*, 2023a; Jain *et al.*, 2024). This research could also facilitate the development of biotechnological approaches to optimize the cultivation of this *D. nobile*, ensuring a sustainable supply of its medicinal components. Pharmacological studies are essential for elucidating the therapeutic effects of *D. nobile*. Investigating its antioxidant, anti-cancer and immune-regulating activities can provide insights into its potential applications in modern medicine. Future research should focus on clinical trials to validate these properties and assess the efficacy of *D. nobile* extracts in various health conditions. The integration

*Corresponding author: shijs@zmu.edu.cn

© Jingshan Shi, Archana Jain, Surendra Sarsaiya, Qin Wu and Qihai Gong 2025. Dendrobium nobile: *Research and Therapeutic Applications* (J. Shi, A. Jain, S. Sarsaiya, Q. Wu and Q. Gong)
DOI: 10.1079/9781800627987.0015

of multi-omics approaches presents another significant opportunity for advancing research on *D. nobile*. By combining genomics, transcriptomics, proteomics and metabolomics, researchers can gain a comprehensive understanding of the complex biological systems in *D. nobile* (Chen *et al.*, 2021; Li *et al.*, 2022a). Such integrative studies can reveal the genetic and biochemical bases of its medicinal properties, paving the way for targeted breeding programmes and conservation strategies. Moreover, chromosome-level genome assembly can provide critical insights into the genetic structure and function of *D. nobile*, facilitating genetic diversity analysis (Wu *et al.*, 2023). Understanding the genetic variation within populations is vital for informing conservation efforts and ensuring the resilience of this species in the face of environmental changes. Environmental stress adaptation is another crucial area of research. Investigating how *D. nobile* responds to salinity, drought and temperature fluctuations can inform cultivation practices and enhance its survival in changing climates (Cruz *et al.*, 2023). This knowledge is essential for developing ecological conservation strategies that preserve both the natural habitats and genetic diversity of *D. nobile*. The future research on *D. nobile* should focus on its biosynthesis, pharmacological properties, multi-omics integration, genetic diversity and environmental stress responses. These areas not only promise to enhance our understanding of this valuable species but also contribute to its sustainable use and conservation in the face of global environmental challenges (Guo *et al.*, 2023; Zhang *et al.*, 2023a).

15.1.1 Current state of *D. nobile* research

The current state of research on *D. nobile* is characterized by growing interest and significant advancements in various aspects of this medicinal plant. Researchers are exploring the traditional knowledge and ethnomedicinal uses of *D. nobile* (see Table 15.1) and investigating its historical use, cultural significance and traditional medicinal practices across various cultures (Ramesh *et al.*, 2019). Extensive phytochemical research has identified a range of bioactive compounds in *D. nobile*, including polysaccharides, alkaloids, phenols, flavonoids and other secondary metabolites, that exhibit potential therapeutic properties (Zhang *et al.*, 2023b). Additionally, *D. nobile* has garnered significant interest in the skincare and cosmeceutical industries because of its promising antioxidant, anti-inflammatory, moisturizing and anti-ageing effects, as well as its potential to enhance skin tone and texture (Cruz *et al.*, 2023). Biotechnological techniques such as tissue culture and micropropagation have been employed to achieve large-scale production of *D. nobile*, with ongoing research aimed at optimizing these methods to enhance plant growth, multiplication and yield (Zhang *et al.*, 2022a). Pharmacological investigations have evaluated the therapeutic potential of *D. nobile* in various health conditions, including its antimicrobial, anti-cancer, immunomodulatory, neuroprotective and anti-diabetes activities (Wang *et al.*, 2023). Researchers have also elucidated the molecular mechanisms underlying its biological activities by exploring its interactions with cellular pathways, receptor binding and gene expression modulation (Cakova *et al.*, 2017). Additionally, formulation studies have focused on developing innovative delivery systems and product formulations, such as nanoencapsulation and liposomal delivery, to improve the bioavailability, stability and efficacy of *D. nobile* extracts (Wei *et al.*, 2024).

Efforts are underway to establish quality control parameters and standardization methods for *D. nobile* products, including developing analytical methods to ensure the consistency and authenticity of herbal preparations and extracts (Luo *et al.*, 2024). Clinical trials are evaluating the efficacy, safety and dosage regimens of *D. nobile* in human subjects with the aim of generating evidence-based data to support its therapeutic use in various health conditions (Wu *et al.*, 2023). Future research on *D. nobile* appears promising, with potential advancements in exploring additional therapeutic applications, optimizing cultivation techniques, investigating novel bioactive compounds and elucidating their mechanisms of action (Fan *et al.*, 2023). Overall, the current state of *D. nobile* research is dynamic and multidisciplinary, encompassing traditional knowledge, phytochemistry, biotechnology, pharmacology, formulation development and clinical studies. This research paves the way for the development of evidence-based therapeutic applications and commercial products derived from *D. nobile* (Table 15.1).

Table 15.1. Current state of *D. nobile* research (authors' own table).

Aspect of research	Key points	Reference
Traditional knowledge	Researchers explore historical use, cultural significance and traditional medicinal practices associated with *D. nobile*.	Ramesh *et al.* (2019)
Phytochemistry and bioactive compounds	Extensive research identifies polysaccharides, alkaloids, phenols, flavonoids and other bioactive compounds with potential therapeutic properties.	Zhang *et al.* (2023b)
Skincare and cosmeceuticals	*D. nobile* shows potential benefits for skin health, including antioxidant, anti-inflammatory, moisturizing and anti-ageing properties. Researchers investigate its effects on skin tone and texture improvement.	Cruz *et al.* (2023)
Tissue culture and biotechnology	Biotechnological techniques like tissue culture and micropropagation are used for large-scale production. Efforts focus on enhancing plant growth, multiplication and yield.	Zhang *et al.* (2022a)
Pharmacological studies	Evaluation of antimicrobial, anti-cancer, immunomodulatory, neuroprotective and anti-diabetes activities reveals therapeutic potential in various health conditions.	Wang *et al.* (2023)
Mechanistic studies	Research elucidates molecular mechanisms underlying therapeutic effects, including interactions with cellular pathways, receptor binding and modulation of gene expression.	Cakova *et al.* (2017)
Formulation development	Focus on developing innovative delivery systems and product formulations to improve bioavailability, stability and efficacy. Techniques include nanoencapsulation and liposomal delivery.	Wei *et al.* (2024)
Quality control and standardization	Efforts to establish quality control parameters and standardization methods to ensure consistency and authenticity of *D. nobile* products.	Luo *et al.* (2024)
Clinical trials	Clinical trials evaluate efficacy, safety and dosage regimens in human subjects, generating evidence-based data to support therapeutic use in various health conditions.	Wu *et al.* (2023)
Future perspectives	Potential for further advancements in exploring additional therapeutic applications, optimizing cultivation techniques, investigating novel bioactive compounds and understanding mechanisms of action.	Fan *et al.* (2023)

15.1.2 Significance and importance of future research

Future research on the orchid *D. nobile*, commonly known as the noble dendrobium, 'shi hu' or 'shi hu gen', has significance and importance in several respects. This section discusses the potential benefits and need for continued investigation in the field of *D. nobile*. *D. nobile* has been widely used in traditional Chinese medicine for centuries owing to its potential medicinal properties. However, further research is needed to fully understand its pharmacological effects, active compounds and mechanisms of action (Sarsaiya *et al.*, 2020). Future research can explore its potential therapeutic applications, such as antioxidant, anti-inflammatory, immunomodulatory and anti-cancer properties, which can contribute to the development of new drugs or complementary therapies. *D. nobile* is a threatened species because of its overharvesting and habitat loss. Future research can contribute to conservation efforts by studying its ecology, distribution and population dynamics (Hinsley *et al.*, 2018). Additionally, research on sustainable cultivation techniques, such as tissue culture, micropropagation and optimized cultivation conditions, can help reduce the dependence on wild populations and promote commercial production while conserving natural resources (Laishram *et al.*, 2019). Understanding the genetic diversity of *D. nobile* populations is crucial for conservation and breeding programmes (He *et al.*, 2022). Future research can utilize molecular techniques such as DNA markers and genomics to assess genetic variability, population structure and gene flow patterns within and among populations (Bhattacharyya and Kumaria, 2015). This knowledge can aid in the development of breeding strategies to enhance desirable traits such as flower colour, size, fragrance and growth characteristics (He *et al.*, 2022).

D. nobile cultivation is an important economic activity in many regions. Future research should focus on optimizing agronomic practices, such as substrate selection, nutrient management, irrigation techniques, and pest and disease control, to enhance the yield and quality of *D. nobile* plants (Ketsa and Warrington, 2023). This study can provide valuable guidance to farmers, growers and horticulturists for sustainable production and commercial success. Advancements in molecular biology tools and techniques have opened up new avenues for future research on *D. nobile* (Sarsaiya *et al.*, 2024). Functional genomics and metabolomics approaches can unravel the genes, enzymes and metabolic pathways involved in key physiological processes and biosynthesis of bioactive compounds. This knowledge can facilitate targeted genetic manipulation, metabolic engineering and biotechnological applications in *D. nobile* research (Tiwari *et al.*, 2022). In conclusion, future research on *D. nobile* will offer significant potential benefits in the areas of medicinal properties, bioactive compounds, conservation, genetic diversity, agronomic practices and functional genomics. Continued investigations in these areas will contribute to the understanding, utilization and sustainable management of *D. nobile*, thereby benefiting certain industries, conservation efforts, and human health and well-being.

15.2 Promising Areas for Future Research

D. nobile, an orchid species with a long history of traditional use, presents several promising areas for future research (Fig. 15.1). Exploring these areas will deepen our understanding of the therapeutic potential of *D. nobile* and expand its applications in various fields. Further research is required to understand the pharmacokinetic properties of *D. nobile* and its bioactive compounds in the human body (Fu *et al.*, 2023). Investigating

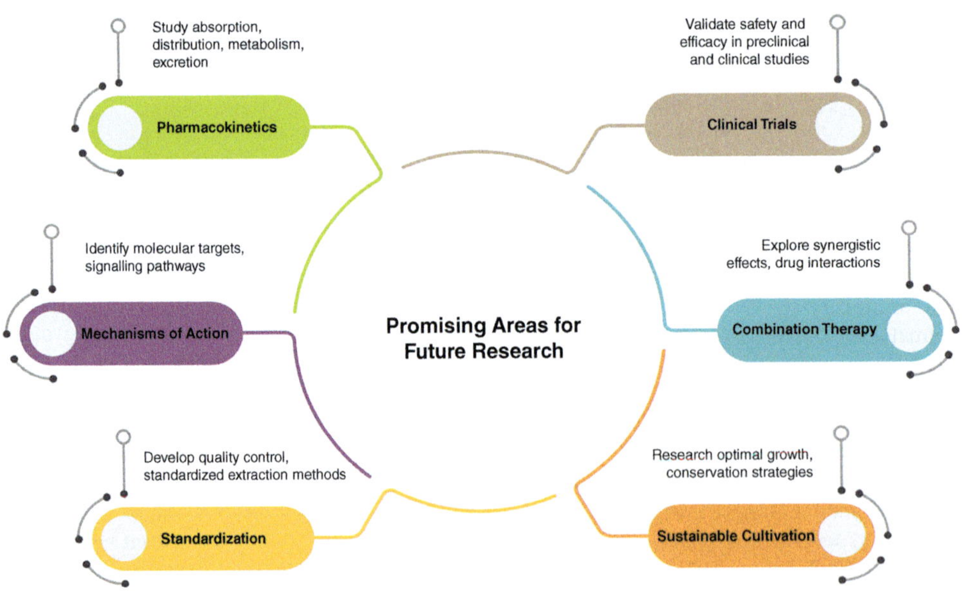

Fig. 15.1. Promising areas for future research (authors' own figure).

the absorption, distribution, metabolism and excretion of *D. nobile* will provide valuable insights into optimizing dosage regimens and developing appropriate formulations for therapeutic use (Wu *et al.*, 2023). Although several bioactive compounds have been identified in *D. nobile*, their specific mechanisms of action remain largely unknown (Li *et al.*, 2024). Future research should focus on elucidating the molecular targets and signalling pathways involved in the therapeutic effects of *D. nobile*, providing a foundation for targeted drug development (Xu *et al.*, 2023). The standardization of *D. nobile* products is essential to ensure their safety, efficacy and reproducibility. Future research should focus on establishing quality control parameters, including the identification and quantification of key bioactive compounds, and developing standardized extraction and purification methods. Preclinical and clinical trials will provide scientific evidence regarding the efficacy and safety of *D. nobile* in various therapeutic applications. These studies can validate its traditional uses and explore new indications, paving the way for evidence-based clinical practice (Li *et al.*, 2022b). *D. nobile* has the potential to be used in combination with other herbal medicines or conventional drugs to enhance therapeutic outcomes. Future research should explore the synergistic effects and potential drug interactions when *D. nobile* is used in combination therapy, leading to the development of more effective treatment strategies (Zhang *et al.*, 2023c). As *D. nobile* is a wild-harvested plant, there is a need for research focused on sustainable cultivation practices and conservation efforts. Investigating optimal growth conditions, cultivation techniques and conservation strategies will contribute to the long-term availability and sustainable use of *D. nobile* (Ninawe and Swapna, 2017). By addressing these promising areas for future research, it can uncover the full potential of *D. nobile* and advance its applications in medicine and healthcare.

15.3 Advanced Experimental Methods and Techniques

The advancement of experimental methods and techniques has played a crucial role in developing our understanding of *D. nobile* and in exploring its therapeutic potential. By employing cutting-edge approaches, researchers have been able to delve deeper into the pharmacological and biochemical aspects of this orchid. Metabolomics allows for comprehensive analysis of small molecules in biological samples. Using techniques such as gas chromatography-mass spectrometry (GC-MS) and liquid chromatography-mass spectrometry (LC-MS), researchers can identify and quantify the diverse metabolites present in *D. nobile* (Jie *et al.*, 2023). Metabolomic studies have facilitated the identification of bioactive compounds and provided insights into the metabolic pathways involved in the therapeutic effects of *D. nobile*. Proteomics enables large-scale study of proteins and their functions within a biological system. Through techniques such as two-dimensional gel electrophoresis (2-DE) and liquid chromatography coupled with tandem mass spectrometry (LC-MS/MS), researchers can identify and quantify the proteins present in *D. nobile* (Lao *et al.*, 2014). Proteomics studies have shed light on protein composition, post-translational modifications and protein–protein interactions, contributing to a comprehensive understanding of the therapeutic mechanisms of *D. nobile* (Yang *et al.*, 2023). Transcriptomics involves high-throughput analysis of RNA transcripts to examine gene expression patterns. Techniques such as RNA sequencing (RNA-seq) have been used to study the transcriptome of *D. nobile* (Li *et al.*, 2022b). Transcriptomic studies have revealed the expression profiles of genes involved in various biological processes and have provided insights into the molecular mechanisms underlying the therapeutic properties of *D. nobile* (Qiu *et al.*, 2023).

Bioactivity-guided fractionation is used to isolate bioactive compounds from complex mixtures. By subjecting *D. nobile* extracts to various separation techniques, such as column chromatography and high-performance liquid chromatography (HPLC), researchers can isolate and identify specific compounds responsible for the observed therapeutic effects (Whaley *et al.*, 2023). This approach allows for the identification of active constituents and determination of their contribution to overall pharmacological activity. Molecular docking and modelling techniques provide insights into the interactions between bioactive compounds and their target proteins. By employing computational methods, researchers can predict the binding affinity and mode of interaction between *D. nobile* compounds and target molecules of interest (Jiaranaikulwanitch *et al.*, 2022). Molecular docking and modelling studies have helped to elucidate the molecular basis of the therapeutic effects of *D. nobile* and facilitate the design of more potent and selective compounds (Zhang *et al.*, 2022b). Plant tissue culture techniques, such as micropropagation and somatic embryogenesis, have been employed to mass produce *D. nobile* plants with desirable traits (Teixeira da Silva and Ng, 2017). Additionally, genetic engineering approaches, including gene transformation and editing, have been utilized to enhance the production of bioactive compounds and improve the overall quality of *D. nobile* (Wang *et al.*, 2020). Utilization of these advanced experimental methods and techniques has significantly advanced our understanding of *D. nobile* and its therapeutic potential. By combining these approaches, researchers can unravel intricate biological processes, identify key molecular targets and develop innovative strategies to harness the therapeutic benefits of *D. nobile* (Fig. 15.2).

15.3.1 Molecular and cellular approaches in *D. nobile*

Molecular and cellular approaches have been instrumental in unravelling the intricate biological mechanisms underlying the therapeutic effects of *D. nobile*. Using these approaches, researchers have gained insights into the molecular pathways, cellular interactions and gene expression profiles associated with the pharmacological effects of this orchid species. Gene expression analysis techniques, such as quantitative real-time polymerase chain reaction (qRT-PCR) and RNA-seq, have been employed to investigate the transcriptional changes induced by *D. nobile* treatment (Gong *et al.*, 2023). By examining the differential expression of genes in response to *D. nobile* extracts or bioactive compounds, researchers can identify key genes and pathways involved in their therapeutic effects (Qiu *et al.*, 2023). Signalling pathways play a crucial role in mediating cellular responses to *D. nobile* treatment (Hsu *et al.*, 2022). Molecular approaches, such as western blotting and immunohistochemistry, have been utilized to investigate the activation or inhibition of specific signalling pathways, including mitogen-activated protein kinase (MAPK), phosphoinositol 3-kinase (PI3K)/Akt and nuclear factor kappa B (NF-κB), in response to *D. nobile* treatment (Hong *et al.*, 2022). These studies provide insights into the molecular mechanisms underlying the therapeutic effects of *D. nobile* (Xu *et al.*, 2022). Cellular imaging techniques, such as confocal microscopy and fluorescence labelling, enable the visualization and localization of *D. nobile* compounds within cells and tissues (Pavlova *et al.*, 2017). These techniques help researchers to understand the cellular uptake, distribution and intracellular localization of bioactive compounds from *D. nobile*. They also provide insights into the

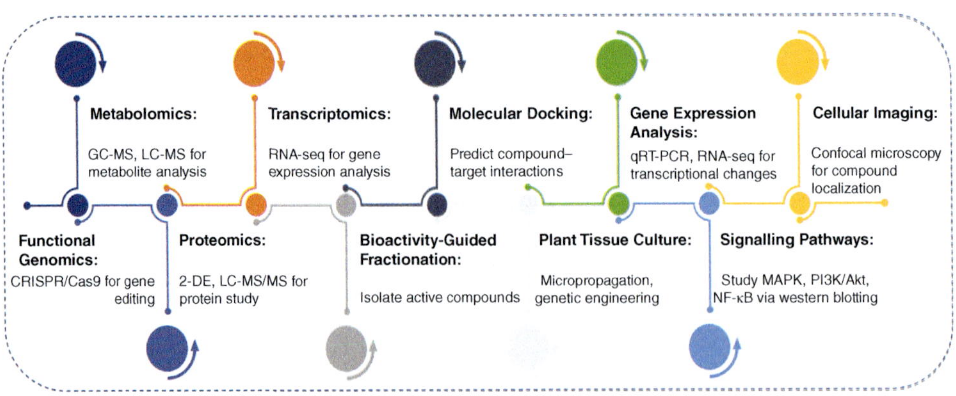

Fig. 15.2. Advanced experimental methods and techniques (authors' own figure). MAPK, mitogen-activated protein kinase; NF-κB, nuclear factor kappa B; PI3K, phosphoinositol 3-kinase; qRT-PCR, quantitative real-time polymerase chain reaction.

interactions between *D. nobile* compounds and cellular components (Zhang *et al.*, 2021a). *In vitro* cell culture models, including primary cell cultures and cell lines, have been used to investigate the cellular effects of *D. nobile* and its bioactive compounds (Favre-Godal *et al.*, 2020). These models allow researchers to study cell viability, proliferation, apoptosis and differentiation in response to *D. nobile* treatment. They also provide a platform for exploring the molecular mechanisms underlying the therapeutic properties of *D. nobile* (Zhang *et al.*, 2023c). Functional genomic approaches, such as gene-knockout or -knockdown techniques, have been used to investigate the functional roles of specific genes or gene families in *D. nobile* (Horvath *et al.*, 2023). These techniques, including CRISPR/Cas9-mediated gene editing, have enabled researchers to elucidate the contributions of specific genes to the therapeutic effects of *D. nobile* and uncover novel targets for intervention (Chen *et al.*, 2021). Metabolic profiling techniques, such as GC-MS and LC-MS, have been employed to study the metabolic changes induced by *D. nobile* treatment (Nongthombam *et al.*, 2024). These approaches enable the identification and quantification of the metabolites involved in the therapeutic effects of *D. nobile*, providing insights into the metabolic pathways and bioactive compounds responsible for its pharmacological properties (Zhang *et al.*, 2023c). By employing these molecular and cellular approaches, researchers have made significant efforts to unravel the molecular mechanisms and cellular interactions underlying the therapeutic potential of *D. nobile*.

15.3.2 Animal models for disease studies

Animal models play a crucial role in investigating the therapeutic potential of *D. nobile* for various diseases. These models provide valuable insights into the efficacy, safety and underlying mechanisms of *D. nobile* treatment. Tumour xenograft models, typically created by implanting human tumour cells into immunodeficient mice, have been used to evaluate the anti-cancer effects of *D. nobile* (Luo *et al.*, 2023). These models enable researchers to assess tumour growth inhibition, metastasis suppression and the potential molecular mechanisms underlying the antitumour activity of *D. nobile*. Animal models of inflammatory diseases such as colitis and arthritis have been employed to study the anti-inflammatory effects of *D. nobile* (Hu *et al.*, 2021). These models, including chemically induced colitis models or collagen-induced arthritis models, allow researchers to evaluate the impact of *D. nobile* treatment on inflammatory markers, histopathological changes and immune responses (Fan *et al.*, 2022). Animal models of neurodegenerative diseases, including Alzheimer's and Parkinson's disease, have been used to investigate the neuroprotective effects of *D. nobile* (He *et al.*, 2020). These models, often induced by neurotoxic agents or genetic manipulations, enable researchers to assess cognitive function, neuronal loss and molecular changes associated with *D. nobile* treatment (Xiong *et al.*, 2021). Animal models of diabetes, such as streptozotocin-induced diabetic rats or mice, have been employed to explore the anti-diabetes effects of *D. nobile* (Hei *et al.*, 2021). These models allow researchers to assess blood glucose levels, insulin resistance and pancreatic beta-cell function following *D. nobile* treatment (Chen *et al.*, 2023). Animal models of wound healing, such as excisional or incisional wound models, have been used to study the wound-healing properties of *D. nobile* (Hong *et al.*, 2022). These models enable researchers to evaluate the rate of wound closure, collagen deposition, angiogenesis and inflammatory responses following *D. nobile* treatment (Xie *et al.*, 2016). Using these animal models, researchers can gain valuable insights into the therapeutic effects and mechanisms of action of *D. nobile* in various diseases.

15.3.3 *In vitro* and *in vivo* pharmacological screening

In vitro and *in vivo* pharmacological screening play a crucial role in evaluating the therapeutic potential of *D. nobile*. These screenings provide valuable insights into the pharmacological activities, safety profiles and underlying mechanisms of *D. nobile* extracts and bioactive compounds. The *in vitro* pharmacological screening of *D. nobile* involves various assays to evaluate its biological effects. Cell viability and cytotoxicity assays, such as the 3-(4,5-dimethylthiazol-2-yl)-2,5-diphenyltetrazolium bromide (MTT) and lactate dehydrogenase release assays, have been used to assess the effects of *D. nobile* on cell viability and cytotoxicity, providing information on its potential toxic effects and helping to determine safe and effective concentrations for further studies (Sut *et al.*, 2017; Xu *et al.*, 2023). Antioxidant activity was evaluated using *in vitro* assays, such as total antioxidant capacity, the 2,2-diphenyl-1-picrylhydrazyl (DPPH) scavenging assay and the superoxide dismutase activity assay, to assess the ability of *D. nobile* extracts or compounds to scavenge free radicals and reduce oxidative stress (Zhang *et al.*, 2020; Rao *et al.*, 2023). Additionally, enzyme inhibition assays, such as acetylcholinesterase and α-glucosidase inhibition assays, have been employed to investigate the inhibitory effects of *D. nobile* on specific enzymes, aiding the identification of potential therapeutic targets and mechanisms of action (Zhai *et al.*, 2022; Li *et al.*, 2023).

The *in vivo* pharmacological screening of *D. nobile* has involved several key studies. Acute toxicity studies in animal models, such as mice or rats, were conducted to evaluate the safety profiles of *D. nobile* extracts or bioactive compounds by determining the maximum tolerated dose and assessing any adverse effects or mortality (Ma *et al.*, 2019). Pharmacokinetic studies in animal models have investigated the absorption, distribution, metabolism and excretion of *D. nobile* compounds, providing insights into their pharmacokinetic properties and bioavailability (Zhang *et al.*, 2021b). Therapeutic efficacy studies use animal models of specific diseases to evaluate the therapeutic potential of *D. nobile* by assessing parameters such as disease progression, biochemical markers, histopathological changes and overall survival (Zhang *et al.*, 2022c). By employing *in vitro* and *in vivo* pharmacological screening techniques, researchers have gained valuable insights into the pharmacological activities and mechanisms of action of *D. nobile*.

15.4 Clinical Trials and Translational Research

Clinical trials and translational research play a crucial role in evaluating the safety, efficacy and potential therapeutic applications of *D. nobile* in humans. These studies bridge the gap between preclinical research and clinical practice by providing valuable insights into the clinical relevance and practical utility of *D. nobile* (see Table 15.2). Li *et al.* (2022c) conducted a randomized controlled trial to evaluate the effect of *D. nobile* extract in improving cognitive function in elderly individuals with mild cognitive impairment. The study demonstrated that *D. nobile* extract significantly improved cognitive performance compared to the placebo group. Wu *et al.* (2023) investigated the effects of *D. nobile* polysaccharides on immune function in patients undergoing colorectal cancer surgery. The study showed that *D. nobile* polysaccharides enhanced immune parameters and reduced postoperative complications, suggesting potential immunomodulatory benefits. Rabidas *et al.* (2023) conducted a prospective clinical study to assess the efficacy of *D. nobile* alkaloids in treating patients with refractory epilepsy. The study revealed that *D. nobile* alkaloids significantly reduced seizure frequency and improved seizure control in patients, indicating their potential as an adjunctive therapy for refractory epilepsy. Long *et al.* (2023) conducted a translational research study to investigate the anti-inflammatory effects of *D. nobile* extract in a mouse model of rheumatoid arthritis. The study demonstrated that *D. nobile* extract effectively suppressed inflammation and attenuated joint damage, suggesting its potential as a therapeutic option for rheumatoid arthritis (Duan *et al.*, 2022). Zhang, *et al.* (2022) conducted a dose–response study to evaluate the effect of different doses of *D. nobile* extract on cognitive function in a rat model of Alzheimer's disease. The study demonstrated a dose-dependent improvement in cognitive performance, with higher doses of *D. nobile* extract leading to greater cognitive benefits. Fu *et al.* (2023) investigated the optimal treatment duration of *D. nobile* polysaccharides in patients with chronic gastritis. The study compared different treatment regimens and found that a 6-week treatment with *D. nobile* polysaccharides resulted in significant improvement in gastritis symptoms and gastric mucosal health.

Zhang *et al.* (2022c) conducted a dose-escalation study to determine the optimal dosage of *D. nobile* alkaloids in patients with rheumatoid arthritis. The study revealed that higher doses of *D. nobile* alkaloids were associated with better control of disease activity and reduction in inflammatory markers. He *et al.* (2023) investigated the treatment regimens of *D. nobile* extract in patients with diabetic retinopathy. The study compared different dosing frequencies and found that a twice-daily regimen of *D. nobile* extract resulted in significant improvement in visual acuity and reduction in retinal lesions. Xiong *et al.* (2021) conducted a prospective study to investigate the association between genetic variations and treatment response to *D. nobile* extract in patients with depression. This study identified specific genetic markers that were predictive of treatment response, allowing for personalized treatment selection and optimization. Peng *et al.* (2022) performed a retrospective analysis of patient data to identify clinical factors associated with favourable responses to *D. nobile* polysaccharides in patients with allergic rhinitis. The study found that patients with a specific phenotype, such as a history of atopy or elevated serum IgE levels, showed better treatment responses to *D. nobile* polysaccharides. Hong *et al.* (2022) conducted a multicentre study to develop a predictive model for therapeutic response to *D. nobile* alkaloids in patients with inflammatory bowel disease. This study combined clinical and molecular factors to create a personalized medicine approach, enabling the identification of patients who are likely to respond favourably to *D. nobile* alkaloids. Wu *et al.* (2023) investigated the role of gut microbiota in mediating the response to *D. nobile* extract in patients with irritable bowel syndrome. This study identified specific gut microbial signatures associated with treatment response, highlighting the potential for microbiome-based patient stratification and personalized treatment approaches. These studies highlight the importance of patient stratification and personalized medicine in maximizing the therapeutic benefits of *D. nobile* by targeting specific patient subgroups or individuals (Table 15.2).

15.5 Future Challenges and Directions

Although research on the therapeutic potential of *D. nobile* has made significant progress, there are still several challenges and areas for further exploration. Addressing these challenges can enhance our understanding of the efficacy, safety and mechanisms of action of *D. nobile*. Clinical trials and human studies are being conducted to evaluate the therapeutic efficacy and safety of *D. nobile* across various disease conditions, exploring the optimal dosage, treatment duration and potential drug interactions. Concurrently, mechanistic studies have investigated the underlying molecular mechanisms of *D. nobile*'s therapeutic effects, elucidating the signalling pathways and cellular targets involved, and exploring interactions with specific disease pathways. Quality control and standardization efforts focus on developing standardized protocols for the cultivation, harvesting and processing of *D. nobile* to ensure consistent quality of herbal products, along with establishing measures to identify and quantify bioactive compounds in *D. nobile* extracts and formulations. The efficacy and safety of different *Dendrobium* species and varieties have been compared to identify the most potent and suitable candidates for therapeutic applications. Studies have focused on assessing the differences in bioactive compound profiles and pharmacological activities among various *Dendrobium* species, with a particular emphasis on *D. nobile*. They explore innovative delivery systems and formulations to enhance the bioavailability and therapeutic efficacy of *D. nobile* compounds while investigating its synergistic effects when combined with other herbal medicines or pharmaceutical agents to achieve improved therapeutic outcomes. Additionally, studies highlight the importance of promoting sustainable cultivation and harvesting practices to ensure the long-term availability of *D. nobile* for research and medicinal applications. Conservation efforts to protect natural populations and maintain biodiversity are also emphasized as critical components for supporting its future use. Addressing these challenges and directions will contribute to further development of *D. nobile* as a potential therapeutic agent. Collaborative research efforts, interdisciplinary approaches and the integration of traditional knowledge with modern scientific methods will be instrumental in advancing our understanding of *D. nobile*'s therapeutic potential.

15.5.1 Regulatory and safety considerations

Similar to any herbal medicine or botanical product, regulatory and safety considerations

Table 15.2. Clinical trials and translational research (authors' own table).

Study design	Patient population	Intervention	Key findings	Reference
Randomized controlled trial	Elderly individuals with mild cognitive impairment	*D. nobile* extract	Improved cognitive performance compared to placebo	Li *et al.* (2022c)
Interventional study	Patients undergoing colorectal cancer surgery	*D. nobile* polysaccharides	Enhanced immune parameters, reduced postoperative complications	Wu *et al.* (2023)
Prospective clinical study	Patients with refractory epilepsy	*D. nobile* alkaloids	Reduced seizure frequency, improved seizure control	Rabidas *et al.* (2023)
Translational research study	Mouse model of rheumatoid arthritis	*D. nobile* extract	Suppressed inflammation, attenuated joint damage	Long *et al.* (2023)
Dose–response study	Rat model of Alzheimer's disease	*D. nobile* extract	Dose-dependent improvement in cognitive performance	Zhang *et al.* (2022c)
Interventional study	Patients with chronic gastritis	*D. nobile* polysaccharides	Significant improvement in gastritis symptoms, gastric mucosal health	Fu *et al.* (2023)
Dose-escalation study	Patients with rheumatoid arthritis	*D. nobile* alkaloids	Better control of disease activity, reduction in inflammatory markers with higher doses	Zhang *et al.* (2022c)
Interventional study	Patients with diabetic retinopathy	*D. nobile* extract	Significant improvement in visual acuity, reduction in retinal lesions with twice-daily regimen	He *et al.* (2023)
Prospective study	Patients with depression	*D. nobile* extract	Identified genetic markers predictive of treatment response	Xiong *et al.* (2021)
Retrospective analysis	Patients with allergic rhinitis	*D. nobile* polysaccharides	Patients with atopy or elevated serum IgE levels showed better treatment responses	Peng *et al.* (2022)
Multicentre study	Patients with inflammatory bowel disease	*D. nobile* alkaloids	Developed predictive model for therapeutic response, enabling personalized medicine approach	Hong *et al.* (2022)

are crucial when exploring the therapeutic potential of *D. nobile*. Local regulations and guidelines governing the research, development and commercialization of herbal medicines and natural products must be followed in all cases. Compliance with good manufacturing practices (GMP) and quality control standards should be adhered to during the production and formulation of *D. nobile* products. Comprehensive toxicological studies were conducted to evaluate the safety profile of *D. nobile* extracts, compounds and formulations. The research should emphasize the comprehensive safety evaluation of *D. nobile* through the assessment of its acute and chronic toxicity, genotoxicity and potential adverse effects using preclinical *in vitro* and *in vivo* studies. It must investigate potential drug interactions with commonly used pharmaceuticals to identify contraindications or adverse effects, as well as the impact of *D. nobile* on drug metabolism enzymes and transporters to understand its influence on drug pharmacokinetics. The allergenic potential of *D. nobile* extracts and products needs to be evaluated through studies such as skin sensitization tests and allergen profiling, aiming to identify allergenic compounds and develop strategies to minimize allergic reactions. Long-term safety studies should be conducted to assess the cumulative effects and safety of prolonged use, complemented by post-marketing surveillance systems to monitor adverse events and ensure ongoing safety evaluation. Furthermore, robust quality control measures will need to be implemented to guarantee the authenticity, purity and consistency of *D. nobile* products. Appropriate analytical techniques, such as DNA barcoding, chemical fingerprinting and marker compound analysis, should be used for product authentication and quality assurance. Collaboration with regulatory agencies, toxicologists, pharmacologists and other experts in the field is essential to navigate regulatory requirements and address safety considerations associated with *D. nobile* research and development.

15.5.2 Collaboration and Interdisciplinary research

Collaboration and interdisciplinary research play a vital role in advancing our knowledge and understanding of the therapeutic potential of *D. nobile*. By bringing together experts from different fields, such as botany, pharmacology, chemistry, clinical medicine and traditional medicine, valuable insights can be gained and comprehensive studies can be conducted. Collaborations with botanists and taxonomists will ensure accurate identification and classification of *D. nobile* variants. Working together to study plant growth conditions, cultivation techniques and conservation strategies can ensure sustainable sourcing. Pharmacologists and chemists need to work together to analyse and isolate bioactive compounds, elucidating their mechanisms of action and identifying potential therapeutic constituents. Clinical researchers will need to design and conduct trials to evaluate the efficacy, safety and optimal dosing of *D. nobile* for specific disease conditions, while collecting and analysing clinical data to assess its benefits and side effects in human subjects. Partnerships with traditional medicine practitioners will help scientists incorporate deep knowledge of *D. nobile*'s historical uses, blending traditional wisdom with modern science to develop evidence-based applications. Biomedical scientists will have the ability to investigate the cellular and molecular mechanisms of its therapeutic effects, exploring interactions with specific cellular targets and signalling pathways. Additionally, experts in data analysis and bioinformatics can contribute to interpreting large-scale omics data generated from *D. nobile* research, ensuring a comprehensive understanding of its medicinal potential. Computational approaches can be applied to identify potential drug targets, predict compound interactions and analyse complex biological networks. Effective collaboration and interdisciplinary research foster a holistic approach towards understanding the therapeutic potential of *D. nobile*, from its botanical aspects to its clinical applications. By combining diverse expertise and resources, researchers can leverage each other's strengths and accelerate scientific research in this field.

15.5.3 Funding and resources

Obtaining adequate funding and utilizing available resources are essential for conducting research on the therapeutic potential of *D. nobile*. To advance *D. nobile* research, it is essential to explore government-funded research grants at both national and international levels, particularly those supporting botanical and natural product research. Funding opportunities focused on herbal medicines, traditional medicine, or botanical therapeutics should be prioritized. Collaborations with research institutions, universities, non-profit organizations, and pharmaceutical companies can provide financial support, specialized facilities, and expertise in formulation development or clinical trials. Utilizing existing research infrastructure, such as laboratories, botanical gardens, and herbaria, along with partnerships granting access to advanced analytical instruments, molecular biology tools, and animal testing facilities, can significantly enhance research efforts. Engaging with other national and international research groups fosters resource-sharing and interdisciplinary expertise. Participation in scientific conferences, workshops, and seminars facilitates networking with researchers and funding agencies, while joining professional societies or associations provides access to valuable funding information and collaboration opportunities. Seeking support from organizations or institutions can provide in-kind contributions such as access to research databases, technical expertise or shared resources. Exploring partnerships with botanical gardens, conservation organizations or herbal medicine companies can provide plant materials or expertise in cultivation and harvesting. Crowdfunding platforms or public engagement initiatives can be considered to raise funds for specific research projects or to create awareness about the importance of *D. nobile* research. Engaging with the public through outreach programmes, educational materials or media coverage promotes the value of research on *D. nobile*. It is crucial to develop a comprehensive funding strategy and effectively utilize available resources to support the research activities, infrastructure and collaborations necessary to explore the therapeutic potential of *D. nobile*.

15.5.4 Ethical and cultural perspectives

Exploring the therapeutic potential of *D. nobile* should be conducted with careful consideration of the ethical and cultural perspectives, prioritizing the well-being and safety of research participants by obtaining informed consent and ensuring that their rights and privacy are protected. The potential benefits, risks and uncertainties associated with *D. nobile* research will need to be communicated to participants in a culturally sensitive manner. Local communities and traditional medicine practitioners need to be engaged with to foster understanding and collaboration, respecting their cultural practices and beliefs. Intellectual property rights must be respected and and the contributions of traditional medicine practitioners, local communities and indigenous knowledge holders acknowledged. Collaborations with traditional medicine practitioners are crucial to ensure that their knowledge and practices related to *D. nobile* are recognized, respected and protected. Sustainable practices must be promoted for the collection, cultivation and utilization of *D. nobile* to preserve its natural habitats and ensure long-term availability. Collaborations with conservation organizations and local communities are also needed to implement sustainable harvesting methods and promote the conservation of *D. nobile*. Research must be conducted in a culturally sensitive manner, taking into account cultural beliefs, practices and rituals associated with *D. nobile*, and guidance and input sought from local communities and cultural experts to ensure that research protocols and methodologies are respectful and appropriate. Mechanisms for benefit-sharing with local communities, particularly when commercializing products derived from *D. nobile*, should be put in place, and community development initiatives and capacity-building programmes engaged with to ensure that local communities benefit from research outcomes. The necessary ethical approval must be obtained from relevant research ethics committees and regulatory bodies to ensure compliance with ethical guidelines and regulations. Good research practices, including data integrity, transparency and responsible reporting of research findings, need to be adhered to. By considering ethical and cultural perspectives throughout the research process, researchers can foster respectful and collaborative relationships with communities and stakeholders, ensuring that the exploration of the therapeutic potential of *D. nobile* aligns with ethical principles and respects cultural values.

15.6 Conclusion and Outlook

In conclusion, research on *D. nobile* has revealed its significant therapeutic potential and has contributed to a better understanding of its pharmacological properties. The rich bioactive compounds found in *D. nobile* have shown promising effects in various diseases, including neurological disorders, metabolic disorders, cardiovascular health, immune modulation and inflammatory diseases. However, there is still much to explore and uncover regarding the full potential of *D. nobile*. Research on *D. nobile* aims to uncover its precise modes of action by investigating the interactions of its compounds with cellular receptors and enzymes. Well-designed clinical trials are essential to evaluate its efficacy, safety and optimal dosage regimens for specific disease conditions. Additionally, the potential of *D. nobile* as an adjunct or complementary therapy alongside conventional treatments is being explored. Understanding interindividual variability in response to *D. nobile* treatment also opens avenues for personalized medicine approaches, ensuring tailored and effective therapeutic applications. Identification of biomarkers or genetic markers can predict individual responses to *D. nobile* and guide treatment decisions. Developing standardized extracts or formulations of *D. nobile* will ensure consistent quality, potency and bioavailability. Exploring innovative delivery systems such as nanoparticles or liposomes will enhance the targeted delivery and efficacy of *D. nobile* compounds. Implementation of sustainable practices for the collection and cultivation of *D. nobile* will ensure its availability for research and traditional use. Exploring the potential of tissue culture techniques or synthetic biology approaches will help to produce bioactive compounds from *D. nobile* in a sustainable and controlled manner. Fostering collaboration between traditional medicine practitioners and scientific researchers will lead to the combination of traditional knowledge with modern scientific approaches. Documenting and preserving traditional knowledge related to the use of *D. nobile* will ensure its continued transmission and the safeguarding of indigenous practices. By pursuing these future directions, researchers can further uncover the therapeutic potential of *D. nobile*, leading to the development of innovative treatments and improved health outcomes in various diseases. It is essential to continue interdisciplinary collaboration, secure adequate funding and uphold ethical and cultural considerations to ensure the progress and success of research on *D. nobile*.

15.6.1 Summary of current findings and knowledge gaps

Current research on *D. nobile* has revealed its significant therapeutic potential and demonstrated its efficacy in various diseases. *D. nobile* exhibits a wide range of pharmacological activities, including antioxidant, anti-inflammatory, immunomodulatory, neuroprotective and antitumour effects. Numerous bioactive compounds, including alkaloids, phenols, polysaccharides and flavonoids, have been identified in *D. nobile*. These compounds also contribute to therapeutic effects. Research has shown the beneficial effects of *D. nobile* in neurological disorders, metabolic disorders, cardiovascular health, immune modulation and inflammatory diseases. Several mechanisms of action have been proposed, including antioxidant activity, modulation of cellular signalling pathways, regulation of gene expression and interaction with specific receptors.

Despite the progress made, significant knowledge gaps remain to be addressed. The complex chemical composition of *D. nobile* requires further investigation to identify additional bioactive compounds and their specific contributions to its therapeutic effects. More detailed studies are needed to elucidate the precise molecular mechanisms through which *D. nobile* exerts its therapeutic effects under different disease conditions. Although preclinical studies have shown promising results, there is a need for well-designed clinical trials to evaluate the efficacy, safety and optimal dosage regimens of *D. nobile* in humans. Understanding the factors that contribute to interindividual variability in response to *D. nobile* treatment and identifying biomarkers for personalized medicine approaches remains an important area for future research. Standardization of *D. nobile* extracts and the development of novel delivery systems are necessary to ensure consistent quality, potency and bioavailability. More research is needed to develop sustainable sourcing and cultivation practices for *D. nobile* to ensure its long-term availability while preserving its natural habitat. Addressing these knowledge gaps will contribute to a more comprehensive understanding of the therapeutic potential of *D. nobile* and will facilitate its translation into clinical practice.

15.6.2 Implications for future research and clinical practice

The current findings regarding the therapeutic potential of *D. nobile* have several implications for future research and clinical practice. Future research should focus on elucidating the precise molecular mechanisms of the action of *D. nobile*. This will enhance our understanding of its therapeutic effects and provide a scientific basis for clinical application. Well-designed clinical trials are needed to evaluate the efficacy, safety and optimal dosing regimens of *D. nobile* for specific disease conditions. Robust clinical evidence will enable evidence-based practice and support the integration of *D. nobile* into mainstream healthcare. It is essential to investigate the potential synergistic effects of *D. nobile* with conventional therapies or other traditional medicines. This may lead to the development of novel combination treatments with enhanced therapeutic outcomes. Developing standardized protocols for the extraction, formulation and quality control of *D. nobile* products is crucial. This will ensure consistency in product quality, potency and safety, thereby facilitating regulatory approval and widespread clinical use. Exploring the interindividual variability in response to *D. nobile* treatment and identifying biomarkers or genetic markers that predict individual responses can guide personalized medicine approaches. Tailoring treatments for specific patient characteristics can enhance efficacy and optimize therapeutic outcomes.

Comprehensive safety studies are needed to evaluate the long-term use of *D. nobile* and monitor any potential adverse effects. This will contribute to the establishment of appropriate safety guidelines and ensure the well-being of patients using *D. nobile* products. Effective knowledge-translation strategies are necessary to bridge the gap between the research findings and clinical practice. Education and awareness programmes for healthcare professionals, patients and the general public can facilitate the appropriate and informed use of *D. nobile*-based therapies. By addressing these implications, future research on *D. nobile* could advance its integration into clinical practice, leading to improved healthcare outcomes and enhanced patient well-being.

15.6.3 Recommendations for further studies

Based on the current state of research on *D. nobile*, several areas warrant further investigation. In-depth mechanistic studies should be conducted to unravel the precise molecular mechanisms by which *D. nobile* exerts its therapeutic effects. This can involve exploring specific cellular signalling pathways, target interactions and modulation of gene expression. Well-controlled clinical trials are needed to evaluate the efficacy, safety and optimal dosage regimens of *D. nobile* under specific disease conditions. This will provide robust evidence to support its use in clinical practice. Long-term safety assessments are needed to evaluate the potential adverse effects and safety profile of *D. nobile*. This is crucial for establishing the safety of prolonged use and ensuring patient well-being. Pharmacokinetic studies must be conducted to understand the absorption, distribution, metabolism and elimination of *D. nobile* compounds in the body. This information will guide dosage regimens and optimize their therapeutic use. The potential synergistic effects of *D. nobile* with other traditional medicines or conventional therapies should also be investigated. This could lead to the development of combination treatments with enhanced efficacy and improved patient outcomes.

Developing standardized protocols for the extraction, formulation and quality control of *D. nobile* products will ensure consistency in product quality, potency and safety and facilitate regulatory approvals. Exploring sustainable sourcing methods and cultivation practices for *D. nobile* will ensure its long-term availability while minimizing environmental impact. This includes investigation of tissue culture techniques and sustainable harvesting practices. Comparative studies should be conducted to compare different varieties or species of *Dendrobium* and evaluate their therapeutic potential and chemical composition. This will provide a comprehensive understanding of variations within the genus. Such studies would provide insights into its mechanism of action and potential clinical applications. Fostering collaborations between researchers, clinicians and traditional medicine practitioners will facilitate the translation of *D. nobile* research findings into clinical practice and bridge the gap between the traditional and modern scientific approaches. By addressing these research recommendations, a deeper understanding of *D. nobile*'s therapeutic potential, mechanisms of action, safety profile and optimal use can be achieved, ultimately benefiting patients and advancing the field of natural medicine.

Acknowledgements

The authors express their gratitude for the financial support received through the Distinguished High-Level Talents Research Grant from the Guizhou Science and Technology Corporation Platform Talents Fund (grant nos. [2017]5733-001 and CK-1130-002), the National Natural Science Foundation of China (U1812403, 82373981 and 82060750), 2011 Collaborative Innovation Centre of Traditional Chinese Medicine in Guizhou Province (no. [2022]026) and the support provided by the Zunyi Medical University, China. Special appreciation is extended to all laboratory colleagues and research staff members for their valuable insights, constructive guidance and assistance with the contents of this chapter.

References

Bhattacharyya, P. and Kumaria, S. (2015) Molecular characterization of *Dendrobium nobile* Lindl., an endangered medicinal orchid, based on randomly amplified polymorphic DNA. *Plant Systematics and Evolution* 301(1), 201–210.

Cakova, V., Bonte, F. and Lobstein, A. (2017) *Dendrobium*: sources of active ingredients to treat age-related pathologies. *Aging and Disease* 8(6), 827–849.

Chen, D., Ji, Q., Liu, J., Cheng, F., Zheng, J. *et al.* (2023) MicroRNAs in the regulation of RIG-I-like receptor signalling pathway: possible strategy for viral infection and cancer. *Biomolecules* 13(9), 1344.

Chen, X.L., Sun, M.C., Chong, S.L., Si, J.P. and Wu, L.S. (2021) Transcriptomic and metabolomic approaches deepen our knowledge of plant-endophyte interactions. *Frontiers in Plant Science* 12, 700200.

Cruz, A.M., Gonçalves, M.C., Marques, M.S., Veiga, F., Paiva-Santos, A.C. *et al.* (2023) In vitro models for anti-aging efficacy assessment: a critical update in dermocosmetic research. *Cosmetics* 10(2), 1–21.

Duan, F.P., Li, Y.S., Hu, T.Y., Pan, X.Q. and Ma, F. (2022) *Dendrobium nobile* protects against ovalbumin-induced allergic rhinitis by regulating intestinal flora and suppressing lung inflammation. *Chinese Journal of Natural Medicines* 20(6), 443–457.

Fan, C., Sun, X., Wang, X. and Yu, H. (2023) Therapeutic potential of the chemical composition of *Dendrobium nobile* Lindl. *Frontiers in Pharmacology* 14, 1163830.

Fan, K.-J., Li, Y.-W., Wu, J., Wang, Q.-S., Xu, B.-X. *et al.* (2022) Pharmacology and molecular docking study of cartilage protection of Chinese herbal medicine Fufang Shatai Heji (STHJ) by inhibiting the expression of MMPs in collagen-induced arthritis mice. *Annals of Palliative Medicine* 11(2), 466–479.

Favre-Godal, Q., Gourguillon, L., Lordel-Madeleine, S., Gindro, K. and Choisy, P. (2020) Orchids and their mycorrhizal fungi: an insufficiently explored relationship. *Mycorrhiza* 30(1), 5–22.

Fu, X., Chen, S., Xian, S., Wu, Q., Shi, J. *et al.* (2023) *Dendrobium* and its active ingredients: emerging role in liver protection. *Biomedicine & Pharmacothorapy* 157, 114043.

Gong, D., Li, B., Wu, B., Fu, D. and Li, Z. (2023) The integration of the metabolome and transcriptome for *Dendrobium nobile* Lindl. in response to methyl Jasmonate. *Molecules* 28(23), 7892.

Guo, Y., Kong, Y., Sun, J., Jiao, Y., Hong, Y. *et al.* (2023) Alleviation of ultraviolet-B radiation-induced photoaging using *Saussurea medusa* maxim polysaccharide. *Photochemistry and Photobiology* 100(3), 622–632.

He, L., Su, Q., Bai, L., Li, M., Liu, J. *et al.* (2020) Recent research progress on natural small molecule bibenzyls and its derivatives in *Dendrobium* species. *European Journal of Medicinal Chemistry* 204, 112530.

He, Q., Xiao, L., Shi, Y., Li, W. and Xin, X. (2023) Natural products: protective effects against ischemia-induced retinal injury. *Frontiers in Pharmacology* 14, 1149708.

He, T., Ye, C., Zeng, Q., Fan, X. and Huang, T. (2022) Genetic diversity and population structure of cultivated *Dendrobium nobile* Lindl. in southwest of China based on genotyping-by-sequencing. *Genetic Resources and Crop Evolution* 69(8), 2803–2818.

Hinsley, A., de Boer, H.J., Fay, M.F., Gale, S.W., Gardiner, L.M. *et al.* (2018) A review of the trade in orchids and its implications for conservation. *Botanical Journal of the Linnean Society* 186(4), 435–455.

Hong, S., Kim, E.-Y., Lim, S.-E., Kim, J.-H., Sohn, Y. *et al.* (2022) *Dendrobium nobile* Lindley administration attenuates atopic dermatitis-like lesions by modulating immune cells. *International Journal of Molecular Sciences* 23(8), 4470.

Horváth, E., Sólyom, Á., Székely, J., Nagy, E.E. and Popoviciu, H. (2023) Inflammatory and metabolic signaling interfaces of the hypertrophic and senescent chondrocyte phenotypes associated with osteoarthritis. *International Journal of Molecular Sciences* 24(22), 16468.

Hsu, W.-H., Chung, C.-P., Wang, Y.-Y., Kuo, Y.-H., Yeh, C.-H. *et al.* (2022) *Dendrobium nobile* protects retinal cells from UV-induced oxidative stress damage via nrf2/HO-1 and MAPK pathways. *Journal of Ethnopharmacology* 288, 114886.

Hu, Y., Ye, Z., Wu, M., She, Y., Li, L. *et al.* (2021) The communication between intestinal microbiota and ulcerative colitis: an exploration of pathogenesis, animal models, and potential therapeutic strategies. *Frontiers in Medicine* 8, 766126.

Jain, A., Sarsaiya, S., Gong, Q., Wu, Q. and Shi, J. (2024) Pharmacological and Therapeutic Biofunction of *Dendrobium nobile*: A Critical Review. *Natural Product Communications* 19(11), doi:10.1177/1934578X241300010.

Jiaranaikulwanitch, J., Yooin, W., Chutiwitoonchai, N., Thitikornpong, W., Sritularak, B. *et al.* (2022) Discovery of natural lead compound from *Dendrobium* sp. against SARS-cov-2 infection. *Pharmaceuticals* 15(5), 620.

Jie, X., Feng, Y., Jiahao, F., Ganggui, L., Jiani, Y. *et al.* (2023) Comprehensive chemical profiling of two *Dendrobium* species and identification of anti-hepatoma active constituents from *Dendrobium* chrysotoxum by network pharmacology. *BMC Complementary Medicine and Therapies* 23(1), 217.

Ketsa, S. and Warrington, I.J. (2023) The *Dendrobium* orchid: botany, horticulture, and utilization. *Crop Science* 63(4), 1829–1888.

Laishram, H., Rocky, T., Sachin, S., Lakshmi, H. and Preema, D.M. (2019) In vitro conservation and asymbiotic propagation of *Dendrobium nobile* Lindl.: a rare threatened orchid of northeast India. *Research Journal of Biotechnology* 14(2), 16–20.

Lao, Y., Wang, X., Xu, N., Zhang, H. and Xu, H. (2014) Application of proteomics to determine the mechanism of action of traditional Chinese medicine remedies. *Journal of Ethnopharmacology* 155(1), 1–8.

Li, D.-D., Fan, H.-X., Yang, R., Li, Y.-Y., Zhang, F. *et al.* (2022a) *Dendrobium nobile* Lindl. alkaloid suppresses NLRP3-mediated pyroptosis to alleviate LPS-induced neurotoxicity. *Frontiers in Pharmacology* 13, 846541.

Li, D.D., Zheng, C.Q., Zhang, F. and Shi, J.S. (2022b) Potential neuroprotection by *Dendrobium nobile* Lindl alkaloid in Alzheimer's disease models. *Neural Regeneration Research* 17(5), 972–977.

Li, Q., Liu, C., Huang, C., Wang, M., Long, T. *et al.* (2022c) Transcriptome and metabonomics analysis revealed the molecular mechanism of differential metabolite production of *Dendrobium nobile* under different epiphytic patterns. *Frontiers in Plant Science* 13, 868472.

Li, L., Chen, H., Huang, G., Lv, Y., Yao, L. *et al.* (2024) Structure of polysaccharide from *Dendrobium nobile* Lindl. and its mode of action on TLR4 to exert immunomodulatory effects. *Foods* 13(9), 1356.

Li, M., Trapika, I.G.S.C., Tang, S.Y.S., Cho, J.-L., Qi, Y. *et al.* (2021) Mechanisms and active compounds polysaccharides and bibenzyls of medicinal *Dendrobiums* for diabetes management. *Frontiers in Nutrition* 8, 811870.

Li, Z., Zeng, M., Geng, K., Lai, D., Xu, Z. *et al.* (2023) Chemical constituents and hypoglycemic mechanisms of *Dendrobium nobile* in treatment of type 2 diabetic rats by UPLC-ESI-Q-orbitrap, network pharmacology and in vivo experimental verification. *Molecules* 28(6), 2683.

Long, Y., Wang, W., Zhang, Y., Du, F., Zhang, S. *et al.* (2023) Photoprotective effects of *Dendrobium nobile* Lindl. polysaccharides against UVB-induced oxidative stress and apoptosis in hacat cells. *International Journal of Molecular Sciences* 24(7), 1–17.

Luo, Y., Liu, G. and Hou, P. (2023) Synergism effect of dendrobine on cisplatin in treatment of H1299 by modulating the balance of Treg/Th17. *Anti-Cancer Agents in Medicinal Chemistry* 23(1), 105–112.

Luo, Y., Yang, D., Xu, Y., Wu, D., Tan, D. *et al.* (2024) Hypoglycemic effects and quality marker screening of *Dendrobium nobile* Lindl. at different growth years. *Molecules* 29(3), 699.

Ma, R.F., Wang, S.G. and Tian, X.J. (2019) Study on the safety of *Dendrobium huoshanense* oral solution. *Journal of Food Safety and Quality* 10(24), 8462–8469.

Ninawe, A.S. and Swapna, T.S. (2017) Orchid diversity of northeast India–traditional knowledge and strategic plan for conservation. *Journal of Orchid Society of India* 31, 41–56.

Nongthombam, K.S., Mutum, S.S. and Pandey, R.R. (2024) Secondary metabolite profiling of the endophytic fungus *Trichoderma longibrachiatum* L2D2 isolated from *Anaphalis contorta*. *Annual Research & Review in Biology* 39(2), 1–15.

Pavlova, A.S., Leontieva, M.R., Smirnova, T.A., Kolomeitseva, G.L., Netrusov, A.I. *et al.* (2017) Colonization strategy of the endophytic plant growth-promoting strains of *Pseudomonas fluorescens* and *Klebsiella oxytoca* on the seeds, seedlings and roots of the epiphytic orchid, *Dendrobium nobile* Lindl. *Journal of Applied Microbiology* 123(1), 217–232.

Qiu, Y., Cai, C., Mo, X., Zhao, X., Wu, L. *et al.* (2023) Transcriptome and metabolome analysis reveals the effect of flavonoids on flower color variation in *Dendrobium nobile* Lindl. *Frontiers in Plant Science* 14, 1220507.

Rabidas, S.S., Prakash, C., Tyagi, J., Suryavanshi, J., Kumar, P. *et al.* (2023) A comprehensive review on anti-inflammatory response of flavonoids in experimentally-induced epileptic seizures. *Brain Sciences* 13(1), 102.

Ramesh, T., Koperuncholan, M., Praveena, R., Ganeshkumari, K., Vanithamani, J. *et al.* (2019) Medicinal properties of some *Dendrobium* orchids – A review. *Journal of Applied and Advanced Research* 4(4), 119–128.

Rao, D., Zhao, R., Hu, Y., Li, H., Chun, Z. *et al.* (2023) Revealing of intracellular antioxidants in *Dendrobium nobile* by high performance liquid chromatography-tandem mass spectrometry. *Metabolites* 13(6), 702.

Sarsaiya, S., Jain, A., Fan, X., Jia, Q., Xu, Q. *et al.* (2020) New insights into detection of a dendrobine compound from a novel endophytic *Trichoderma longibrachiatum* strain and its toxicity against phytopathogenic bacteria. *Frontiers in Microbiology* 11, 337.

Sarsaiya, S., Jain, A., Shu, F., Yang, M., Pu, M. *et al.* (2024) Enhancing dendrobine production in *Dendrobium nobile* through mono-culturing of endophytic fungi, *Trichoderma longibrachiatum* (MD33) in a temporary immersion bioreactor system. *Frontiers in Plant Science* 15, 1302817.

Sut, S., Maggi, F. and Dall'Acqua, S. (2017) Bioactive secondary metabolites from orchids (Orchidaceae). *Chemistry & Biodiversity* 14(11), e1700172.

Teixeira da Silva, J.A. and Ng, T.B. (2017) The medicinal and pharmaceutical importance of *Dendrobium* species. *Applied Microbiology and Biotechnology* 101(6), 2227–2239.

Tiwari, P., Sharma, A., Bose, S.K. and Gautam, A. (2022) Biotechnological interventions in orchids: recent updates, translational success, and commercial outcomes. *Research Square* 1(2), 1–38.

Wan Hei, J., Hung, W., Yanbo, Z. and Zhang, Z.-J. (2021) *Dendrobium* in diabetes: a comprehensive review of its phytochemistry, pharmacology, and safety. *Journal of Translational Science* 7(6), 1–10.

Wang, Y., Zhang, Y., Cong, H., Li, C., Wu, J. *et al.* (2023) Cultivable endophyte resources in medicinal plants and effects on hosts. *Life* 13(8), 1695.

Wang, Z., Zhao, M., Cui, H., Li, J. and Wang, M. (2020) Transcriptomic landscape of medicinal *Dendrobium* reveals genes associated with the biosynthesis of bioactive components. *Frontiers in Plant Science* 11, 391.

Wei, X., Liu, J., Xu, Z., Wang, D., Zhu, Q. *et al.* (2024) Research progress on the pharmacological mechanism, in vivo metabolism and structural modification of Erianin. *Biomedicine & Pharmacotherapy* 173, 116295.

Whaloy, A.K., Minakov, D.A., Orlova, A.A., Ponkratova, A.O., Fock, E. *et al.* (2023) Analysis of *Empetrum nigrum* L. lipophilic secondary metabolites, thoir metabolomic profiles and antioxidant activity. *Journal of Essential Oil Research* 35(3), 310–323.

Wu, W., Lin, Y., Farag, M.A., Li, Z. and Shao, P. (2023) *Dendrobium* as A new natural source of bioactive for the prevention and treatment of digestive tract diseases: a comprehensive review with future perspectives. *Phytomedicine* 114, 154784.

Xie, J.-H., Jin, M.-L., Morris, G.A., Zha, X.-Q., Chen, H.-Q. *et al.* (2016) Advances on bioactive polysaccharides from medicinal plants. *Critical Reviews in Food Science and Nutrition* 56(1), 60–84.

Xiong, T.-W., Liu, B., Wu, Q., Xu, Y.-Y., Liu, P. *et al.* (2021) Beneficial effects of *Dendrobium nobile* Lindl. alkaloids (DNLA) on anxiety and depression induced by chronic unpredictable stress in rats. *Brain Research* 1771, 147647.

Xu, Q., Niu, S.-C., Li, K.-L., Zheng, P.-J., Zhang, X.-J. *et al.* (2022) Chromosome-scale assembly of the *Dendrobium nobile* genome provides insights into the molecular mechanism of the biosynthesis of the medicinal active ingredient of *Dendrobium*. *Frontiers in Genetics* 13, 844622.

Xu, X., Yu, Y., Yang, L., Wang, B., Fan, Y. *et al.* (2023) Integrated analysis of *Dendrobium nobile* extract dendrobin A against pancreatic ductal adenocarcinoma based on network pharmacology, bioinformatics, and validation experiments. *Frontiers in Pharmacology* 14, 1079539.

Yang, L., Yang, Y., Huang, L., Cui, X. and Liu, Y. (2023) From single- to multi-omics: future research trends in medicinal plants. *Briefings in Bioinformatics* 24(1), bbac485.

Zhai, D., Lv, X., Chen, J., Peng, M. and Cai, J. (2022) Recent research progress on natural stilbenes in *Dendrobium* species. *Molecules* 27(21), 7233.

Zhang, X., Zhao, R., Zheng, S., Chun, Z. and Hu, Y. (2020) *Dendrobium* liquor eliminates free radicals and suppresses cellular proteins expression disorder to protect cells from oxidant damage. *Journal of Food Biochemistry* 44(12), e13509.

Zhang, J., Xu, H.X., Zhao, Z.L., Xian, Y.F. and Lin, Z.X. (2021a) *Dendrobium nobile* Lindl: a review on its chemical constituents and pharmacological effects. *Chinese Medicine and Culture* 4(4), 235–242.

Zhang, S., Hu, R., Geng, Y., Chen, K., Wang, L. *et al.* (2021b) The regulatory effects and the signaling pathways of natural bioactive compounds on ferroptosis. *Foods* 10(12), 2952.

Zhang, B., Niu, Z., Li, C., Hou, Z., Xue, Q. *et al.* (2022a) Improving large-scale biomass and total alkaloid production of *Dendrobium nobile* Lindl. using a temporary immersion bioreactor system and MeJA elicitation. *Plant Methods* 18(1), 10.

Zhang, H., Liu, M.M., Liu, X.N., Li, Z.Y., Zhao, L.L. *et al.* (2022b) Impact of endophytic microorganisms on the pharmaco-active compounds production in medicinal plants: a review. *Biotechnology Bulletin* 38(8), 41.

Zhang, W., Zhang, M., Wu, Q. and Shi, J. (2022c) *Dendrobium nobile* Lindl. alkaloids ameliorate Aβ25-35-induced synaptic deficits by targeting Wnt/β-catenin pathway in Alzheimer's disease models. *Journal of Alzheimer's Disease* 86(1), 297–313.

Zhang, Q., Fu, X., Liu, H., Chen, Y., Chen, S. *et al.* (2023a) A systematical review on ethnobotanical, phytochemical and pharmacological aspects of *Dendrobium nobile* Lindl. *Phytochemistry Reviews* 22(3), 743–780.

Zhang, Y., Lu, J., Li, H. and Song, H. (2023b) Advances in dietary polysaccharides as hypoglycemic agents: mechanisms, structural characteristics, and innovative applications. *Critical Reviews in Food Science and Nutrition* 64(18), 1–21.

Zhang, Y.W., Shi, Y.C. and Zhang, S.B. (2023c) Metabolic and transcriptomic analyses elucidate a novel insight into the network for biosynthesis of carbohydrate and secondary metabolites in the stems of a medicinal orchid *Dendrobium nobile*. *Plant Diversity* 45(3), 326–336.

Zheng, J., Zhao, Y., Qi, Q., Gao, M. and Yu, J. (2022) Exploring the molecular mechanism of Sishen decoction in the treatment of rheumatoid arthritis. *Annals of Translational Medicine* 10(18), 977.

Index

Note: Page numbers in **bold** type refer to **figures**; Page numbers in *italic* type refer to *tables*

This book is published by **CABI**, an international not-for-profit organisation that improves people's lives worldwide by providing information and applying scientific expertise to solve problems in agriculture and the environment.

CABI is also a global publisher producing key scientific publications, including world renowned databases, as well as compendia, books, ebooks and full text electronic resources. We publish content in a wide range of subject areas including: agriculture and crop science / animal and veterinary sciences / ecology and conservation / environmental science / horticulture and plant sciences / human health, food science and nutrition / international development / leisure and tourism.

The profits from CABI's publishing activities enable us to work with farming communities around the world, supporting them as they battle with poor soil, invasive species and pests and diseases, to improve their livelihoods and help provide food for an ever growing population.

CABI is an international intergovernmental organisation, and we gratefully acknowledge the core financial support from our member countries (and lead agencies) including:

UKaid from the British people

Ministry of Agriculture People's Republic of China

Australian Government Australian Centre for International Agricultural Research

Agriculture and Agri-Food Canada

Ministry of Foreign Affairs of the Netherlands

Schweizerische Eidgenossenschaft Confédération suisse Confederazione Svizzera Confederaziun svizra Swiss Agency for Development and Cooperation SDC

Discover more

To read more about CABI's work, please visit: **www.cabi.org**

Browse our books or explore our online products at **https://www.cabidigitallibrary.org**

Interested in writing for CABI? Find our author guidelines here: **www.cabi.org/publishing-products/information-for-authors/**

Printed and bound by CPI Group (UK) Ltd, Croydon, CR0 4YY

13/09/2026

14961085-0002